VIII International Scientific Conference "Issues of Physics and Technology in Science, Industry and Medicine"

AA006781

IOP Conference Series: Materials Science and Engineering
Volume 135

Tomsk, Russia
1 – 3 June 2016

Editor:

Potylitsyn Alexander

ISBN: 978-1-5108-2818-6
ISSN: 1757-8981

Printed from e-media with permission by:

Curran Associates, Inc.
57 Morehouse Lane
Red Hook, NY 12571

Some format issues inherent in the e-media version may also appear in this print version.

Printed by Curran Associates, Inc. (2016)

For permission requests, please contact the Institute of Physics
at the address below.

Institute of Physics
Dirac House, Temple Back
Bristol BS1 6BE UK

Phone: 44 1 17 929 7481
Fax: 44 1 17 920 0979

techtracking@iop.org

Additional copies of this publication are available from:

Curran Associates, Inc.
57 Morehouse Lane
Red Hook, NY 12571 USA
Phone: 845-758-0400
Fax: 845-758-2633
Email: curran@proceedings.com
Web: www.proceedings.com

Table of contents

Volume 135

VIII International Scientific Conference "Issues of Physics and Technology in Science, Industry and Medicine"
1–3 June 2016, Tomsk, Russia

Accepted papers received: 16 June 2016
Published online: 02 August 2016

Preface

Papers

IOP Conf. Series: Materials Science and Engineering **135** (2016) 011001 doi:10.1088/1757-899X/135/1/011001

VIII International Scientific Conference "Issues of Physics and Technology in Science, Industry and Medicine" was held at Tomsk Polytechnic University, 1-3 June, 2016.

The Organizing Committee included representatives from Karlsruhe Institute of Technology, Tokyo Metropolitan University, Academy of Sciences of the Czech Republic, ROSENERGOATOM Electric Power Division of ROSATOM, FSUE Mining and Chemical Combine.

The Conference brought together specialists in industry, teachers, students and young scientists conducting research in the field of Power Engineering and Nuclear Technology.

CONFERENCE SESSIONS:

1 Physical and power plants and electrophysical installations

2 Innovative materials and technologies for nuclear and 'green' energy

3 Mathematical modeling in basic and applied research

4 Physical/chemical and isotope technologies in science, industry and medicine

5 Radiation and beam-plasma technologies in science, engineering and medicine

6 Topical issues of nuclear non-proliferation, safety and ecology of the nuclear industry

7 Information technologies, automation and control systems

More than 450 papers of young and experienced researchers from Russia and abroad were submitted which covered the topics of the conference. 55 selected papers were recommended for publication in *IOP Conference Series: Materials Science and Engineering.*

We also appreciate the financial support from the Russian Foundation for Basic Research

The Organizing Committee of Issues of Physics and Technology in Science, Industry and Medicine 2016

Dr. Potylitsyn Alexander, TPU

IOP Conf. Series: Materials Science and Engineering **135** (2016) 011002 doi:10.1088/1757-899X/135/1/011002

Peer review statement

All papers published in this volume of *IOP Conference Series: Materials Science and Engineering* have been peer reviewed through processes administered by the proceedings Editors. Reviews were conducted by expert referees to the professional and scientific standards expected of a proceedings journal published by IOP Publishing.

IOP Conf. Series: Materials Science and Engineering **135** (2016) 012001 doi:10.1088/1757-899X/135/1/012001

Lithium and magnesium isotopes fractionation by zone melting

D V Akimov, N B Egorov, A N Dyachenko, M P Pustovalova, I R Podoinikov

National Research Tomsk Polytechnic University, Tomsk, Russia

E-mail: egorov@tpu.ru

Abstract. The process of changing isotopic composition of the lithium and magnesium salts was studied by using the process of zone melting. It was founded in the paper that the process of separation of the lithium isotopes is more effective than for magnesium isotopes when the conditions of process were the same. The coefficients of isotopes separation were calculated and have the next value: $\alpha = 1.006$ for ^{26}Mg isotope and $\alpha = 1.0022$ for ^{6}Li isotope.

1. Introduction

Naturally occurring lithium is composed of two stable isotopes ^{6}Li (7.589 %) and ^{7}Li (92.411 %) [1]. Each of them is very important for nuclear energetics. The isotope of ^{7}Li is used as additive in the coolant of the first loop pressurized water reactor and also for production chemical reagents for nuclear energetics [2].

Naturally occurring magnesium is composed of tree stable isotopes ^{24}Mg (78.992 %), ^{25}Mg (10.003 %) and ^{24}Mg (11.005 %) [3]. The isotope of ^{24}Mg is used for production radioisotope ^{22}Na [4]. Other isotopes ^{25}Mg and ^{26}Mg are used for biological labeling during the process of study the absorption and metabolism of magnesium in the human body.

Isotopes of lithium and magnesium are nonvolatile compounds. For this reason for separation of these isotopes we use the next methods: electromagnetic [5-6], crown ethers extraction [7-11], ion exchange [12-14], vacuum distillation [15], amalgam-lithium and amalgam-magnesium systems [16, 17] during precipitation from aqueous solutions [18, 19].

The process of zone melting is widely used for deep cleaning the different substances and for their production in single crystal form [20-23]. Also, this process can be used for separation isotopes such as H_2O and D_2O [24-26], the isotopes of boron, zinc, magnesium, lead and tin [27-32].

The purpose of this work is to investigate the process of changing isotope composition of lithium and magnesium during the process of zone melting of their salts.

2. Experimental part

2.1. Reagents

$MgCl_2 \cdot 6H_2O$ (analytical grade), $LiClO_4 \cdot 3H_2O$ and $LiClO_4$ were used. $LiOH \cdot H_2O$ (pure) and $HClO_4$ (pure) were used for synthesis $LiClO_4 \cdot 3H_2O$. The anhydrous lithium perchlorate $LiClO_4$ was obtained by dehydration of the $LiClO_4 \cdot 3H_2O$ at 200 °C for 3 hours. According to the results of X-ray diffraction analysis, the obtained salts were correspond to $LiClO_4 \cdot 3H_2O$ (PDF 00-008-0156) and $LiClO_4$ (PDF 30-0751).

2.2. Instrumentation

Investigations were conducted on samples with a length of 100 mm and a diameter of 3 mm. For preparing samples $MgCl_2 \cdot 6H_2O$, $LiClO_4 \cdot 3H_2O$ or $LiClO_4$ were heated until salt melted. After that a part of the melt was placed into a glass tube which was sealed at one end. When the process of zone melting was done, the glass tube was cut into samples, 10 mm in length. Then the samples were removed and dissolved in distilled water. After this, in the obtained solutions the isotope composition of the lithium or magnesium was determined. The analysis of isotope compositions was carried out on the mass spectrometer MX-1301T.

2.3. The process of zone melting

All experiments were carried out on the installation, which has five heating elements and five cooling elements (Figure 1). The main part of the installation was a platform with the heating elements placed at equal distance in the form of a ring, the inner diameter of which corresponds to the outer diameter of the glass tube. Power supply GW INSTEK PSH -10100 was used for regulating the temperature of the heating elements. Temperature was measured by using a chromel-alumel thermocouple. The cooling elements were installed between the heating elements.

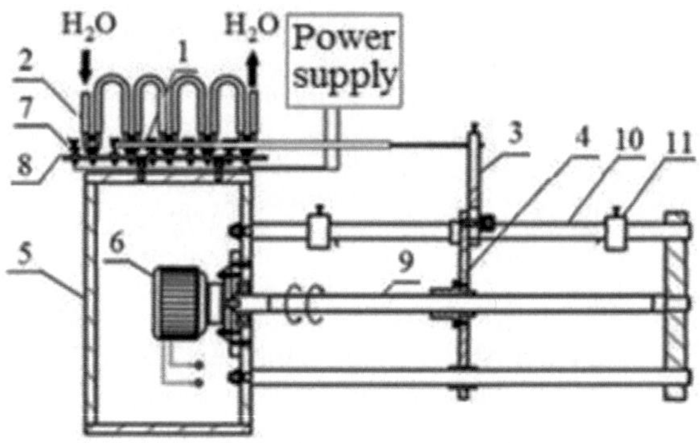

Figure 1. Installation for zone recrystallization (1 – mixture of REEs; 2 – cooling element; 3 – rod; 4 – carriage; 5 – housing; 6 – stepper motor; 7 – heating element; 8 – platform; 9 – shaft; 10 – guides; 11 – sensor).

Nichrome wire was used as the heating element. In Table 1 the temperatures of the heating elements for different work substances are presented. The length of the melt of the ingot was from 7 to 8 mm.

Table 1. The melting point of the work substances.

The work substance	The melting point, °C	Temperature of the heating elements, °C
$LiClO_4$	236	285±15
$LiClO_4 \cdot 3H_2O$	95	115±5
$MgCl_2 \cdot 6H_2O$	120	140±5

3. Results and discussions

In the experiments with lithium the parameters of the process of zone melting were the following: the speed of the tube was 3 cm/h, the number of zone processes was equal to 30. The results of isotopic analysis show that the isotopic composition of lithium in the work substances has change. The process of isotopes fractionation is more effective if we use $LiClO_4 \cdot 3H_2O$ as the work substance. The results of isotopic analysis of the lithium isotopes after the process of zone melting are shown in the Table 2.The coefficients of separation of the lithium isotopes are shown in Table 3.

Table 2. The isotopic composition of $MgCl_2 \cdot 6H_2O$ in the initial zone of crystallization.

The work substances	The isotope content, %			
	^{6}Li		^{7}Li	
	the original composition	after the process of zone melting	the original composition	after the process of zone melting
$LiClO_4$	6.950±0.01	7.32±0.03	93.045±0.01	92.68±0.04
$LiClO_4 \cdot 3H_2O$	6.955±0.01	7.15±0.02	93.050±0.02	92.85±0.03

Table 3. The coefficients of separation of the lithium isotopes.

The work substances	The coefficients of separation	
	$\alpha(^6Li)$	$\alpha(^7Li)$
$LiClO_4$	1.0019	0.9981
$LiClO_4 \cdot 3H_2O$	1.0022	0.9978

The parameters of the process of zone melting of $MgCl_2 \cdot 6H_2O$ were the same. The results of isotopic analysis of the magnesium isotopes after the process of zone melting are shown in the Table 4.

Table 4. The isotopic composition of $MgCl_2 \cdot 6H_2O$ in the initial zone of crystallization.

The isotope content, %					
^{24}Mg		^{25}Mg		^{26}Mg	
the original composition	after the process of zone melting	the original composition	after the process of zone melting	the original composition	after the process of zone melting
79.58±0.03	79.35±0.04	9.8±0.02	9.87±0.04	10.62±0.02	10.78±0.04

The data obtained from the Table 4 shows that the isotopes ^{25}Mg and ^{26}Mg concentrate in the initial zone of crystallization and, also, the isotope ^{24}Mg moves with the melt zone and concentrates in the end of zone crystallization. The coefficients of separation of the magnesium isotopes are presented in Table 5.

Table 5. The coefficients of separation of the magnesium isotopes.

The coefficients of separation		
$\alpha(^{24}Mg)$	$\alpha(^{25}Mg)$	$\alpha(^{26}Mg)$
0.9995	1.0002	1.0006

These results can be explained as follows. Burton, Prim and Slichter obtained the formula for the effective distribution coefficient [33]:

$$K = \frac{1}{1+\left[\left(\frac{1}{K_0}\right)-1\right]e^{-f\delta/D}}, \quad (1)$$

where K_0 – the equilibrium distribution coefficient, f – the crystal growth rate, δ – the thickness of the diffusion layer in front of the crystallization front, D – the diffusion coefficient.

From this equation we can conclude that differences between the effective distribution coefficient K and the equilibrium distribution coefficient K_0 will be more significant if the value of the diffusion coefficient D is higher and the thickness of the diffusion layer in front of the crystallization front is smaller.

4. Conclusion

The distribution of lithium and magnesium isotopes was studied during the process of zone melting of their salts. During the process of zone melting the isotopes ^{25}Mg and ^{26}Mg concentrate in the initial zone of crystallization and, also, the isotope ^{24}Mg moves with the melt zone and concentrates in the end of zone crystallization. The coefficients of isotopes separation were calculated and have the next value: α = 1.006 for ^{26}Mg isotope and α = 1.0022 for ^{6}Li isotope. In the next work effectiveness of separation isotopes during the process of zone melting can be changed by using the other work substances, such as $LiClO_4 \cdot H_2O$ [34].

References

[1] Qi H P, Tailor P D P, De Bievre P 1997 Calibrated Measurements of the Isotopic Composition and Atomic Weight of the Natural Li Isotopic Reference Material IRMM-016 *International Journal of Mass Spectrometry and Ion Processes* **171** 263–268

[2] Report to the Ranking Member, Subcommittee on Oversight, Committee on Science, Space, and Technology, House of Representatives 2013 Stewardship of Lithium-7 Is Needed to Ensure a Stable Supply United States Government Accountability Office № GAO-13-716 31

[3] Sabatier M, Arnaud M J, Kastenmayer P, Rytz A, Barclay D V 2002 Meal effect on magnesium bioavailability from mineral water in healthy women *Am. J. Clin. Nutr.* **75** 65–71

[4] Wälti M K, Zimmermann M B, Walczyk T, Spinas G A, Hurrell R F 2003 Measurement of magnesium absorption and retention in type 2 diabetic patients with the use of stable isotopes *Am. J. Clin. Nutr.* **78** 448–453

[5] Gusev V M, Chkuaseli D V, Guseva M I 1957 Separation of germanium and magnesium isotopes in a small electromagnetic separator *The Soviet Journal of Atomic Energy* **3 3** 999–1005

[6] Keim C P 1952 Electromagnetic separation of stable isotopes *Annu. Rev. Nucl. Sci.* **1** 263–292

[7] Nishizawa K, Nishida T, Miki T, Yamamoto T, Hosoe M 1996 Extractive behavior of magnesium chloride and isotopic enrichment of magnesium by a crown ether *Separation Science and Technology* **31 5** 643–654

[8] Kim D W 2001 Enrichment of magnesium isotopes by monoazacrown bonded merrifield peptide resin *Bull. Korean Chem. Soc.* **22 6** 570–574

[9] Symons E A 1985 Lithium Isotope Separation: A Review of Possible Techniques *Separation Science and Technology* **20** 633–651

[10] Kim D W, Kang B M 2001 Enrichment of magnesium isotopes by 2'-aminomethyl-15-crown-5 bonded Merrifield peptide resin *Journal of Radioanalytical and Nuclear Chemistry* **250** 291–294

[11] Levkin A V, Basmanov V V, Demin S V, Tsivadze A Yu 1990 Segregation of magnesium isotopes at extraction with crown ethers *Russ. J. Phys. Chem.* **64 5** 1376–1377

[12] Oi T, Yanase S, Kakihana H 1987 Magnesium Isotope Fractionation in Cation-Exchange Chromatography *Separation Science and Technology* **22 11** 2203–2215

[13] Kim D W 2002 Separation of lithium and magnesium isotopes by hydrous manganese(IV) oxide *Journal of Radioanalytical and Nuclear Chemistry* **252 3** 559–563

[14] Isupov V P, Mitrofanova R P, Chupahina L E 2001 Partition coefficients of lithium isotopes by chemical isotopic exchange *Chemistry for Sustainable Development* **9** 183–198

[15] Pugachev Yu I, Rylov V S 1963 Segregation of magnesium isotopes at vacuum distillation

Russ. J. Phys. Chem. **37** **3** 691–693
[16] Allred V D, Cutler I B 1953 *Chemical methods for the separation of lithium isotopes* (Oak Ridge National Laboratory) 177
[17] Chen D, Chang Z, Nomura M, Fujii T 1997 Isotope effects of magnesium in amalgam/organic solution systems *J. Chem. Soc. Faraday Trans.* **93** **14** 2395–2398
[18] Mavromatis V, Pearce C R, Shirokova L S, Bundeleva I A, Pokrovsky O S, Benezeth P, Oelkers E H 2012 Magnesium isotope fractionation during hydrous magnesium carbonate precipitation with and without cyanobacteria *Geochimica et Cosmochimica Acta* **76** 161–174
[19] Li W, Chakraborty S, Beard B L, Romanek C S, Johnson C M 2012 Magnesium isotope fractionation during precipitation of inorganic calcite under laboratory conditions *Earth and Planetary Science Letters* **333-334** 304–316
[20] Yang G, Guan Y T, Jian F Y 2015 Zone Refinement of Germanium *Crystals Journal of Physics: Conference Series* **606** 012014
[21] Sun D Y, He L W, Lian J Q 2015 Process for solar grade silicon production by molten salt electrolysis and zone melting *Journal of Materials Engineering* **43** **4** 73–78
[22] Yang G, Govani J, Mei H 2014 Investigation of influential factors on the purification of zone-refined germanium ingot *Crystal Research and Technology* **49** **4** 269–275
[23] Akimov D V, Egorov N B, Obmuch K V 2013 Separation of rare earth elements by zone recrystallization *Fundamental Research* **8** **3** 529–533
[24] Smith H A, Thomas C O 1959 The separation of mixtures of ordinary and heavy water by zone refining *J. Phys. Chem.* **63** **3** 445–447
[25] Herington E F G 1959 Zone melting with some comments on its analytical potentialities *Analyst* **84** 680–689
[26] Kuhn W, Thürkauf M 1958 Isotopentrennung beim gefrieren von wasser und diffusionskonstanten von D und ^{18}O im Eis. Mit diskussion der moglichkeit einer multiplikation der beim gefrieren auftretenden Isotopentrennung in einer haarnadelgegenstromvorrichtung *Helvetica Chimica Acta* **41** **4** 938–970
[27] Makarov A V, Koreckay T V, Panchenkov G M 1962 About boron isotope separation zone melting *Russ. J. Phys. Chem.* **36** 391–393
[28] Troitskii O A 1978 Fractionation of stable isotopes of metals by combined electric migration and zone melting *Phys. Stat. Sol. (a)* **48** 229–234
[29] Troitskii O A, Spitsyn V I, Moiseenko M M, Tsiganov A D 1979 Fractionation of stable isotopes and impurities during zone recrystallization of metals and preparation of single crystals rich in individual isotopes *Phys. Stat. Sol. (a)* **54** 651–654
[30] Akimov D V 2012 Preparation of isotopically enriched lead suitable for use in fast reactors *Russian Physics Journal* **55** **2/2** 123–129
[31] Egorov N B, Akimov D V Gerin I I, Obmuch K V 2014 Fractionation of magnesium isotopes by using zone recrystallization *Russian Physics Journal* **57** **2/2** 249–254
[32] Andrienko O S, Egorov N B, Gerin I I 2007 Magnesium isotopes selection at recrystallization of $MgCl_2 \cdot 6H_2O$ *Bulletin of the Tomsk Polytechnic University* **311** **3** 67–69
[33] Pfann V 1970 *Zone melting* (M: Mir)
[34] Ivlev S I, Akimov D V, Egorov N B, Kraus F 2016 Syntesys and charaxterization of $LiClO4 \cdot H_2O$ *Monatshfte fur Chemie* **147** 279–288

Development of an oximeter for neurology

A Aleinik[1], Z Serikbekova[1], N Zhukova[2], I Zhukova[2], M Nikitina[2]

[1] National Research Tomsk Polytechnic University, Tomsk, Russia
[2]Siberian State Medical University, Department of Neurology Tomsk, Russia

E-mail: aleinik@tpu.ru

Abstract. Cerebral desaturation can occur during surgery manipulation, whereas other parameters vary insignificantly. Prolonged intervals of cerebral anoxia can cause serious damage to the nervous system. Commonly used method for measurement of cerebral blood flow uses invasive catheters. Other techniques include single photon emission computed tomography (SPECT), positron emission tomography (PET), magnetic resonance imaging (MRI). Tomographic methods frequently use isotope administration, that may result in anaphylactic reactions to contrast media and associated nerve diseases. Moreover, the high cost and the need for continuous monitoring make it difficult to apply these techniques in clinical practice. Cerebral oximetry is a method for measuring oxygen saturation using infrared spectrometry. Moreover reflection pulse oximetry can detect sudden changes in sympathetic tone. For this purpose the reflectance pulse oximeter for use in neurology is developed. Reflectance oximeter has a definite advantage as it can be used to measure oxygen saturation in any part of the body. Preliminary results indicate that the device has a good resolution and high reliability. Modern applied schematics have improved device characteristics compared with existing ones.

1. Introduction

During the surgery a reduction of oxygen saturation of the brain can occur, with the other controlled parameters change slightly. Prolonged intervals of cerebral anoxia may damage the central nervous system. General anesthesia has become safer in recent years due to the emergence of more reliable methods of control [1]. These technologies make it possible to identify early onset of adverse events and significantly reduce morbidity and mortality from the use of anesthetics [2, 3]. Quantitative assessment of physiological parameters permits to reduce the effects of the factors associated with a poor outcome. Pulse oximetry is a fast, noninvasive method for monitoring the oxygen saturation of a patient's blood. Pulse oximeter provides to diagnose problems in the respiratory and circulatory systems. Commonly used method for measurement of cerebral blood flow is the Kety-Schmidt method [4]. In this case, the patient inhales a mixture of O_2 and N_2O, and blood is sampled using invasive catheters.

Stimulation of the sympathetic nervous system leads to vasoconstriction and increased activity of the heart, while vagal activity to the heart is inhibited. At the same time some changes occur simultaneously: arterial pressure increases first, because of arterial constriction; then venous constriction displaces blood towards the heart and increases cardiac output and as a consequence stimulated heart increases the heart rate and the force of contraction. In the case of anesthesia the patient can not feel anything, but surgical intervention causes the typical sympathetic system reaction to pain.

Heart rate variability (HRV) may be used as an effective indicator of sympathetic tone and autonomic nervous system activation. In this case very low frequency in the HRV signal corresponds to both sympathetic and parasympathetic activity. At the same time higher frequency components correspond to only parasympathetic activation [5]. The aim of this study is to investigate the use of the reflectance pulse oximeter for remotely detecting a sudden sympathetic system changes.

2. Materials and experimental

To measure changes in sympathetic activity the reflectance pulse oximeter is developed. The block-diagram of the device is shown in Figure 1.

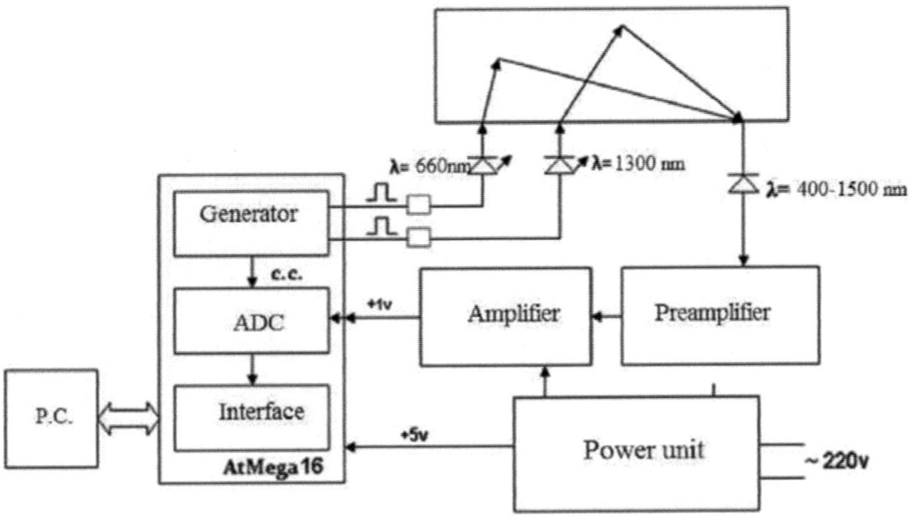

Figure 1. The block-diagram of the device

The main control element is a microcontroller Atmega16. The signal is processed by10 bit ADC. Each LED is controlled via a multiplexer. The main objective is to extract the AC signal with the exception of its constant part. This task is accomplished via an operational amplifier. The main contribution to the noise is ambient light with a frequency of 50 Hz and 100 Hz. Necessary information contained in the signal lies in the range of 0-30 Hz. Typically, LEDs are directly controlled by voltage. The disadvantage of this method is the dependence of the amplitude on the temperature of the LED. To eliminate his shortcoming the current source is designed. To improve the signal/noise ratio filtering is needed. As smoothing filter the moving window is selected. Such a filter is easily accomplished with macros.

3. Results and discussion

The device was attached to the forehead of volunteer by using the elastic band. Then the measurements of oxygen saturation during normal breathing and breath delay for 30s were made. The results are shown in Figure 2. As can be seen from the graph a breath delay leads to a reduction in oxygen saturation.

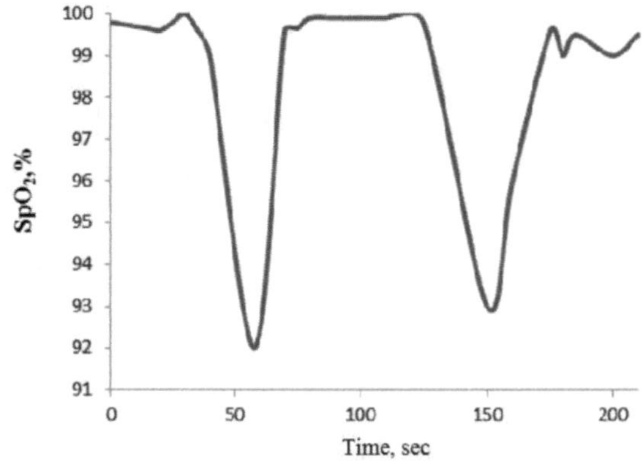

Figure 2. Typical view of hypoxia presence

A rapid change in sympathetic tone can be simulated by a sudden decrease in arterial pressure. The classic supine-standing experiment was performed. In this experiment, a person initially is in the supine position and then stands up suddenly. This rapid change in body position causes a sudden drop in arterial pressure due to the effects of gravity pulling blood into the lower extremities. This process activates the baroreceptor reflex, and as a consequence sympathetic arterial constriction and increased heart rate. The results are shown in Figures 3 and 4. It can be seen that the device detects the change in sympathetic tone confidently. After changing the body position the heart rate first decreases and then increases with subsequent return to the normal value.

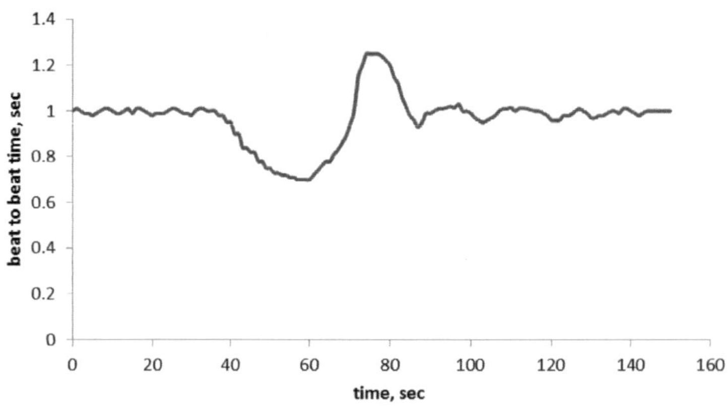

Figure 3. Beat-to-beat interval

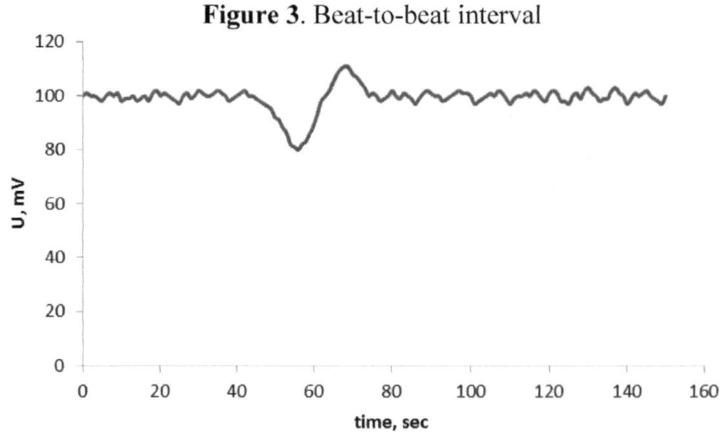

Figure 4. Peak value of PPG for each pulse

Changing the PPG is delayed, which is necessary for the blood flow to reach the point of measurement.

4. Summary

The reflection pulse oximeter to detect sudden changes in sympathetic tone is developed. The device has a definite advantage as it can be used for measurement in any part of the body. Preliminary results indicate that the device reliably detects changes in sympathetic tone, has a good resolution and high reliability. Modern applied schematics have improved device characteristics compared with existing ones.

Acknowledgements

This work was partly supported by Competitiveness Enhancement Program of Tomsk Polytechnic University.

References

[1] Olsen K, Svendsen L, Larsen F. 1996 Validation of transcranial near-infrared spectroscopy for evaluation of cerebral blood flow autoregulation. *J. Neurosurg. Anesthesiol.* **8** pp 280-285.

[2] Kim M, Durduran T, Frangos S, Edlow B, Buckley E, Moss H, Zhou C, Yu G, Choe R, Maloney Wilensky E, Wolf R, Grady M, Greenberg J, Levine J, Yodh A, Detre J, Kofke W 2010 Noninvasive measurement of cerebral blood flow and blood oxygenation using near-infrared and diffuse correlation spectroscopies in critically brain-injured adults. *Neurocrit Care* **12** pp 173-180.

[3] Henson L, Calalang C, Temp J, Ward D 1988 Accuracy of a cerebral oximeter in healthy volunteers under conditions of isocapnic hypoxia. *Anesthesiology* **88** pp 58-65.

[4] Taudorf S, Berg R, Bailey D, MÃller K. 2009 Cerebral blood flow and oxygen metabolism measured with the Kety-Schmidt method using nitrous oxide. *Acta Anaesthesiol Scand* **53** pp 159-16.

[5] Bianchi A and et al. 1993 Time-variant power spectrum analysis for the detection of transient episodes in hrv signal. *IEEE Transactions on Biomedical Engineering*, **40** pp 136–144.

IOP Conf. Series: Materials Science and Engineering **135** (2016) 012003 doi:10.1088/1757-899X/135/1/012003

Communication research between working capacity of hard-alloy cutting tools and fractal dimension of their wear

K Arefiev, V Nesterenko, N Daneykina

National Research Tomsk Polytechnic University, Tomsk, Russia

E-mail: nest2004@ bk.ru

Abstract. The results of communication research between the wear resistance of the K applicability hard-alloy cutting tools and the fractal dimension of the wear surface, which is formed on a back side of the cutting edge when processing the materials showing high adhesive activity are presented in the paper. It has been established that the wear resistance of tested cutting tools samples increases according to a fractal dimension increase of their wear surface.

1. Introduction

To produce details and knots of nuclear reactors the steels of an austenitic class are used, usually processed with cutting tools of the K applicability groups. High working capacity of these tools can provide high standards and safety of production. Therefore, it is important to forecast wear resistance of the cutting tools at their use.

The structure of firm alloys of the K applicability group at their production and operation is exposed to saturation by various gas elements of the surrounding atmosphere [1]. The interaction intensity of firm alloy components with atoms and molecules of the gas environment is defined by the technology of their preparation, impurity existence in their composition, and the modes of composite agglomeration in general and the conditions of their operation. The covalent, ionic and hydrogen bonds are formed along with metal communication between atoms in components structure of a firm alloy. Thereof, the destruction of a hard-alloy surface contacting to the processed material, due to the intensive intermolecular interaction caused by adhesive processes will be essentially heterogeneous. At the zone of hard-alloy material wear, as a rule, there are the signs of fragile, viscous and mixed types of the destruction, arising through separation of elementary fragments of carbide and cobalt phases. Primary the character of this or that type of destruction assumes the formation of some microrelief of the wear and its fractal dimension. A preliminary radiation exposure of the hard-alloy cutting plates by gamma rays influences components interaction with gas elements appearing as a result mainly at high temperatures, and predetermines some change in the destruction nature of a surface and the microrelief formation with fractality of a bit different size in comparison with control samples.

2. Experiment technique

The research of wear resistance of the hard-alloy cutting plates of the K applicability group was carried out on a thread-cutting lathe. The steel 12X18H10T causing an intensive adhesive wear was used as the processed material. The processing was made by the hard-alloy cutting plates of the industrial BK6 brand, relating to the above specified group K. Test cutting speed was 70 m/mines and, approximately, it was an optimum one (providing the minimum intensity of wear). The speed of the longitudinal movement of a cutter – 0.23 mm/turn (I), the cutting depth – 1.5 mm. Wear resistance was estimated in minutes (T) of operating time of the cutting tool to the established criterion of a tool dulling (extent of loss of the cutting properties) – wear on a back surface of the cutting wedge equal to 0.8 mm.

The gamma ray exposure of the cutting tools was made in a special radiation unit with a fluency near of 10^{17} gamma-quant/cm^2. The morphology of a wear surface was investigated using the contactless three-dimensional MICROMEASURE 3DSTATION profile meter of the French firm STIL. Investigation of a wear surface was made by the pulse light bunch generated by a halogen lamp with a frequency of 100 Hz. This frequency provided the most accurate reproduction of a profile of wear. The sizes of ledges and hollows of a microrelief were defined from a difference of the intensity of falling and reflected light. The roughness of a profile of a surface was estimated based on the results of its measurement on the piece equal 0.5 mm. The received profiles processing was made with a special program, being a part of a profile meter. The size of fractal dimension was defined in the automatic mode, based on the obtained data on a roughness of the measured profile. Roughness control and the determination of fractal dimension were made on the wear facet located on a cutting plate back surface in its central zone. The places of diagnosing were constants for all workout cutting plates subjected to coordinate control.

3. Results and discussion

Processing of the materials causing an intensive adhesive wear is carried out, as a rule, with the K applicability group hard-alloy cutting tools, which are exposed to intensive oxidation by atoms and molecules of oxygen of the surrounding gas environment. The formed superficial oxide structures carry out a role of solid greasing when cutting and shield intermolecular interactions. They also have a great influence on process of formation of development degree of a contact surfaces roughness of a cutting tool. Processes of adsorption of oxygen and other gas elements have a great influence on formation of oxide films, their destruction because of adhesion and formation of a roughness on contact surfaces of the cutting tool. As measurements and control showed, a microrelief of a formed wear surface possesses fractal properties [2]. It is shown that the area of rough surface is not aspiring to its final limit, but increases sedately at a reduction of measurement scale. The main characteristic of fractal surfaces is their fractal dimension. The degree of development of rough structure of a wear surface - fractal dimension is closely connected with mechanochemical and electro physical properties of a surface and near-surface layer.

Fig. 1 shows dependence of time change of the cutting plates functioning on the established criterion of a tool dulling, depending on size of fractal dimension created on their facet of wear. It follows that the increase in roughness development of a wear facet, fractal dimension leads to an increase of wear resistance of cutting plates.

The initial degree of dispersion of carbide grains in structure of a firm alloy, their distribution by sizes, composition of carbide grains, structure and properties of a cobalt sheaf, structure and degree of deficiency of interphase borders, availability of impurity, sorption properties of carbide grains and the formed microrelief as well as electric and thermal properties of a surface and volume have their impact on the formed fractal dimension on a wear surface. The vacancy deficiency of the workout material structure has also its influence on the size of the formed development of microrelief, and its specific surface also. The deficiency increase promotes adsorption intensification with a wear surface of atoms and molecules of surrounding gas environment.

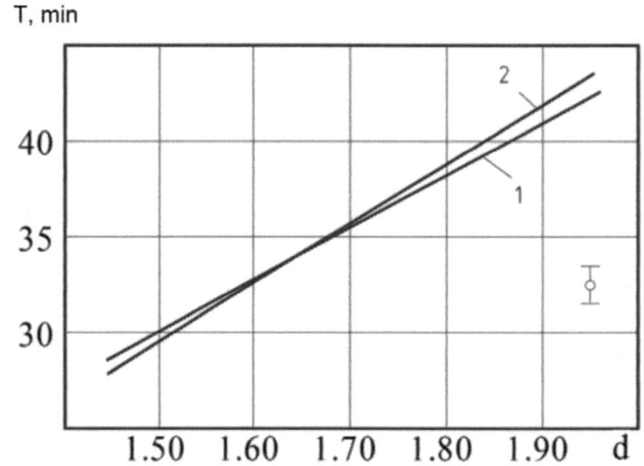

Figure 1. Change of wear resistance of the cutting plates from a firm alloy of BK8 when cutting steel 12X18H10T depending on the size of fractal dimension of a facet of wear: 1 – control plates, 2 – control plates previously processed by gamma rays.

The formation process of a surface microrelief, the wear facet of hard-alloy groups of the K applicability of the cutting plates can be compared to the process of formation of a microrelief on a surface with the phenomena of self-organization of a matter in areas of coverage of high temperatures and pressure. Self-organization process, formation of a microrelief, necessary for degree of development, assumes, in turn, adsorption of oxygen, necessary for intensity, and sufficient oxidation of contact surfaces of the cutting plate [2]. The formed oxide films thus shield processes of adhesion, transfer heat, weight and are solid greasing. Owing to realization of an order of self-regulation when machining materials increase in coefficient of friction, components of forces of cutting decline sharply in quality of the processed surface coming at higher values of the parameters of characterizing process of destruction of the cutting wedge, and the general wear resistance of the cutting tools increases.

The process of adsorption of molecules of oxygen contact surfaces of the hard-alloy cutting tools of the K applicability group consists of two stages [3]. The first stage includes physical adsorption. Its intensity is connected with extent of polarization of molecules of oxygen by a wear surface on the cutting wedge. In turn, the polarizing ability of a surface depends on development of a microrelief. With an increase in a microrelief development of a wear facet degree of the concentration of oxygen on its nominal surface wear unity increases. Under certain conditions, the physical adsorption of oxygen on surfaces transforms into a chemical one. The second stage of adsorption includes a chemical interaction of oxygen atoms with atoms of contact surfaces. As a result, chemical compounds of oxygen with carbide of tungsten and cobalt are formed on its surface. Superficial oxide film structures are formed on the basis of covalent or ionic types of communications formed between atoms.

Many internal and external factors influence the formation of a certain type of chemical bonds in oxide structures. With growth of free carbon as a part of a firm alloy, and some impurity alkaline and the alkaline earth metals (potassium, sodium, calcium), a probability of complex oxide formation compounds increases, which are formed due to interaction of several oxides [4]. In this case, superficial and near-surface structures of oxides are formed, for example, on the basis of tungsten, cobalt and other elements. The share of covalent communication between elements increases and their destruction happens in the zone of contact of the cutting and processed materials. Then formation of the increased development of a microrelief of a wear facet occurs. Oxide destruction has the subsequent disperse nature; oxide carbide particles and conglomerates of a surface also promote their more effective role as solid greasing in contact zones.

Thus, the development degree of the formed microrelief of wear platforms on contact surfaces will depend on a character of the formed chemical bonds between elements of a surface of carbide grain and oxygen, and the intensity of the fragile destruction happening in micro volumes of superficial and near-surface structure of a firm alloy. In addition, both factors, a character, and the process of formation of chemical bonds, and the subsequent destruction of the created superficial and near-surface structures are closely connected and have an essential mutual impact. A low viscosity coefficient of destruction of carbide grains has a great influence on a degradation of a surface layer and the formation of the developed microrelief, which is characterized by high fractal dimension.

It provides, on one hand, a low crack resistance of oxide and oxide carbide layers on a carbide component, and on the other- an effective branching of the formed superficial cracks.

Big ability to cracks branching, formed at destruction of near-surface structure, is provided by high degree of vacancy deficiency of a firm alloys structure. Thereof there is also a considerable concentration of vacancies, small and large pores in superficial oxide and the oxicarbide structures, which are formed on contact surfaces of the cutting tool. The main part of dot defects in oxidic and oxicarbide tungsten films is made by cationic vacancies, and in oxidic films of cobalt by anion vacancies [4]. With increase in concentration of both types of vacancies, at simultaneous decrease in their associations (small and large pores), development of a microrelief at destruction of superficial structures increases. Intensity of capture of atoms and molecules of oxygen a contact tool surface (contact happens to the processed material) increases with pressure decrease in intercontact area (space between contact surfaces). The pressure in intercontact area is lower; the capture is stronger, than the atoms get into this area and molecules of oxygen. Process of capture happens in the conditions of the high competition between various active centers - the centers of capture of a surface.

The most probable centers of capture are the following structural defects: anion vacancy in a combination with an electron in structure of an oxidic film of tungsten; cationic vacancy in a combination with a hole in structure of an oxidic film of cobalt; an impurity ion and cationic vacancy in a combination with anion vacancy in structure of oxide of tungsten; an impurity ion and anion vacancy in a combination with cationic vacancy in structure of oxide of cobalt, the atomic centers in crystals with monovalent cationic impurity; ions of one and bivalent impurity in anion vacancy in structure of oxide of tungsten, periodic process of transition of an impurity ion of a cationic sub lattice in anion and vice versa and process of deformation hashing of cationic and anion sub lattices [5]. The considered static and dynamic active centers have the lowered energy of ionization and actively interact with oxygen molecules.

Process of superficial diffusion of the captured atoms and molecules of oxygen is one of the main stages in realization of effective adsorption of oxygen. In the course of diffusion atoms and molecules of oxygen find optimum in the thermodynamic relation of situation with the purpose of formation of connection with metal.

Formation of oxides leads to decrease in superficial energy and adhesive activity in relation to the processed material. Thus, the probability of local destructions on a surface from action of adhesion decreases and, respectively, the size of fractal dimension of a microrelief is stabilized.

The size of the fractal dimension formed at a wear surface depends little on the time of cutting and significantly depends on the cutting speed temperature.

The most developed rough surface on a back side of the cutting wedge (a wear facet) is formed at some optimum temperature, corresponding to optimum speed of cutting. The development of a microrelief of a surface of wear is lower below and above the optimum temperature of operating in a contact zone on a back surface of the cutting tool, and the size of fractal dimension is less, than that which turns out at an optimum speed of cutting. At temperatures, below optimum, low development of a microrelief of a surface of wear and small size of fractal dimension are caused by insufficiently high intensity of oxidation of near-surface area of a firm alloy. It leads to the formation of oxidic films of poor quality – with insufficiently high shielding ability to resist to adhesive manifestations and to carry out lubricant function. The inefficiency of oxidic structures is a consequence of low level of their stoichiometry. The oxygen deficiency is formed owing to insignificant activity of the adsorptive processes and including their second stage.

At temperatures, (speeds) above optimum, low development of a microrelief of a wear surface and small size of fractal dimension are caused by intensive formation of oxidic films and their intensive sublimation. It occurs owing to high activity of the adsorptive processes, high speed of oxidation of a surface and high speed of evaporation of the created oxides. Both in the first and in the second cases on contact surfaces of the cutting wedge, owing to intensive adhesion the facet surface of wear which is characterized by the low size of fractal dimension is formed.

A preliminary radiation processing of the cutting plates with gamma rays changes somewhat the size of fractal dimension of a wear surface. Its influence depends probably on a condition of internal structure of firm alloys: availability of impurity in components, initial extent of internal oxidation of structure, character and degree of deficiency of components. There are some options of samples with slight increase and decrease of fractal dimension in a party of tested cutting plates in relation with samples in a control party.

4. Conclusion

It is possible to assume that radiation has its impact on the sorption processes, which in turn are generally a function of a structure condition of firm alloy components. On the basis of stated above it is possible to draw conclusions that the wear resistance of the hard-alloy cutting plates of the K applicability group is defined by of a degree of superficial and near-surface oxide and the oxicarbonic structures formed. With the increase in fractal dimension of a wear surface the duration of trouble-free functioning of the cutting plates has a tendency to growth.

References

[1] GOST 3882 – 74
[2] Roldugin V I 2008 *Fisikochimija poverchnosti* (M.: Intellekt) p 563
[3] Krilov O V 1981 *Adsorbzija i katalis na perechodnich metallach i ich ocsidach* (M.: Chimija) p 286
[4] West A 1988 *Chimija tverdogo tela* (M.: Mir) p 588
[5] Parfenovich I A 1977 *Elektronnie zentri okraski v ionnich cristallach* (Irkutsk) p 208
[6] Tretykov Yu D 1978 *Tverdofasnii reakzii* (M.: Chimija) p 359

IOP Conf. Series: Materials Science and Engineering **135** (2016) 012004 doi:10.1088/1757-899X/135/1/012004

Analysis constants for database of neutron nuclear data

S V Bedenko[1], J Joseph Jeremiah[2], V V Knyshev[1] and I V Shamanin[1]
[1]National Research Tomsk Polytechnic University, Tomsk, Russia
[2]Madras Christian College, Chennai, Indian

E-mail: bedenko@tpu.ru

Abstract. At present there is a variety of experimental and calculation nuclear data which are rather entirely presented in the following evaluated nuclear data libraries: ENDF (USA), JEFF (Europe), JENDL (Japan), TENDL (Russian Federation), ROSFOND (Russian Federation). Libraries of nuclear data, used for neutron-physics calculations in programs: Scale (Origen-Arp), MCNP, WIMS, MCU, and others. Nevertheless all existing nuclear data bases, including evaluated ones, contain practically no information about threshold neutron reactions on ^{232}Th nuclei; available values of outputs and cross-sections significantly differ by orders. The work shows necessity of nuclear constants corrections which are used in the calculations of grids and thorium storage systems. The results of numerical experiments lattices and storage systems with thorium.

1. Introduction

Accuracy of cross-sections evaluation of neutron interaction in fuel compositions is very important for reliable evaluation of uranium-thorium nuclear fuel cycle (NFC). At present there is a variety of experimental and calculation nuclear data which are rather entirely presented in the following evaluated nuclear data libraries: ENDF (USA), JEFF (Europe), JENDL (Japan), TENDL (Russian Federation), ROSFOND (Russian Federation).

One of the most valuable sources of information is the international library ENDSF (Evaluated Nuclear Structure Data File), which provides recommended data as well as initial experimental data's. It should be noted that the libraries of such nuclear data are used for neutron-physical calculations in such precision programs as Scale (Origen-Arp), MCNP, WIMS, MCU and other.

Nevertheless all existing nuclear data bases, including evaluated ones, contain practically no information about threshold neutron reactions on ^{232}Th nuclei; available values of outputs and cross-sections significantly differ by orders. If we consider nuclear data base JENDL-4.0 as an example, in which during neutron data preparation for thermal and epithermal the energies of ^{232}Th fission cross-sections, were prepared based on simple physical consumptions, which give the dependence of $\sim 1/\upsilon$ type. There is no sufficient data necessary for calculation of threshold neutron reactions on ^{232}Th nuclei in the nuclear data base ENDF/B-VIII.0, which is recommended by the majority of computing engineers. Thus it is difficult to speak about practical certainty of existing data for threshold reactions on ^{232}Th nuclei in the range of thermal energies and resolved resonances.

Current situation concerning cross-sections of radiation capture is perfectly demonstrated by the researches carried out in Japan (Kyoto university, 2004) [1]. The comparison of the values of cross-sections radiation capture on ^{232}Th nuclei was performed using SRAC system code from evaluated nuclear data libraries (JENDL-3.3, JENDL-4.0, ENDF/B-VIII, JEFF 3.1.1). Result shows that there is significant gap (up to ~40%) between the libraries.

Considering the importance of thorium data and concerning about the accuracy of ^{232}Th cross section libraries, a series of experiments on thorium critical core with different neutron spectra has been implemented at Kyoto University Critical Assembly [2]. Calculations of sensitivity coefficient of ^{232}Th reactivity were conducted and its result shows the more contribution of ^{232}Th cross section uncertainty at the thermal range to the calculation error. It suggests that ^{232}Th capture cross section is needed to be adjusted at thermal energy range to obtain more reliable calculated data.

It is obvious that plurality of information sources, their imperfection and disagreement can influence the results of critical calculations of the fuel-rod arrays and systems with thorium, mainly because in calculation of k_{eff} the required functionals have dependences $\sigma_i(E,r,\Omega)$.

In researches carried out for multiplying thorium-containing systems {(m%U,n%Th)O_2, (m%Pu,n%Th)O_2} operating on thermal neutrons the accuracy evaluation of available methods of resonant effects description was made. The calculation values showed that formalism describing the processes of neutrons interaction with nuclei, including absorption and scattering did not provide satisfactory agreement with the experimental data (Germany, Institute ISR-2 of the Forschungszentrum Julich, 1999) [3, 4].

Therefore there is vital necessity have precise nuclear data for calculating resonant absorption parameters in fuel-rod arrays ad systems containing thorium.

During numerical experiments on the study of reactor system physics, the most important is the constant provision state. Constant provision to calculation of slow nuclide kinetics even for uranium-plutonium fuel cycle is far from perfect. In the case of reactor systems with thorium nuclear fuel calculations is necessary to take into account imperfection of constant provision [5, 6].

It should be noted that the purpose of the work is to attract attention to the problem of using thorium in nuclear fuel cycle during execution of the new technology platform of nuclear power industry in Russia.

2. Analytical model of nuclear-physical processes in thorium-containing systems

The neutron interaction mechanism in the region of allowed and not allowed resonances, resonant interaction effects belong to the class of wave processes and can be considered as the product of the effective neutron "size" $\pi(\lambda_n/2\pi)^2$ and the transmissivity p of the phase surface formed by [7, 8]:

$$\sigma_c = \pi(\lambda_n / 2\pi)^2 \cdot p, \qquad (1)$$

where the permeability $p = |J_1| / |J_2| = 4k_1 \cdot k_2/(k_1 + k_2)^2$ can be found from the Schrodinger equation, $|J_1|$ and $|J_2|$ are the probability density vectors of the transmitted and incident waves, and k_1 and k_2 are
the corresponding wave numbers of neutron.

At a stage of calculated estimates, we used the Fermi gas model within which the potential $V_c(r)$ should be presented as

$$V_c(r) = V_r(r) + iV_m(r), \qquad (2)$$

where the real and imaginary parts of expression (2) are responsible for elastic and inelastic channels ofnuclear interactions, respectively.

One of such potentials is the Fermi pseudopotential

$$V_c(r) = \frac{2\pi\hbar^2}{\mu} \cdot b \cdot \delta(r) = U \cdot \delta(r), \qquad (3)$$

where $\delta(r)$ is the Dirac delta function, b is the scattering length, and the parameter $b = x + iy$ is a complex quantity in the general case.

The parameter b in Eq. (3) was selected so that the calculated potential well depth U would correspond to the data of [7].

IOP Conf. Series: Materials Science and Engineering **135** (2016) 012004 doi:10.1088/1757-899X/135/1/012004

Taking into account relation (1) and the wavenumbers k_1 and k_2 found in solving the Schrödinger equation, the cross sections for ^{238}U and ^{232}Th were estimated in the energy range from 10^{-3} to 24 eV.

The found cross sections were compared with the data given in the ENDF/B-VII, JENDL-4.0, JEFF-3.1, and TENDL-2013 libraries. The results of comparisons are shown in Fig. 1 and 2.

The uncertainty in fission (σ_f) and radiative capture (σ_γ) cross sections for ^{238}U in the energy regions to 2.9 eV and 1.75 eV, respectively, does not exceed 30% (see Fig. 1). For ^{232}Th, the cross sections $\sigma\gamma$ with an accuracy of 30% can be found in the energy range from 10^{-3} to 0.8 eV (see Fig. 2).

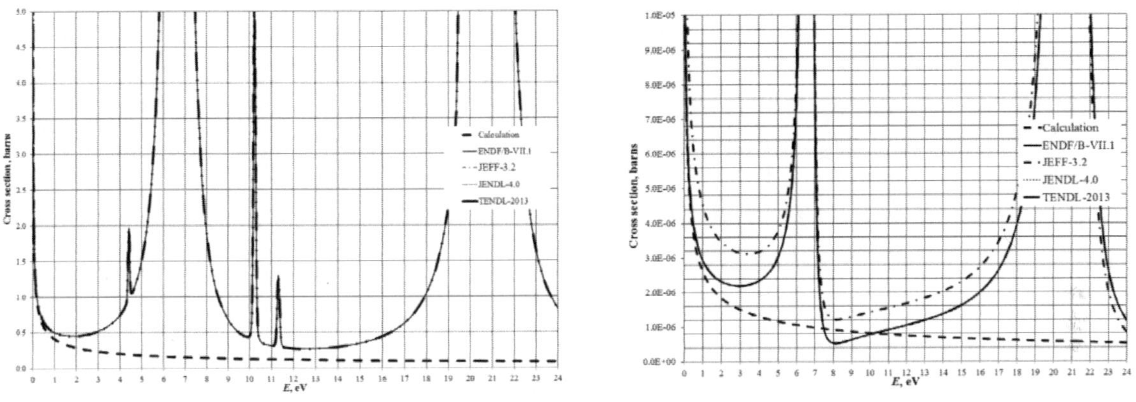

(a) (b)

Figure 1. The results of calculated ^{238}U cross-sections are compared with evaluated nuclear data (ENDF/B, JEFF-3.2, JENDL-4.0, TENDL-2013) (a) Dependence of cross-section σ_γ of ^{238}U nucleus in the neutron energy range from 10^{-3} to 24 eV (b) Dependence of cross-section σ_f of ^{238}U nucleus in the neutron energy range from 10^{-3} to 24 eV.

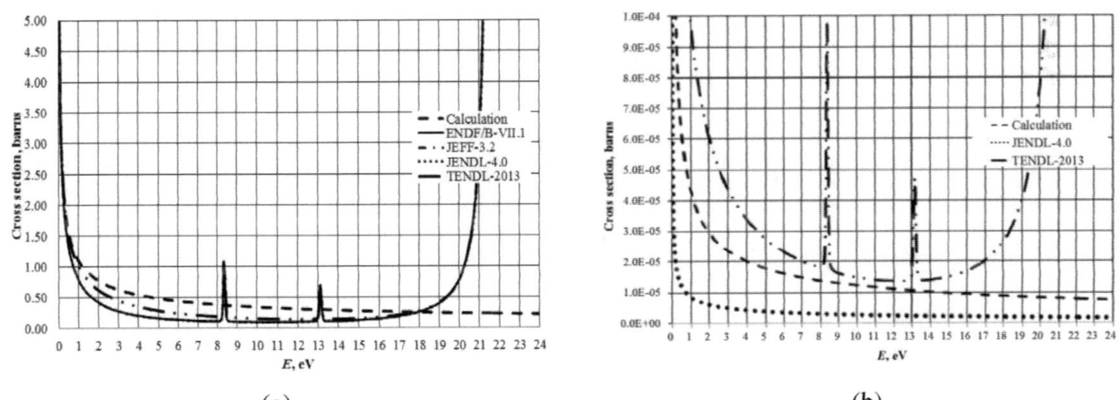

(a) (b)

Figure 2. The results of calculated values of ^{232}Th cross-sections are compared with evaluated nuclear data (JENDL-4.0, TENDL-2013): (a) Dependence σ_γ of ^{232}Th nucleus in the neutron energy range form 10^{-3} to 24 eV (b) Dependence of ^{232}Th nucleus cross-section in the energy range from 10^{-3} to 24 eV.

We note that the cross sections σf in the energy region to 4 eV, given in the TENDL for ^{232}Th differ significantly from those in the JENDL-4.0 (see Fig. 2), and the presence of the first two allowed resonances in the energy range from 6 to 14 eV seems doubtful.

In general, the structure of nuclear ^{232}Th absorption resonant region is enough in detail explored in the works [1, 2, 7–11]. Table №1 shows nuclear ^{232}Th absorption resonant region parameters convolution results for nuclear ^{232}Th in the range from 4 eV to 2,15 keV [9].

Table 1. Characteristics of ^{232}Th resonances.

Number group	The boundaries of the energy interval, eV	Number of resonance	E_r(eV)	σ_r,(b)	0,5Γ (eV)	$\overline{\sigma_r}$ (b)	$\overline{E_r}$ (eV)	$\overline{\Gamma}$ (eV)
21	4.65…10	–	–	–	–	–	–	–
20	10...21.5	–	–	–	–	–	–	–
19	21.5…46.5	1	22.5	1000	2	1000	34	4
18	46.5…100	2	60	1000	1.25	1000	73.3	4.25
		3	68	1000	3			
17	100…2150	4	114	425	1	405	157.5	2.2
		5	121.5	600	1			
		6	129	200	0.5			
		7	171	600	1.5			
		8	193	330	1.25			
		9	199	275	1.35			

Further nuclear-physical researches were carried out considering analytically corrected nuclear data and MCU5 program code [12].

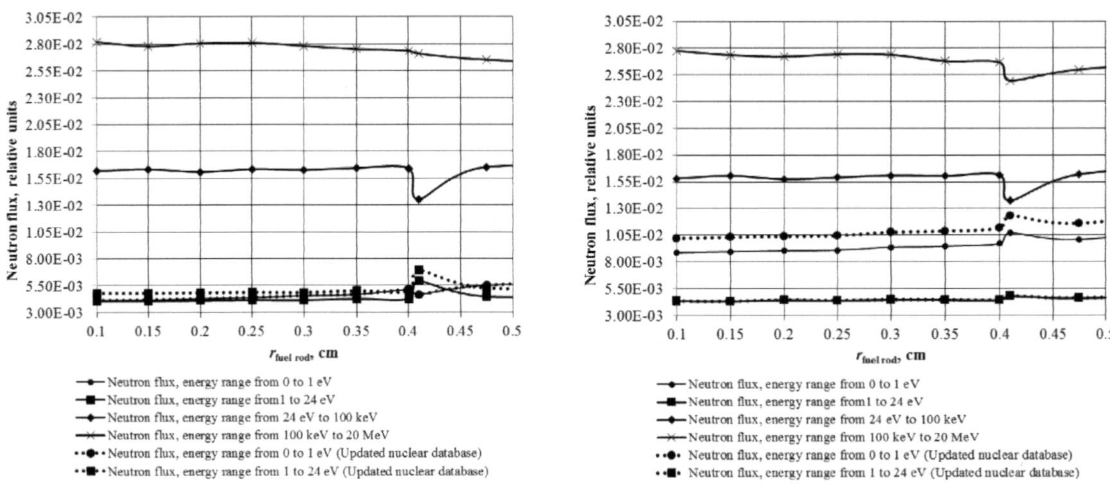

(a) (b)

Figure 3. Space-energy distribution of neutron flux density in fuel element semi-array in IGR (a) Neutron flux distribution in fuel-rod arrays with thorium (5%Pu, 95%,Th)O_2; (b) Neutron flux distribution in fuel-rod arrays with thorium– (5%U, 95%Th)O_2.

Using the calculation code of MCU5 and evaluated data library files (ENDF/B and JENDL-4.0) and analytically calculated cross-sections values the nuclear constants used for determination of k_{eff} and evaluation of necessary functionals (σ_i(E_i,r_i,Ω) и φ(ΔE_i,r_i)) were found. During preparation of the nuclear constants the technology developed in Nuclear Safety Institute, Russian Academy of Sciences

[13] was used. It is worth noting that macro constants for the spectrum part from 18 eV to 100 keV were determined using the MCU5 program code without correction of the nuclear data.

Such kind of approach allows not only to conduct correction of nuclear constants but to take into account fine structure of neutron absorption by nuclear ^{232}Th resonant region.

So we used the maximum opportunities theoretical analysis and nuclear data, which have long been received, evaluated and do not cause doubts. The results of calculations for multiplying systems with thorium are presented graphically in figure 3.

Neutron-physical researches of storage systems containing thorium were also carried out by the authors considering analytically corrected nuclear data. The calculation of k_{eff} and $\varphi(\Delta E_i, r_i)$ was made in the multigroup approximation using subgroup parameters during calculation of group coefficients in the decomposed Boltzmann equation (Scale 4.4) with the used of calculation methods [13, 14].

Figure 4 shows one of the DSSNF loading layouts allowing a decrease in the neutron radiation level at critical storage points to maximum allowable values (NRB-99/2009) [6]. The effective multiplication factor of the simulated system is $k_{eff} = 0.2148 \pm 0.0003$.

	1	2	3	4	5	6	7	8	9	10	11	12	13	14	15	16	17	18	19	20	21	22
11	5	5	5	5	5	5	5	5	5	5	5	5	5	5	5	5	5	5	5	5	5	5
10	39	1	1	1	1	1	1	1	1	1	1	1	1	1	1	1	1	1	1	1	1	5
9	40	1	1	1	1	2	2	2	2	2	2	2	2	2	2	2	2	2	1	1	1	5
8	41	1	1	1	2	2	2	3	3	3	3	3	3	3	3	2	2	2	1	1	1	5
7	42	1	1	1	2	2	3	3	3	3	3	3	3	3	3	3	2	2	1	1	1	5
6	43	1	1	1	2	3	3	3	3	3	3	3	3	3	3	3	3	2	1	1	1	5
5	44	1	1	1	2	2	3	3	3	3	3	3	3	3	3	3	2	2	1	1	1	5
4	45	1	1	1	2	2	2	3	3	3	3	3	3	3	3	2	2	2	1	1	1	5
3	46	1	1	1	1	2	2	2	2	2	2	2	2	2	2	2	2	2	1	1	1	5
2	47	1	1	1	1	1	1	1	1	1	1	1	1	1	1	1	1	1	1	1	1	5
1	48	49	50	51	52	53	54	55	56	57	58	59	60	61	62	63	64	65	66	67	68	69

(a)

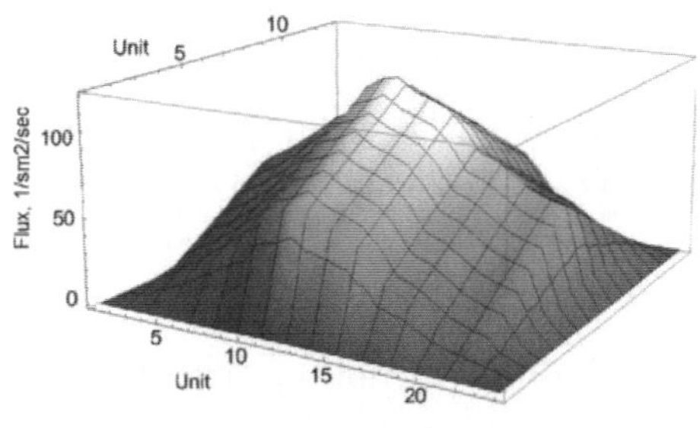

(b)

Figure 4. Results of calculations of k_{eff} and $\varphi(\Delta E_i, r_i)$ of DSSNF systems with modified UGR fuel: (a) array 22×11 in size. Units 1, 2, 3 are seats with canisters, burnup fractions are 20, 25, 30 GW·day/tHM, respectively; Units 5, 39-69 are empty seats; (b) fast neutron flux as a function of unit layout in the array; INF storage is 10 years.

3. Conclusion

The work shows necessity of nuclear constants corrections which are used in the calculations of grids and thorium storage systems. The results of numerical experiments lattices and storage systems with thorium.

The performed theoretical studies and numerical experiments will make it possible to increase ecological, nuclear, and radiation safety of dry storage and transport systems for ceramic irradiated nuclear fuel of next generation reactor facilities.

Furthermore, the performed studies will allow the development of technical and regulating solutions for handling of promising irradiated nuclear fuel.

Acknowledgments

The study was supported by the state. GK 14411.9990019.05.110 of 11.21.2014. Topic: Possibility justification and program development of thorium-bearing fuel materials used in nuclear power systems.

References

[1] Shiroya S and Kudo K 2004 Criticality Analysis of Highly Enriched Uranium/Thorium Fueled Thermal Spectrum Cores of Kyoto University Critical Assembly *PHYSOR 2004.*

[2] Vu T M, Fujii T, Wada K, et al 2013 Accuracy of thorium cross section of JENDL-4.0library in thorium based fuel core evaluation *Annals of Nucl. Energy* **57** 173–178.

[3] Shamanin I V, Ukhov A A, Rutten H J, and et al 1999 The Use of (Th,U,Pu)O_2 Fuel in a Water Water Energy Reactor: Physics and Fuel Cycle Simulation by means of the V.S.O.P. (97) Computer Code FZJ **1** 40.

[4] Shamanin I V, Ukhov A A, Rutten H J, et al 2000 Rezultaty modelirovaniya parametrov toplivnogo cikla dlya vodo-vodyanogo energeticheskogo reaktora WWER-1000 s zagruzkoj torijuranplutonievyx toplivnyx kompozicij *Izv. Vyssh. Uchebn. Zaved. Yadernaya Energetika* **4** 53–64 (in Russian).

[5] Shamanin I, Chertkov Y, Bedenko S, Gubaydulin I 2015 Gas-Cooled Thorium Reactor with Fuel Block of the Unified Design *Adv. in Mater. Science and Eng.* **2015** 1−8 doi:10.1155/2015/392721.

[6] Plevaka M N, Bedenko S V, Gubaydulin I M, Knyshev V V 2015 Neutron-physical studies of dry storage systems of promising fuel compositions *Bull. Lebedev Phys. Inst.* **42(8)** 240–243.

[7] Mengoni A, Nakajima Y Fermi-gas model parametrization of nuclear level density 1994 *J. Nucl. Sci. Tech.* **31(2)** 151–162.

[8] Koning A J, Delaroche J P Local and global nucleon optical models from 1 keV to 200 MeV 2003 *Nucl. Phys. A* **713(3-4)** 231–310.

[9] Shamanin I V 2006 Sledstvija, vyzvannye otlichijami struktur rezonansnoj oblasti pogloshhenija jader ^{238}U i ^{232}Th *Al'ternativ. jenerg. jekol.* **11(43)** 47–53 (in Russian).

[10] Joseph Jeremiah J, Jyrwa B M Pöschl Teller model for the total cross-section of neutron scattering from ^{232}Th 2012 *Indian J. Pure Ap. Phys.* **50(5)** 289–294.

[11] Capote R, Chiba S, Quesada J M, et al 2008 A global dispersive coupled-channel optical model potential for actinides *J. Nucl. Sci. Tech.* **45** 333–340.

[12] Oleynik D S, Shkarovskiy D A, Gomin E A, et al 2012 The status of MCU-5 *Phys. of Atomic Nuclei* **75(13)** 1634–1646.

[13] Blokhin D A, Mitenkova E F, Blokhin F I 2011 Podgotovka polnyh bibliotek jadernyh dannyh v potochechnom predstavlenii na osnove fajlov ocenennyh dannyh ENDF/B-VII.0, JEFF-3.1.1, JENDL-4.0 (***Preprint*** № IBRAE-2011-08).

[14] Abramovich S N, Gorelov V P, Gorshikhin A A, et al 1996 Baza dannyx dlya chislennogo modelirovaniya nakopleniya oskolkov v toplive energeticheskix reaktorov *Vopr. Atom. Nauki Tekhn. Ser. Mat. Model. Fiz. Prots.* **3** 40–61 (in Russian).

IOP Conf. Series: Materials Science and Engineering **135** (2016) 012005 doi:10.1088/1757-899X/135/1/012005

Fluorine distribution during thermal treatment of sulfuric acid leaching products of complex beryllium raw materials at JSC UMP

A N Borsuk[1], I I Zherin[2], G N Amelina[2] and I Pivovarov [2]

[1] JSC Ulba Metallurgical Plant, 102 Abay Ave., Ust-Kamenogorsk, 070005, Republic of Kazakhstan
[2] National Research Tomsk Polytechnic University, Tomsk, Russia

E-mail: amelina@tpu.ru

Abstract. X-ray diffraction, infrared spectroscopy, thermogravimetry and mass spectrometry were employed to study the behavior and distribution of fluorine in sulfuric acid treatment of the melt of beryl-bertrandite-phenakite-fluorite concentrates with alkali fluxes. It was established that fluorine in a solid phase is predominantly present in the form of sodium and calcium fluorosilicates and in a gaseous phase it is evolved in the form of silicon tetrafluoride and fluorosulfonic acid.

1. Introduction

Currently at JSC Ulba Metallurgical Plant the production of beryllium constitutes the processing of complex beryllium raw materials consisting of beryl concentrates $Be_3Al_2[Si_6O_{18}]$ and collective concentrates of bertrandite $Be_4[Si_2O_7](OH)_2$, phenakite $Be_2[SiO_4]$ and fluorite CaF_2. Hydrometallurgical technology of their processing consists in preliminary melting with alkali fluxes (Na_2CO_3 and $CaCO_3$) at 1400 °C, water melt granulation and subsequent sulfuric acid (60%) leaching of the baked melt (hereinafter referred to as the melt). The resulting product is sulfate-silicate mass (SSM) which is subjected to aqueous leaching of soluble sulfates, including beryllium. After phase separation sulfate-silicate sludge (SSS) is sent for disposal, and sulfate solution is sent for alkaline precipitation of beryllium hydroxide. The presence of significant amounts of fluorine in the system leads to the formation of stable, highly soluble and hardly hydrolysable complex ions [1, 2]. In terms of hydrolysis resistance, complex ions can be arranged in a sequence: $BeF_4^{2-} > BeF_3^- > BeF_2 > BeF^+$. Therefore, to decompose complexes increasing pH up to 12 is required which causes conjoint precipitation of impurities and incomplete beryllium precipitation. For these reasons, fluorine removal before the precipitation stage of $Be(OH)_2$ is an urgent task.

In [3] the results of the research in thermal treatment of sulfate-silicate masses for fluorine removal are discussed. The nature of this process lies in leaching of the melt with 60% sulfuric acid and subsequent heating of sulfate-silicate masses up to 300 °C. Herein due to water evaporation the concentration of sulfuric acid occurs, i.e. leaching ends up with sulfatization with possible removal of fluorine in the form of HF and SiF_4. This method is rather attractive because it does not require any alteration of the existing technology.

It has been found that SSM calcination allows reducing fluorine content by 40-50% in solutions after the leaching of sulfates from SSM. However, fluorine behavior and the forms of its condition in these processes have not been established. Consequently, addressing these issues is the objective of the present paper.

2. Materials and methods

A representative sample of the crushed melt containing beryllium and fluorine, 2.90 and 2.85% by weight respectively was used for the present research. Complete elemental composition is given in [3].

The research methodology consisted in leaching crushed melt with 60% H_2SO_4 at up to 140 °C followed by thermogram analysis up to 900 °C.

To determine the nature of element bonds in the initial melt and silicate-sulfate masses IR research on IR-Fourier spectrometer Nicolet 5700 in the range of 400-4000 cm^{-1} wavenumbers was conducted. OMNIC database [4], as well as data from [5] were used to interpret spectra. Thermogravimetric research was carried out on synchronous thermal analyzer «SDT Q600». X-ray diffraction research was conducted on an automatic diffractometer Shimadzu XRD 6000 within scanning range of 10-70°. Mass spectrometer «PROLAB» by «Thermo» was involved for the analysis of gaseous products in the mass range of 1-150 a. m. u., simultaneously the identification of gaseous products was made with the help of mass spectra library NIST 08 [6].

3. Results and Discussion

3.1 Melt analysis

The preliminary research of the phase composition of the melt showed no diffraction signals, indicating that it is X-ray amorphous [3].

Infrared absorption spectrum showed that concentrate melting followed by water quenching leads to partial destruction of Si-O-Si bonds, as indicated by the shift of the maximum of Si-O bond in a silicon-oxygen tetrahedron towards low frequency range – 948.5 cm^{-1}. The absorption bands in the 850-1150 cm^{-1} and 600-850 cm^{-1} regions indicate the violation of a distant order in the crystalline structure which is typical of amorphous solids. The break of Si-O-Me bonds and the formation of relatively simple silicates occur in a disordered structure. Also aluminates are formed, as evidenced by an intense absorption band at 463.4 cm^{-1}, characteristic of the octahedrally coordinated aluminum. At 912.9 cm^{-1}, 721.0 cm^{-1}, 615.7 cm^{-1} and ~ 520 cm^{-1} oscillation frequencies of Al-OH appear in the tetrahedrally coordinated anion $[Al(OH)_4]^-$ that proves the presence of aluminum in tetrahedral coordination.

A broad peak with a maximum at 3413.8 cm^{-1} (valence vibrations of crystallization water in silicates) and 1625.2 cm^{-1} (deformation vibrations of OH in H_2O) indicate that water molecules are linked to the granulate crystal lattice. In the investigated melt spectrum fluorosilicate absorption bands are detected. At 753.3 cm^{-1} the band corresponding to valence vibrations of SiF_6^{2-} ion is present for calcium fluorosilicate and sodium fluorosilicate. Furthermore, in the 500 cm^{-1} region sodium fluorosilicate spectrum reveals a set of bands at 482.4; 492.2; and 523.5 $cм^{-1}$ which is observed in the present spectrum. Calcium fluorosilicate spectrum is characterized by the presence of the absorption band at 571.2 $cм^{-1}$.

Thus, X-ray and infrared investigations suggest that the melt is amorphous depolymerized partially hydrated aluminum silicate. The melt contains fluorine mostly in the form of sodium and calcium fluorosilicates.

3.2. Interaction of the melt with sulfuric acid solution

Melt samples were mixed with sulfuric acid solution and left at the temperature of 100-120 °C for 2 hours. Interaction of the melt with sulfuric acid solution may lead to the destruction of silicates and aluminates forming soluble sulfates, hydrosulfates and crystalline hydrates. Infrared spectroscopy,

IOP Conf. Series: Materials Science and Engineering **135** (2016) 012005 doi:10.1088/1757-899X/135/1/012005

thermal analysis and mass spectrometry were employed to determine the mechanism of the interaction of sulfuric acid solutions with the melt.

The comparison of absorption spectra of the melt and sulfate mass makes it possible to conclude that the interaction between the melt and H_2SO_4 solution leads to the formation of hydrosulfates, absorption bands of which appear at 1321.4 cm^{-1}, 1280.2 cm^{-1}, 1225.5 cm^{-1}, 1063.4 cm^{-1}, 1011.1 cm^{-1}, 878.5 cm^{-1}, 852.4 cm^{-1}.

The absorption band at 1173.8 cm^{-1} can refer to the valence vibrations of $[HSO_4]^-$ ion, as well as to the deformation vibrations of the SiO_4-tetrahedron. The absorption bands at 608.2 cm^{-1} and 584.6 cm^{-1} in the field of deformation vibrations of $[HSO4]^-$ also indicate hydrosulfate formation. Alongside hydrosulfate ion absorption bands, absorption bands are observed at 1121.9 cm^{-1} and 1063.4 cm^{-1} which are characteristic of fluorosulfonic acid salts. Fluorosulfonic acid formation is possible as a result of the interaction between the formed during sulfatization hydrogen fluoride and excess sulfuric acid as follows:

$$Na_2SiF_6 + 2H_2SO_4 = 2NaHSO_4 + 2HF + SiF_4 \quad (1)$$

$$H_2SO_4 + HF = HSO_3F + H_2O \quad (2)$$

A diffused band in the frequency range of 650-800 cm^{-1}, as in case of the melt, reveals the presence of SiF_6^{2-}-ion, although it is partially overlapped by S-O deformation vibrations. Thus, the infrared spectra of the fluorosilicates are characterized by the valence vibration absorption band in the 740-750 cm^{-1} region. The deformation vibration band at 527.8 cm^{-1} points at the presence of sodium fluorosilicate. This band coincides with the deformation vibration of $[HSO_4]^-$ for calcium fluorosilicate.

The cleavage of the absorption bands in the region of deformation vibrations of water (1600-1700 cm^{-1}) demonstrates the presence of several types of nonequivalent crystallochemical water molecules.

A wide band in the region of 3500-2900 cm^{-1} also proves the presence of several crystallization water modes. In the region of valence O-H vibrations the referred modes are difficult to distinguish. It is believed that weakly polarized in the field of cations water usually absorbs in the 3650-3400 cm^{-1} region, more strongly polarized, participating in stronger H-bonds, water usually gives broad diffused absorption bands lying below 3400 cm^{-1}. In addition, valence O-H oscillation frequencies of water shift to a low frequency oscillation spectrum due to strengthening of H-bonds. In the IR spectrum of sulfate mass two bands at 2483.4 cm^{-1} and 2325.1 cm^{-1} corresponding to a composite tone $\delta_{H2O} + \nu_L$ (deformation + libration) are observed. Since libration vibrations occur only with the participation of H_2O molecules in hydrogen bonds, the appearance of composite tone in IR spectrum proves the presence of free unbound water in the substance. The presence of unbound water in the sulfate mass was confirmed by thermal analysis data, represented in Figure 1.

Heating sulfate-silicate mass at 10 °C/min in the atmosphere is accompanied with a series of endothermic processes followed by weight loss. Mass spectra library NIST 08 was employed for the identification of gaseous products; mass spectrum of sulfuric acid decomposition products at atmospheric pressure in the air was determined experimentally and appeared identical to SO_2 spectrum. On the basis of the mass spectra investigated, the identification of water, sulfurous anhydride and silicon tetrafluoride in the reaction products was carried out by the presence of ions with M/Z= 18; 64; 85 in the mass spectra.

The processes with a maximum gas emission rate of 96 °C and 153 °C are caused by the loss of free (96 °C) and bound (153 °C) water , which is confirmed by mass spectrometry analysis. The content of free (unbound) water is about 9%, and the hydrated crystalline water content is 18% of the sample weight.

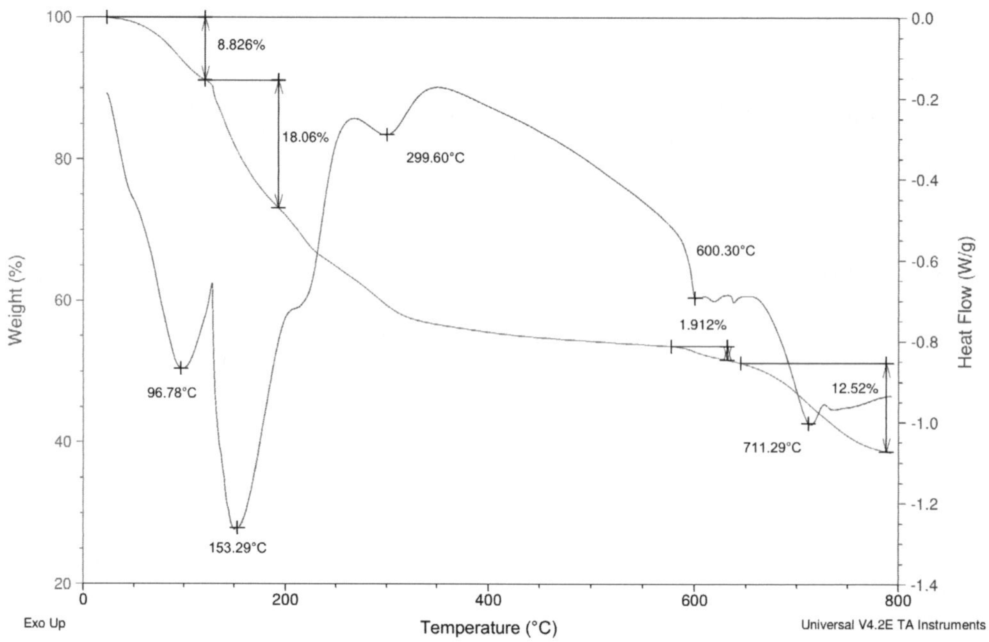

Figure 1. Sulfate-silicate mass thermogram at the rate of 1.6 g of 60% sulfuric acid per 1 g of melt

The endothermic process occurring in the temperature interval of 600-800 °C with 12.5% weight loss and the evolution of gaseous SO_3 indicate decomposition of sulfates. According to the gas phase analysis, an endothermic process with 1.9% mass loss in the temperature range of 580-650 °C is followed by SiF_4 evolution. This confirms the IR analysis data of the presence of fluorosilicates in the sulfate-silicate mass composition. According to mass spectrometry analysis, in the interval of 200-300 °C the occurring in the system process is accompanied with a simultaneous release of water, SiF_4 and SO_3.

The IR spectrum of the sulfate-silicate mass calcinated at 300 °C revealed that the absorption band of sodium (calcium) fluorosilicates disappears in the interval of 740-750 cm^{-1} as a result of thermal treatment at 300 °C. The reduction of the absorption band at ~ 3400 cm^{-1} indicates the removal of crystallization water weakly bound with crystal lattice. The dehydration of hydrosulfate leads to a peak at 1024.3 cm^{-1}, associated with valence vibrations of a sulfate anion. Both the peak at 674.5 cm^{-1}, characteristic of Si-O-Mg(Na) bonds, and a diffused absorption band at 792.9 cm^{-1}, characteristic of Si-O-Si bridging bonds, are observed.

Thermal analysis and IR spectroscopy results demonstrate that the process occurring in the system in the range of 200-300 °C is determined by the interaction between melt sulphatization products according to the reaction scheme:

$$Na_2SiF_6 + 2H_2SO_4 = 2NaHSO_4 + 2HF + SiF_4 \quad (3)$$

$$4HF + SiO_2 = SiF_4 + 2H_2O \quad (4)$$

Thereby, sulfate-silicate mass examination revealed that the main reaction products of the melt and sulfuric acid solution interaction are crystalline hydrates of calcium and sodium hydrosulfates with a portion of water remaining in the unbound state. In the reaction products fluorine is represented in the form of fluorosilicates and fluorosulfonic acid salts.

Further investigation was focused on the study of the characteristics of sulfuric decomposition of fluorosilicates. For this purpose, the influence of sulfuric acid amount on the process of granulate desulfurization was studied. Leaching was done with 60% sulfuric acid solution. The amount of sulfuric acid solution was varied in the interval of 1.35 – 3.24 g per 1 g of the melt. Typical thermogram of the obtained sulfate-silicate masses is represented in Figure 2.

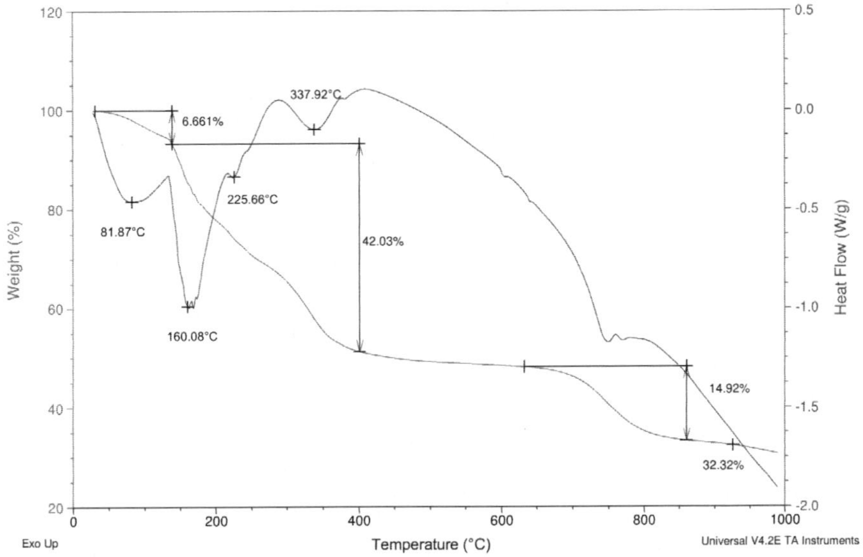

Figure 2. Sulfate-silicate mass thermogram at the rate of 3.24 g of 60% sulfuric acid per 1 g of melt.

It was established that the increase in the amount of sulfuric acid solution does not cause any significant change to the amount of free unbound water and sulfates. The reduction of the residue after calcination is determined by significant differences in the processes in the range of 100-400 ° C indicating the formation of chemical compounds with decomposition temperature from 100 to 400 °C. In this area of the thermograms 3 occurring processes can be allocated, the speed peaks of which are observed at 150-160 °C, 215-240 °C and 300-330 °C. Increasing the amount of acid has a positive effect on the process that takes place in the range of 300-400 °C, it inhibits the process in the range of 200-300 °C and does not affect the process in the range of 130-200 °C.

Mass spectrometry analysis of gaseous reaction products was carried out to identify separate stages. Typical mass-spectrum of gaseous products is shown in Figure 3.

It was depicted above that the process occurring in the interval of 200-300 °C is caused by the interaction between sodium and calcium fluorosilicates and sulfuric acid followed by sodium hydrosulfate decomposition. The product of the process occurring in the range of 300-400 °C is SiF_4. SO_3 and water are not available, what indicates the occurrence of thermal dissociation of calcium fluorosilicate. In the region of about 600 °C thermal dissociation of sodium hexafluorosilicate Na_2SiF_6 proceeds and at temperatures above 700 °C the decomposition of sulfates occurs.

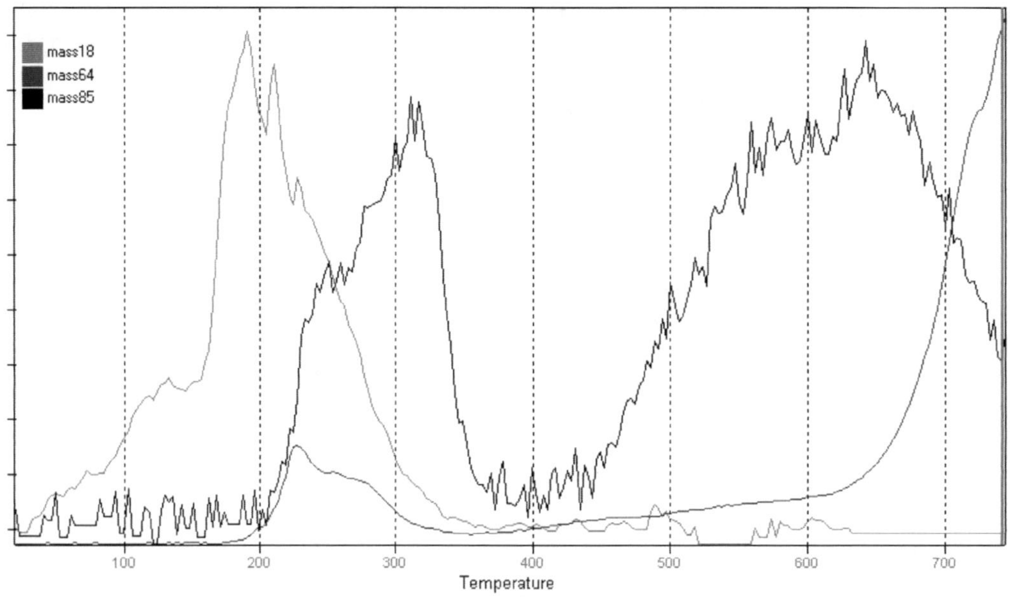

Figure 3. Formation of gaseous products after thermal treatment of sulfate-silicate masses in the course of granulate leaching with 60% sulfuric acid solution at the rate of 3.24 g per 1 g of the melt

Generally sulfate-silicate mass defluorination is possible by thermal dissociation of fluorosilicates at temperatures of 550-650 °C, or by acid leaching of fluorosilicates at 200-300 °C. Defluorination by thermal dissociation can destroy soluble beryllium compounds, resulting in a significant loss of beryllium at the subsequent leaching step. Therefore, acid leaching of sodium and calcium fluorosilicates with concentrated sulfuric acid at 200-300 °C is preferred.

4. Conclusion

X-ray diffraction, infrared spectroscopy, thermogravimetry and mass spectrometry were employed to study the behavior and distribution of fluorine in sulfuric acid treatment of the melt of beryl-bertrandite-phenakite-fluorite concentrates with alkali fluxes. It was established that fluorine in a solid phase is predominantly present in the form of sodium and calcium fluorosilicates, and in a gaseous phase it is evolved in the form of silicon tetrafluoride and fluorosulfonic acid. The preferred method of defluorination is acid decomposition of sodium and calcium fluorosilicates with concentrated sulfuric acid at 200-300 °C. The obtained results are the basis for detailed kinetic studies to determine the optimal technological conditions for fluorine removal process.

References

[1] Hogben M G, Radley R and Reeves L W 1970 *Canadian J. of chemistry* Studies of fluoroberyllate complexes in aqueous solution by 19F nuclear magnetic resonance **48** 2960-65.

[2] Bjerrum J, Schwarzenbach G and Sillen L G 1958 *Stability Constants P. II Inorganic Ligands* (London: The Chemical Society).

[3] Borsuk A N, Amelina G N, Zherin I I, Grigoryeva E O and Ivlev S I 2014 *Procedia Chemistry* Study of the Thermal Method for Removing Fluorine from Products of Sulfuric Acid Leaching of Beryllium Ores at UMP JSC **11** 113-11.

[4] http://www.thermoscientific.com/content/tfs/en/product/omnic-series-software.html.

[5] Smith A 1982 *Applied IR spectroscopy* (Moscow: Mir)

[6] http://www.nist.gov/srd/upload/Ver20Man.pdf.

IOP Conf. Series: Materials Science and Engineering **135** (2016) 012006 doi:10.1088/1757-899X/135/1/012006

Issues of Intruder Analytical Model Applicability for Evaluating an Efficiency of Security Systems

Anton V. Bukovetskiy[1], Vladimir I. Boyko[2], Gennady N. Kolpakov[3], Michael V. Poner[4]

[1] Engineer, Federal State Unitary Enterprise «Mining and Chemical Combine», Zheleznogorsk, Krasnoyarsk region, Russia
[2] Full Doctor, Prof., Department of Physical and Power Plants, Institute of Physics and Technology, National Research Tomsk Polytechnic University, Tomsk, Russia
[3] Doctor, Assoc. Prof., Department of Physical and Power Plants, Institute of Physics and Technology, National Research Tomsk Polytechnic University, Tomsk, Russia
[4] Bachelor of Arts, Department of Physical and Power Plants, Institute of Physics and Technology, National Research Tomsk Polytechnic University, Tomsk, Russia

E-mail: antaresbav@tpu.ru

Abstract. Physical protection system (PPS) is created at a nuclear object to prevent unauthorized acts regarding to nuclear materials, nuclear installations, points of storage of nuclear materials and other items under physical protection. The ability of the PPS to prevent unauthorized actions of intruder was adopted as the main criterion for evaluating the efficiency of the physical protection system. The article considers the possibility of creation of the intruder analytical model to determine time indicators of overcoming boundaries of protection of nuclear object by an intruder with set features.

1. Introduction

Physical protection system integrates people, procedures and equipment for the protection of assets or facilities against theft, sabotage or other malevolent attacks. Among critical facilities, nuclear facilities and nuclear weapon sites require the highest level of PPS. Thus, the International Atomic Energy Agency (IAEA) adopted a convention [1] and published documents outlining requirements for physical protection at nuclear facilities [2]. After the September 11, 2001 terrorist attacks in the U.S.A., the international community, including the IAEA, have made substantial efforts to protect nuclear material and nuclear facilities. These efforts include the Nuclear Security Fund established by the IAEA in 2002 and the Global Initiative to Combat Nuclear Terrorism launched by the USA and Russia in 2006.

Principles of physical protection are realized through administrative and technical measures, including physical barriers. The measures for the physical protection of nuclear material in use and storage and during transport, and of nuclear facilities presented herein are recommended for use by States as required in their physical protection systems. These measures are based on the state of the art in physical protection hardware and systems and on the types of nuclear material and nuclear facilities.

Physical protection system consisting of the physical protection personnel, organizational and technical measures and actions carried out by it, as well as complex of the physical protection engineering & technical facilities, should perform the following tasks: prevention of unauthorized actions; prompt detection of unauthorized actions; hindering (slowing down) the intrusion (advancement) of offenders; responding to unauthorized actions of offenders and neutralization of such in order to prevent the unauthorized actions.

The implementation of the basic principles when designing a physical protection system is aimed at achieving the required level of system efficiency, which is determined by its ability to resist unauthorized actions of intruders regarding to physical protection items. A numerical value characterizing the effectiveness of the physical protection system is a system efficiency index. Its value is determined during the assessment of the effectiveness of PPS subject to certain external and internal threats, the list of vulnerabilities and the model of the intruder.

2. Efficiency index

PPS is a complex system, which includes a variety of structural and functional subsystems. One of the basic properties of PPS as an integrated system is ability to work in conditions of uncertainty. The latter leads to the need of defining efficiency index that has stochastic nature.

The ability of the PPS to prevent unauthorized actions of the intruder is taken as the main criterion. The probability of suppression of the intruder action (P_s) is considered as an efficiency index, since the fact of restraint is a random event. In general, the suppression can be considered as a combination of the following events: intruder detection by technical means of PPS, the delay of the intruder in the progress toward goals, reaction and neutralization of the intruder action [3]. In this approach, the probability of suppression of unauthorized actions in relation to items under physical protection can be described by the formula [4]:

$$P_s = P_{dt} P_{dl} P_n \quad (1)$$

- probability of timely detection of unauthorized actions of the intruder (P_{dt}) is determined by the characteristics of detection equipment and capabilities of the security forces; in the first case it can be derived from the detection equipment technical characteristics, in the second – on the basis of available statistics or as a result of simulation;
- the probability of delay and slow progress of the intruder (P_{dl}) represents the quantile of a random variable of time delay, i.e. the probability that the delay t is not less than a predetermined value (t_{dl}):

$$P_{dl} = P(t \geq t_{dl}) \quad (2)$$

The calculation of P_{dl}, as a rule, is carried out using simulation and requires knowledge of time needed for physical barriers overcoming by different types of intruder, who differs in action tactics, special training, level of technical equipment, etc.

- the probability of neutralization of an intruder by the guard forces (P_n) represents the quantile of a random variable generated at the slice implementation of random functions, defined as "neutralization" in a fixed moment of time *t*.

Simulation is commonly used to determine the value of P_n based on special programs.

Currently, the most prevalent approach to evaluate the effectiveness of security systems in models of the functioning of the PPS is the approach which examines possible scenarios of unauthorized actions and routes of intruder to locations of physical protection items in protected areas on the object. Further, the system's ability to counter the actions of intruders on selected routes is assessed.

In this case, the choice of the efficiency index should meet the following requirements [5]:

- representativeness - strict compliance of the indicator to security system goals;
- sensitivity – a sufficient change of the value of efficiency index when changing the most important parameters and initial data describing the system and affecting its effectiveness;
- simplicity – a lack of mathematical difficulties in the calculation of the efficiency index;
- visibility – clear physical meaning of the defined indicator.

The ability of the PPS to prevent unauthorized actions of intruder was adopted as the main criterion for evaluating the effectiveness of the physical protection system. System efficiency is measured by quantitative indicators, reflecting the probability of restraint of the intruder's actions by the security forces, acting on the alarm.

Efficiency index depends on threats, model of the intruder and vulnerabilities that were adopted in the process of analyzing the nuclear object. Following indicators are used to assess the effectiveness of the physical protection system [6]:

- differential efficiency index takes into account the probability of preventive action against the intruder to one target. When considering multiple scenarios of the intruder action against selected targets, the differential indicator of the effectiveness of the physical protection system of this vulnerability is taken equal to the minimum (worst) value among all the considered scenarios. The scenario of the intruder's actions, corresponding to the minimum value of the probability of preventing actions against selected targets, is adopted as the critical;
- integral indicator represents an average efficiency index of physical protection system for nuclear object with regard to the rank of targets importance.

The realized approach involves using the probability of suppressing the actions of the intruders moving along the route, providing the maximum probability of success with minimal time to achieve the goal, as the primary indicator of the effectiveness of physical protection systems [7]. Elements of the route that intruders have to overcome in different ways depending on intruder chosen tactics are engineering tools (i.e. physical barriers), areas of open terrain, the construction elements of buildings, premises and facilities, equipped with means of detection and control. Overcoming each element of the route is characterized by tactics, method, time of intruder movement and the probability of their detection by technical means of physical protection, site personnel, guard forces. Further measures are taken to intercept, block and neutralize intruders.

3. Intruder model

Intruders can be separated into three classes:

- outsiders;
- insiders;
- outsiders in collusion with insiders.

For each class of intruder, the full range of tactics (i.e., deceit, force, stealth, or any combination of these) is considered. Deceit is the attempted defeat of a security system by using false authorization and identification; force is the overt, forcible attempt to overcome a security system; and stealth is the attempt to defeat the detection system and enter the facility covertly. Adversary capabilities include knowledge of the PPS, level of motivation, skills useful in carrying out the attack (e.g., knowledge of the safety systems), the speed with which the attack is carried out, and the ability to carry and use tools and weapons.

Theft and sabotage may be prevented in two ways: by deterring the adversary or by defeating the adversary. Deterrence occurs by implementing a PPS that is seen by potential adversaries as too difficult to defeat; it makes the facility an unattractive target. In addition, legal ramifications associated with attacking a site may deter some adversaries, although not a determined one. The problem with deterrence is that it is impossible (or, at least, extremely difficult) to measure, and therefore, will not be discussed further. Defeating the adversary refers to the actions taken by the protective or response force to prevent an adversary from accomplishing his goal once he actually begins a malevolent action against a facility. There are three major functions that the PPS must perform. These include: detection, delay and response.

The goal of an adversary is to complete a path with the least likelihood of being stopped by the PPS. To achieve this goal, the adversary may attempt to minimize the time required to complete the path. This strategy involves penetrating barriers with little regard to the probability of being detected. If the adversary completes the path before the guards can respond to interrupt his activities, he is successful. Alternatively, the adversary may attempt to minimize detection with little regard to the time required. If the adversary completes the path without being detected, he is successful.

Selection and description of the intruder model is one of the key issues for creating an effective physical protection system, as the characteristics of the intruder determine the shape, the functionality and the structure of the security system. Characteristics and the parameters of an intruder in its mathematical nature can be divided into three groups [5].

Random with a non-stochastic nature. The values of these parameters for the intruder could not be identified in the analysis of preparation for the implementation of the illegal action or by analytical methods described in the formation of an intruder model. An example of such a parameter may be the following: the intruder group size seeking to realize a threat. The common technique of working with these parameters is to assign probabilistic characteristics to these parameters by the using of various techniques. In many cases modification the mathematical nature of the parameters of this group are carried out through the use of one or simultaneously several methods of expert estimations.

Random with a stochastic nature. The values of these parameters can be obtained statistically or with using specialized computer programs, based on a statistical database and implementing special procedures with them. Examples of such characteristics and parameters may be the following: intruder movement speed on the object site, the time of physical barriers overcoming, the chance to trigger the technical means of protection (detection) under different actions of the intruders and many others. The main difficulty when using parameters of this group is the lack of completeness of source data bases. At the same time, obtaining of any individual parameter of the intruder is time consuming and requires a large number of relevant experiments.

Deterministic. The values of these parameters are constant and depending on the accepted limitations of the performed calculations. Examples of such parameters in the particular case may be the rate of fire of weapons, the power of explosives, the number of physical barriers that intruder must overcome when moving towards the location of the physical protection items at the nuclear object. The proportion of deterministic parameters in the totality of characteristics of an intruder model is small.

Unauthorized actions against nuclear materials, nuclear installations and points of storage of nuclear materials are the set of actions aiming to overcome the boundaries of protection, movement in a protected area of a nuclear object in the direction of the physical protection items placement, participation in the armed clash with security forces. In this case, the time required for intruders to perform the task can be divided into time of overcoming the route elements and time of action with the target object. The time required for moving on a site between the selected boundaries of protection is determined by the speed of the intruder, the nature and length of the surmounted site and security forces actions. The value of the detection probability and the delay time at the turn of protection depends on characteristics of the technical means of physical protection and ways of overcoming the boundaries of engineering means. Analysis of the interactions in the system «intruder – PPS» allows to determine intruder general characteristics (technical equipment, level of physical training, group size) and tactics of action (power, covert, deceptive, or mixed) that affect the value of the efficiency index of PPS.

Nowadays, time of physical barriers overcoming is determined by conducting exercises and by the expert way taking into account selected characteristics and parameters of the intruder. The exercise is difficult in organization, time-consuming and lengthy process, requiring usage of large number human and material resources. Thus, obtained data allow covering only a small sample of events with different initial conditions, and data in most cases are random with the stochastic nature. The characteristics may be determined by the laws of random variables distribution or parameters of these laws.

Data obtained by the expert estimates has low reliability and directly depends on the subjective views and competence of the expert. The original data obtained by the expert estimates is recorded in the database as deterministic parameters and weakly depends on the characteristics and parameters of intruder model. In this case, there are difficulties of taking into account the level of skills and abilities of intruder in overcoming physical barriers due to their random nature. Usually there are three conditional levels of skills of the intruder in specialized programs: high, medium, low [8]. Also, features and specifications (material, dimensions, fences, locations of obstacles) of physical barriers cannot be fully taken into account. All this reduces the reliability of results obtained in the evaluation of the effectiveness of the nuclear object physical protection system.

4. Conclusion

Solving practical issues associated with the evaluation of the effectiveness of the nuclear object security system allows making a conclusion about the necessity for determining more accurate time values of overcoming actually installed physical barriers established within the existing security system. The approach allows to apply for this purpose the algorithm describing the motion of the intruder on the territory of the object and overcoming the boundaries of protection, taking into account the equipment, physical and physiological characteristics of the human body and other factors influencing the possibility of performing unauthorized actions. Its use in the definition of the source data to assess the effectiveness of the physical protection system will require the creation of analytical model of intruder actions. Obtained data allow to take into account the internal and external factors of the interactions in the system «intruder – PPS» that provide the maximum security of nuclear materials at the nuclear facility.

References

[1] IAEA, INFCIRC/274 Convention on the Physical Protection of Nuclear Material (1981)
[2] IAEA, INFCIRC/225/rev.4 The Physical Protection of Nulcear Material and Nuclear Facilities (1999); IAEA, TECDOC-967 (2000); IAEA, TECDOC-1276 (2000).
[3] Garcia M L 2001 *The design and evaluation of physical protection systems* (Oxford: Elsevier Butterworth-Heinemann)
[4] Anton V. Bukovetskiy, Boris P. Stepanov, Denis A. Tatarnikov 2016 *Initial Data Forming for Process Simulation in System "Intruder – Physical Protection System"* (Key Engineering Materials) vol 685 (Switzerland: Trans Tech Publications) pp 148-152
[5] Barnard R L 1988 *Intrusion Detection Systems, 2nd ed.* (Boston: Butterworth-Heinemann)
[6] Fischer R J, Halibozek E, Green G 2008 *Introduction to Security, 8th ed.* (Boston: Butterworth-Heinemann)
[7] Fennelly L 2013 *Effective Physical Security,4 ed.* (Boston: Butterworth-Heinemann)
[8] Fischer R J, Halibozek E, Walters D 2013 *Introduction to Security, 9nd ed.* (Boston: Butterworth-Heinemann)

The influence of the dislocation distribution heterogeneity degree on the formation of a non-misoriented dislocation cell substructures in f.c.c. metals

D N Cherepanov[1,2], O V Selivanikova[1] and M V Matveev[1]
[1]National Research Tomsk Polytechnic University, Tomsk, Russia
[2]Tomsk State University of Architecture and Building, Tomsk, Russia

E-mail: selov@tpu.ru

Abstract. Dislocation loops emitted by Frank-Reed source during crossing dislocations of the non-coplanar slip systems are accumulates jogs on the own dislocation line, resulting in the deceleration of the segments of dislocation loops with high jog density. As a result, bending around of the slowed segments the formation of dynamic dipoles in the shear zone occurs. In the present paper we consider formation mechanism of non-misoriented dislocation cell substructure during plastic deformation of f.c.c. metals and conclude that the increase in the degree heterogeneity of dislocation distribution leads to an increase in the jog density and reduce the mean value of arm dynamic dipoles.

1. Introduction

Non-misoriented dislocation cell substructure (NMDCS) was the first experimentally observed dislocation structure [1] and attracted the attention of researches due to its contribution to the strengthening of the material [2 – 4]. It is observed as bright spots, surrounded by lots of dark stripes that forms cell boundaries. Cell boundaries may contain areas with misorientations of no more than half a degree [4].

According to the modern concept of cell structures, non-misoriented dislocation cell substructures assumed as heterogeneous substructure, dislocations of which mostly located in the walls of finite thickness, consisting of dipoles and multipoles hindering movement of shear-forming dislocations [4].

It is known that f.c.c. metals with high energy (more than 60 MJ/m^2) stacking fault characterized by cellular substructure with blurred cell walls under moderate or impulse loads. With decreasing of this energy the tendency of cell formation become weaker and dislocation plexus become more observable. Structure is dominated by twins and stacking faults at low energies (less than 25 MJ/m^2) [5, 6].

Study of dislocation structure of f.c.c. single crystal copper oriented for single slip [7] showed that crystals under small deformations have free strongly curved primary dislocations as well as dipole and multipole configurations, consisting mainly of screw dislocations. Screw dislocations and dipoles were observed either rarely or not at all, what allows to conclude that they are quickly annihilating by cross slip [8].

In the beginning of strain, the dislocation reactions resulting to formation of dislocation plexus uniformly over the entire sample from which the walls of cells are formed in single crystals oriented for multiple slip [8– 14].

In the copper even at low stresses and strains $\tau \approx 10\,\text{Pa}$ cellular structures with flat rounded cells, lying in the planes perpendicular to the axis of tension, are formed. Average area of cells that are almost freed of dislocations, decreases with increasing of stress measured in MPa according to the law $25.5\tau^{-2}$. Average cell diameter d_C is always smaller, then the length of the slip traces. Since the length of the slip traces has the same order as the diameter of shear zone D, then for single crystal copper at room temperature $D \approx 4d_C$ [9], i.e. four cells are formed in the shear zone. With increasing stress, the dislocation density in the cell walls increases proportionally $\tau^{2,3}$ with the proportionality coefficient that depends on the orientation of the sample [9].

When using the cyclic strain, the cells with more regular shape are formed and their walls are built of more correct and less intertwined dislocation structures [15]. On the basis of the observation was concluded that the plastic strain is concentrated inside the cells, and the saturation state inside the cells there are groups of free dislocations with generally screw orientation, as well as the vacancy dipoles with arm of $3.5\,\text{nm}$ [15].

In copper polycrystals for an average diameter of dislocation cells were found: $d_C = 16\rho^{-1/2}$ and $\tau = 10.5Gb/d_C \approx 0{,}66Gb\rho^{1/2}$ [16]. Inverse proportionality of the shear stress τ to the average diameter of cells was found in other single and polycrystalline f.c.c. metals. For nickel $d_C = (7\ldots12)\rho^{-1/2}$, moreover proportionality coefficient is little affected by temperature and deformation degree, and close to the values of other pure metals [17].

Thus, in metals and alloys cells are characterized by average diameter $d_C = (2\ldots16)\rho^{-1/2}$ and wall width $h_C = (0{,}3\ldots0{,}5)d_C$ $h_C = (6\ldots80)\cdot10^{-5}$ at $\rho = 10^8\,\text{sm}^{-2}$.

Consequently, the diameter of the cells corresponds to the mean free path of screw segments. Therefore, the formation of the cell structure may be linked with the evolution of the dynamic multipole structure. With increasing strain cells are grinding, and their average size is reduced to value of $0.2\,\mu m$.

Formation and evolution of NMDCS studied both experimentally, and by mathematical modeling [18 – 21]. The main quantity characterizing the evolution NMDCS with increasing degree of plastic deformation, is a dislocation density of the dynamic dipole configurations ρ_d. In addition, mathematical models of plastic deformation include NMDCS characteristics as an equation parameters, such as the average value of the dipole arm h_d, the mean free path ℓ_d of a screw dislocation with jogs and the average shear zone area S_D, i.e. the average area swept out by shear-forming dislocations emitted by Frank-Read sources.

For convenience, the dipoles are considered to consist of elementary dipoles with length ℓ_d and for the output of dipoles accumulation intensity number of elementary dipoles n_d is used.

To construct a mathematical model firstly we need to answer the following questions. What is the resistance to movement of the screw orientation of the segments have jogs? How many jogs on the segment with screw orientation is necessary for a significant slowdown of the segment? What is the average value ℓ_d of path that run segments of shear-forming dislocations having jogs with screw orientation before bending around adjacent segments with lower jog density? How many elementary dipoles are formed in the area S_D after the passage of a dislocation loop? What is distribution function for arm of dynamic dipoles? What is the intensity of the dynamic dipole generation? What is the annihilation rate of the thermodynamically stable dynamic dipoles under the influence of flow deformational point defects? What is the intensity of point defects generation due to the dissociation of the thermodynamically unstable dipoles with a small dipole arm? What effect does the reduction of the shear zone area S_D with increasing degree of deformation on the formation of NMDCS?

2. Formation of non-misoriented dislocation cell substructure

In the first papers on the modeling of evolution NMDCS assumed that the deceleration and self-locking of screw segments of dislocation loop after the run for a distance ℓ_d because of the increase in the jog concentration it leads to a bending around the adjacent more mobile segments.

In the process of bending around as a result of run edge segments the new segments with a screw orientation are formed, which also bending around after run of about the same distance ℓ_d. As a result, within the shear zone the system of dipole and multipole dislocation configurations occurs. NMDCS formed due to the transformation of the dipole configuration system and disappears due to the dissociation and annihilation of dipoles. The dislocation density in the cell walls is the same as ρ_d, while the dislocation density inside the cells - is part of ρ_d determined by not transformed dipoles remaining at the time of observation.

Equating work of screw segments on the drawing of jogs $\tau_j b \mathrm{d}S_D^s$ to the total energy of generated point defects $0.5(U_i^f + U_\upsilon^f) c_j \mathrm{d}S_D^s / b$, where U_i^f and U_υ^f - formation energy of interstitial atoms and vacancies, respectively, we get strain $\tau_j = 0.5 c_j (U_i^f + U_\upsilon^f)/b^2 \approx 0.25Gbc_j$ that necessary for nonconservative drawing of jogs [18 – 21].

The value of ℓ_d depends on the jogs concentration c_j, because stopping of the screw segment occurs when strain $\tau_j = 0.25Gbc_j$ that need for drawing of jogs becomes larger than effective strain $\tau_{eff} \approx \alpha_{eff} Gb\rho^{1/2}$ that spreads the loop. Since the concentration of jogs c_j does not exceed the value $\beta_j \xi \rho \ell_d$ obtained by integration of jogs generation rate $c_j \approx \int_b^{\ell_d} \beta_j \xi \rho_m \mathrm{d}x \approx \beta_j \xi \rho_m \ell_d$, then basis on the inequalities $0.4\rho^{1/2} \le 0.25 c_j \le 0.25 \beta_j \xi \rho \ell_d$ it follows that value ℓ_d is not less than $4\alpha_{eff} \beta_j^{-1} \xi^{-1} \rho^{-1/2} \approx (2\ldots8)\rho^{-1/2}$ on condition of $\alpha_{eff} = 0.1\ldots0.4$.

Due to the movement of a dislocation loop $n_d = 2F_d \ell_d^{-2} S_D^s$ elementary dipoles are formed, all the dipoles consist of them. Next apply $n_d = 42$; $F_d \approx 0.3$; $\ell_d = 8\rho^{-1/2}$. In this case on the segment with length ℓ_d there is on average $c_j \ell_d \approx 14$ jogs.

If N of dislocation sources per unit of volume emit portion on the average of Δn dislocations loops during time Δt, then shear plastic deformation a will increase by the amount of $\Delta a = S_D b n N$. Since dipole consists of two dislocation segments, then with increasing shear strain at Δa density of dislocation that form dynamic dipoles during one elementary act of segment-source, increases by the amount of $\Delta\rho_d = \Delta\rho_d^\upsilon + \Delta\rho_d^i = 2n_d \ell_d \Delta n N$.

Consequently, the dislocations in the dynamic dipole configurations of interstitial and vacancy types accumulates with intensity of $G_d = 2n_d \ell_d S_D^{-1} b^{-1}$.

Considering the intensities of dynamic dipole generation with interstitial G_d^i and vacancy G_d^υ types equal between each other we get

$$G_d^i = G_d^\upsilon = 0{,}5 G_d = F_d S_D^s S_D^{-1} \ell_d^{-1} b^{-1} \approx 0{,}125 F_d b^{-1} \rho^{1/2}. \quad (1)$$

Where ρ_d^i, ρ_d^υ - dislocation density in dynamic dipole configurations of interstitial and vacancy types, respectively.

Dislocations that forms vacancy dipoles, annihilates by the climbing as a result of the interstitial atom deposition, whereas dislocations, that form interstitial dipoles, annihilates by climbing as a result of monovacancies and bivacancies deposition.

Annihilation rate A_{id}, $A_{1\upsilon d}$, $A_{2\upsilon d}$ of point defects during their deposition at the edges of extraplanes of dipoles have form of:

$$A_{id} = \rho_d^{\upsilon} D_{id}\, c_i,\ A_{1\upsilon d} = \rho_d^{i} D_{1\upsilon d}\, c_{1\upsilon},\ A_{2\upsilon d} = \rho_d^{i} D_{2\upsilon d}\, c_{2\upsilon}. \quad (2)$$

Where D_{kd} - diffusion coefficients, c_i, $c_{1\upsilon}$, $c_{2\upsilon}$ - concentrations of point defects (interstitial atoms, mono- and bivacancies respectively).

Annihilation rate of the dipoles under the influence of ping defects have form []:

$$A_{di} = w_d^i A_{id} < h >^{-1} b^{-1},\ A_{dk} = w_d^k A_{kd} < h >^{-1} b^{-1},\ k = 1\upsilon,\ 2\upsilon. \quad (3)$$

Where w_d^i, $w_d^{1\upsilon}$, $w_d^{2\upsilon}$ - ratio of total surface area of the tubes with radius b around dislocations, that forms dipoles, to total surface area of all sinks for point defects of every type.

3. Influence of the degree of the heterogeneity of the distribution of dislocations on the NMDCS formation

The value ℓ_d depends on the jog concentration c_j, so the accumulation of dislocations in the dynamic dipole configuration depends significantly on the distribution by dislocation sign of non-coplanar slip systems, crossed by the screw segments.

The dislocations intersection of non-coplanar slip systems leads to the fact that the segments, bending around the decelerating segment of screw orientation, turn out to be in planes different from the original slip plane of the shear-forming dislocation. The distance from the original slip plane where the bending around segments are moved away. It depends on its swept out area and distribution of the jogs-forming dislocations in this area.

The minimum distance equal to the distance $\sqrt{6}b/3$ between adjacent planes of octahedral slip, however it is possible that the enveloping segments will meet on one plane, after that the dipole is not formed or the arm dipole can be considered to be equal to zero.

The maximum distance will be reached, if the enveloping segments cross all the noncoplanar shear zone dislocations, and it will be equal to Δnb of the product of the number of emitted loops per module of the Burgers vector. In the random distribution of the "forest" dislocations the appearance of a dipole with a maximum arm is much less likely than the appearance of a dipole with a minimum and even average lever.

The area occupied by one dipole, is deduced from a ratio $\gamma_d \ell_d^2 \approx S_D^s n_d^{-1} = 0{,}5\ell_d^2 F_d^{-1}$ of the area gripping by the screw segments to the number n_d of elementary dipoles.

Average arm dipole is expressed as the product of the area $< h >= \left|\beta_j^+ - \beta_j^-\right| \xi\rho\, \gamma_d \ell_d^2 \sqrt{6}b/3$ per one dipole, on the difference between the fractions of jog-forming dislocations with a different sign, on the "forest" dislocations density $\xi\rho$ and the distance $\sqrt{6}b/3$ between adjacent slip planes.

In the entire material volume the values β_j^+ and β_j^- are approximately equal, but in the area per one dipole, they may vary. It is clear that $\beta_j^+ + \beta_j^- = \beta_j \approx 0.43$ - the fraction of the "forest" jog-forming dislocations.

If dislocations of only one sign are crossing, the maximum arm dipole of the dipole is: $h_{max} \approx 0.43\xi\rho\, \gamma_d \ell_d^2 \sqrt{6}b/3$.

Suppose that the function graph of distribution function for arm of dynamic dipoles is a straight line cutting-off on the x-axis a segment whose length is h_{max}, then the density function is restored as follows:

$$f_h(x) = 2h_{max}^{-1}(1 - xh_{max}^{-1}), \qquad (4)$$

and the mathematical expectation is: $<h> = h_{max}/3$.

Given that, $\gamma_d = 0.5F_d^{-1}$, $F_d \approx 0.3$, $\ell_d \approx 8\rho^{-1/2}$ for the maximum value of the dipole lever from the dislocation in dynamic dipole configurations and dislocation fragments we get $h_{max} \approx 0.43\xi\rho\,\gamma_d \ell_d^2\sqrt{6}b/3 \approx 18b$, and for average value - $<h> \approx 6b$.

At the intersection of dislocations with the opposite signs the jogs that can be annihilated, moving along the dislocation line are formed on the screw segments. This reduces the deceleration force of shear-forming dislocation, increases ℓ_d and decreases the number of generated elementary dipoles. If ℓ_d is close to the shear zone diameter, then the dynamic dipoles are not formed.

The jogs fraction w_j that can annihilate is determined by their distribution on screw segments by the sign. The value w_j is being decreased with the increasing of the excessive density of dislocations creating a local disorientation. At the intersection of the area with the excessive density of dislocations on a segment by screw segment, the jogs of the same sign are formed, that lead to a significant inhibition of the segments, reduce ℓ_d and increase the number of the generated elementary dipoles. Furthermore, by virtue of the fact that w_j is evidently proportional to $|\beta_j^+ - \beta_j^-|$, then with the excessive dislocation density increasing, the maximum value of the dipole arm is reduced.

Dipole with an arm of a few interatomic distances that are thermodynamically unstable, so a dipole arm decrease leads to a rate dissociation of interstitial dipoles and annihilation of vacancy dipoles under the influence of the forming interstitial atoms. This fact seems to be the reason that with increasing of the strain degree and excessive dislocation density the NMDCS is a small factor in the hardening by the time of fragmented substructure formation is not observed.

4. Summary

It is shown that with increasing degree of heterogeneity of distribution of dislocations due to the formation of excess dislocation density, the following occurs: decreasement in the mean free path of screw segments, increasement in number of elementary dipoles and decreasement in the dipole arm. As a result of the dissociation of the interstitial dipoles with a small arm the annihilation of dipoles becomes more intense and rapid disappearance of the non-misoriented dislocation cell substructure occurs.

References

[1] Howie A. Direct Observations of Imperfections in Crystals, (Interscience Pub., Inc., N. Y., 1961), p.283.

[2] Sevillano J.G., Houtte P., Aernoudt E.// Progr. Mater. Sci.-1981, v. 25, p. 69-412.

[3] Malygin G.A. Dislocation self-organization processes and crystal plasticity. // Phys. Usp.-1999, Vol. 42, pp. 887 – 916.

[4] Koneva N.A., Starenchenko V.A., Lychagin D.V., Trishkina L.I., Popova N.A., Kozlov E.V. Formation of dislocation cell substructure in face-centred cubic metallic solid solutions. // in: Materials Science and Engineering A, Vol. 483-484, № 1-2 C, 2008, pp. 179-183.

[5] Murr L.E., Kuhlmann-Wielsdorf D. Experimental and theoretical observations on the relationship between dislocation cell size, dislocation density, residual hardness, peak pressure and pulse duration in shock-loaded nickel.// Acta Met.- 1978, v. 26, № 5, p. 847-857.

[6] Greulich F., Murr L.E. Effect of grain size, dislocation cell size and deformation twin spacing on the residual strengthening of shock-loaded nickel.// Mater. Sci. and Eng.-1979, v. 39, № 1, p. 81-93.
[7] Mughrabi H. Elektronenmikroskopische Untersuchung der Versetzungsanordnung verformter Kupfereinkristalle im belasteten Zustang.// Phil. Mag.- 1971, v. 23, № 184, p. 869-895.
[8] Essman V., Mughrabi H. Annihilation of dislocations during tensile and cyclic deformation and limits of dislocation densities.// Phil. Mag.- 1979, v. 40, № 6, p. 731-756.
[9] Ambrosi P., Gottler E., Schwink Ch.// Scripta Met.- 1974, v. 8, № 8, p. 1093-1098.
[10] Tabata T., Imanaka Sh., Fujita H. In situ deformation of the single crystals [111] Al observed by high voltage electron microscopy.//Acta met.- 1978, v. 26, p. 405-414.
[11] Kawasaki J., Takeuchi T.// Scripta Met.- 1980, v. 14, № 3, p. 183-188.
[12] Ambrosi P., Homeier W., Schwink Ch.// Scripta Met.- 1980, v. 14, № 3, p. 325-329.
[13] Chow C.K., Nembach E.// Acta met.- 1976, v. 24, p. 453-462.
[14] Luft A.// Phys. stat. sol.- 1970, v. 42, № 1, p. 429-440.
[15] Kocanda S. Fatigue failure of metals.- Springer, Engelska, 2011.
[16] Staker M.R., Holt D.L. The dislocation cell size and dislocation density in copper deformed at temperatures between 25 and 700°C// Acta Met.- 1972, v. 20, № 4, p. 569-579.
[17] Holt D.L. Dislocation Cell Formation in Metals.// J. Appl. Phys.- 1970, v. 41, № 8, p. 3179-3201.
[18] Starenchenko V.A., Cherepanov D.N., Solov'eva Y.V., Popov L.E. Generation and accumulation of point defects in FCC single crystals upon plastic strain // Russian Physics Journal.- 2009.- T. 52, № 4, C. 398-410.
[19] Starenchenko V.A., Cherepanov D.N., Selivanikova O.V. Modeling of plastic deformation of crystalline materials on the basis of the concept of hardening and recovery.// Russian Physics Journal.- 2014, Volume 57, Issue 2, June 2014 (Russian Original So. 2. February. 2014), pp. 139-151.
[20] Starenchenko V.A., Cherepanov D.N., Selivanikova O.V., Barbakova E.A. Formation of Shear Zone's Defect Structure in F.C.C. Metals // Advanced Materials Research.- 2015, Vol. 1084, pp. 26-29.
[21] Starenchenko V.A., Cherepanov D.N., Selivanikova O.V. Generation of interstitial atoms in f.c.c. single crystals // Russian Physics Journal.- 2015, Vol. 58, № 4, pp. 446-453.

Radiopharmaceuticals for SPECT Cancer Detection

V I Chernov[1,2], A A Medvedeva[1,2], R V Zelchan[1,2], I G Sinilkin[1,2], E S Stasyuk[2], L A Larionova[2], E M Slonimskaya[1], E L Choynzonov[1]

[1]Tomsk Cancer Research Institute, Tomsk, Russia
[2] National Research Tomsk Polytechnic University, Tomsk, Russia

E-mail: chernov@oncology.tomsk.ru

Abstract. The purpose of the study was to assess the efficacy of single photon emission computed tomography (SPECT) with ^{199}Tl and ^{99m}Tc-MIBI in the detection of breast, laryngeal and hypopharyngeal cancers. Materials and Methods: a total of 220 patients were included into the study. Of them, there were 120 patients with breast lesions (100 patients with breast cancer and 20 patients with benign breast tumors) and 100 patients with laryngeal/hypopharyngeal diseases (80 patients with laryngeal/hypopharyngeal cancer and 20 patients with benign laryngeal/hypopharyngeal lesions). Results: no abnormal ^{199}Tl uptake was seen in all patients with benign breast and laryngeal lesions, indicating a 100% specificity of ^{199}Tl SPECT. In breast cancer patients, increased ^{199}Tl uptake in the breast was visualized in 94.8% patients, ^{99m}Tc-MIBI in 93.4% patients. Increased ^{199}Tl uptake in axillary lymph nodes was detected in 60% patients and ^{99m}Tc-MIBI in 93.1% patients. In patients with laryngeal/hypopharyngeal cancer, sensitivity of SPECT with ^{199}Tl and ^{99m}Tc-MIBI were 95%. The ^{199}Tl SPECT sensitivity in identification of regional lymph node metastases in patients with laryngeal/hypopharyngeal cancer was 75% and the ^{99m}Tc-MIBI SPECT sensitivity was 17%. Conclusion: the data obtained show that SPECT with ^{199}Tl and ^{99m}Tc-MIBI can be used as one of the additional imaging methods in detection of tumors.

1. Introduction

According to recent WHO data, more than 10 million newly diagnosed cancer cases occur annually in the world. In many countries, cancer ranks the second most common cause of death following cardiovascular diseases [1].

Nuclear medicine techniques have the advantage over X-ray or magnetic resonance imaging in terms of higher specificity with a similar high sensitivity for cancer detection [2, 3 ,4]. Moreover, the use of nuclear medicine imaging in detection of laryngeal/hypopharyngeal tumors allows prevention of unnecessary diagnostic invasive procedures.

^{99m}Tc-methoxy-isobutyl-isonitrile (^{99m}Tc-MIBI) is the most widely used single photon emission computed tomography (SPECT) tracer in detecting malignant disease [5, 6[6]]. However, this tracer with a relatively high sensitivity has a rather low specificity. Therefore, the search for radiopharmaceuticals characterized by a high specificity while maintaining adequate sensitivity is of great importance.

More than 30 years ago, the diagnostic value of ^{201}Tl for detecting malignant tumors was studied. It was shown that ^{201}Tl, a biological potassium analogue, actively accumulated in tumors [6[6]]. Findings from animal studies demonstrated that ^{201}Tl had higher tumor accumulation compared to other radiopharmaceuticals [8, 9[9]]. However, a long half-life (72 hours) of ^{201}Tl limited its widespread use, since there was a risk of increased radiation dose to the patient's body and critical organs, especially during re-examinations, for example during chemotherapy.

It is proposed that ^{199}Tl is a more promising isotope than ^{201}Tl, which is currently used in nuclear medicine. Unlike ^{201}Tl, ^{199}Tl has a more simple and cheaper production method. There are many reports on the use of this isotope for myocardial perfusion scintigraphy. Cellular uptake mechanism of ^{199}Tl is the same as that of ^{201}Tl. However, because of its short half-life (7.4 hours), radiation dose to the patient's whole body and critical organs is significantly reduced. By using ^{199}Tl, the radiation dose to critical organs is 4-15 times lower than that by using ^{201}Tl. It allows ^{199}Tl scintigraphy to be performed 5 times a year for the same patient [10, 11, 12].

The aim of our study was to evaluate the efficacy of SPECT with ^{199}Tl in the detection of breast and laryngeal/hypopharyngeal carcinomas.

2. Experiment

A total of 220 patients were included into the study. There were 120 women with breast disease. Out of them, 39 patients with stage $T_{1\text{-}4}N_{0\text{-}2}M_0$ breast cancer and 20 patients with benign breast lesions were injected with ^{199}Tl, and 61 patients with stage $T_{1\text{-}4}N_{0\text{-}2}M_0$ breast cancer were injected with ^{99m}Tc-MIBI. There were 80 patients with stage $T_{1\text{-}3}N_{0\text{-}2}M_0$ laryngeal and hypopharyngeal cancers (including 20 patients with recurrent laryngeal/hypopharyngeal cancer) and 20 patients with benign diseases, among them, 50 patients were injected with ^{199}Tl and the other 50 patients were injected with ^{99m}Tc-MIBI. None of the patients at the time of the survey showed evidence of distant metastasis.

Besides conventional imaging methods such as computed tomography, ultrasound, fibrolaryngoscopy and biopsy, all patients underwent preoperative SPECT with ^{199}Tl or ^{99m}Tc-MIBI. The patients were subsequently operated on and surgical specimens were examined histologically and immunohistochemically. Patients with benign breast lesions underwent sectoral resection of the breast with subsequent histological examination.

Single photon emission computed tomography was performed using a double-head gamma-camera (E.CAM 180, Siemens) equipped with parallel high energy collimators for ^{199}Tl and parallel low energy collimators for ^{99m}Tc-MIBI. The injection of ^{199}Tl at the dose of 260 MBq or ^{99m}Tc-MIBI at the dose of 740 MBq was made intravenously. In patients with breast disease, injection was made into the antecubital vein contralaterally to the breast lesion or in the pedal vein. Images of the breast were obtained with the patients lying in the supine position with the arms raised over the head and with the chest, both breasts, axilla and the myocardium, included into the field of view. For obtaining SPECT images in patients with laryngeal/hypopharyngeal cancer, patients were also placed in a supine position, and the detector's field of the view encompassed the head and neck region and trachea to the bifurcation. A total of 32 projection images were recorded into a 64×64 matrix (30 seconds per projection), without magnification factor. The scan images were analyzed using the manufacturer software (e.soft, Siemens, Germany). Three-dimensional images of the chest, sagittal, transverse and coronal sections were obtained. Single photon emission computed tomography scans were visually assessed. Images of contralateral areas were compared and asymmetrically increased radiotracer uptake was considered pathological.

3. Results and discussion

Single photon emission computed tomography revealed no abnormal ^{199}Tl uptake in the chest wall and breasts of patients with benign breast lesions. Despite a small group of patients (n=20), the specificity of SPECT with ^{199}Tl was 100%.

In breast cancer patients, the increased ^{199}Tl uptake was visualized in 37 (94.8%) of the 39 patients, of the 6 patients with tumors less than 10 mm, 4 showed abnormal radiotracer uptake (Figure 1). In one case, a large mass detected by mammography and ultrasound appeared as a cluster of microcalcifications. Carrying out of SPECT with ^{99m}Tc-MIBI revealed abnormal formations in the projection of the mammary glands in 59 (93.4%) out of 61 patients (Figure 2). Tumors greater than 10 mm were detected by ^{199}Tl and ^{99m}Tc-MIBI SPECT in 100% of cases.

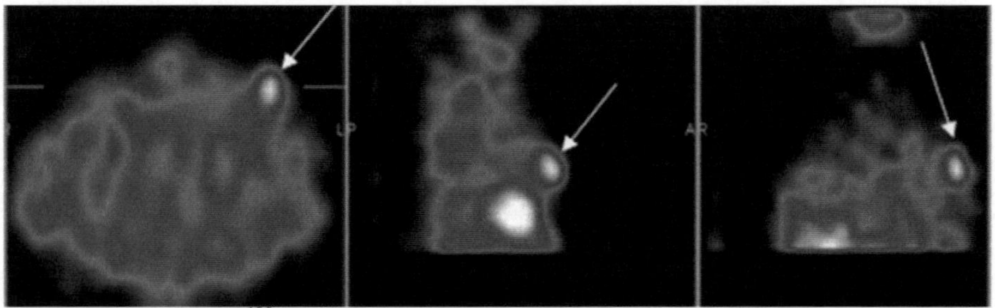

Figure 1. SPECT with ^{199}Tl for the patient with stage T2N0M0 cancer in the left breast.

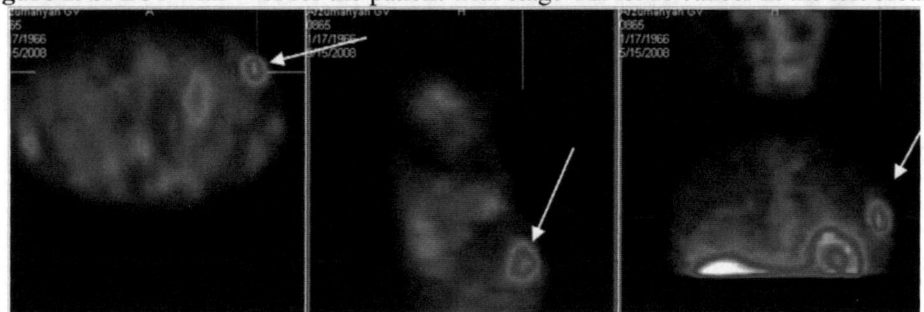

Figure 2. SPECT with ^{99m}Tc-MIBI for the patient with stage T1N0M0 cancer in the left breast.

It is well known that secondary infiltrative ductal breast carcinoma can be difficult to diagnose. In our study, the accuracy in differentiating malignant from benign breast lesions by SPECT with ^{199}Tl was 100%, as well as with ^{99m}Tc-MIBI (Figures 3,4). Identification of multicentric breast cancer is an important factor in diagnosis of breast cancer. In our study, multicentric breast cancer was diagnosed in 66.6% patients during the study with ^{199}Tl and in 46.1% with ^{99m}Tc-MIBI (Figures 5,6).

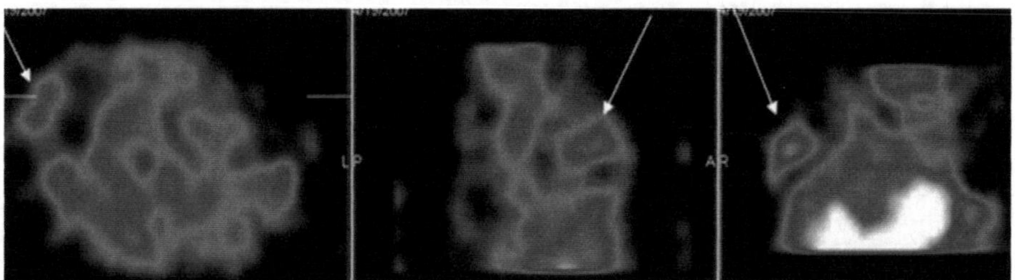

Figure 3. SPECT with ^{199}Tl for the patient with secondary infiltrative ductal breast carcinoma in the right breast.

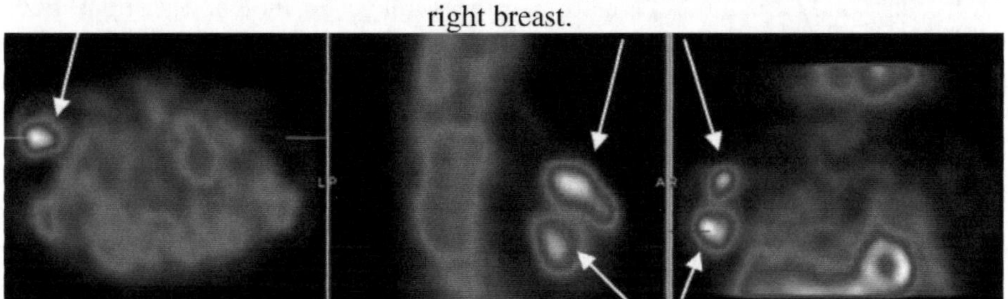

Figure 4. SPECT with ^{99m}Tc-MIBI for the patient with secondary infiltrative ductal breast carcinoma in the right breast.

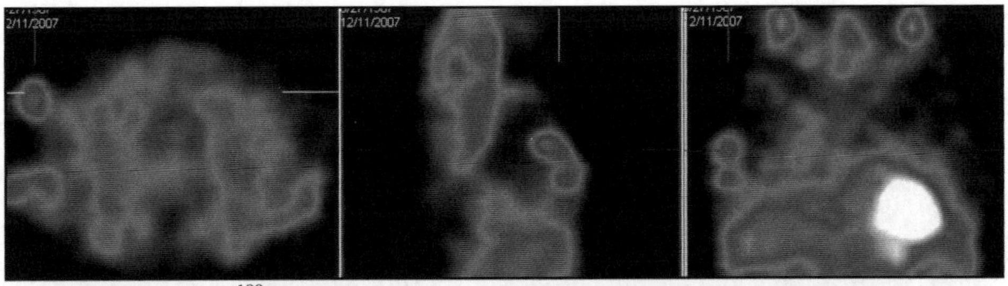

Figure 5. SPECT with ^{199}Tl for the patient with multicentric breast cancer in the right breast.

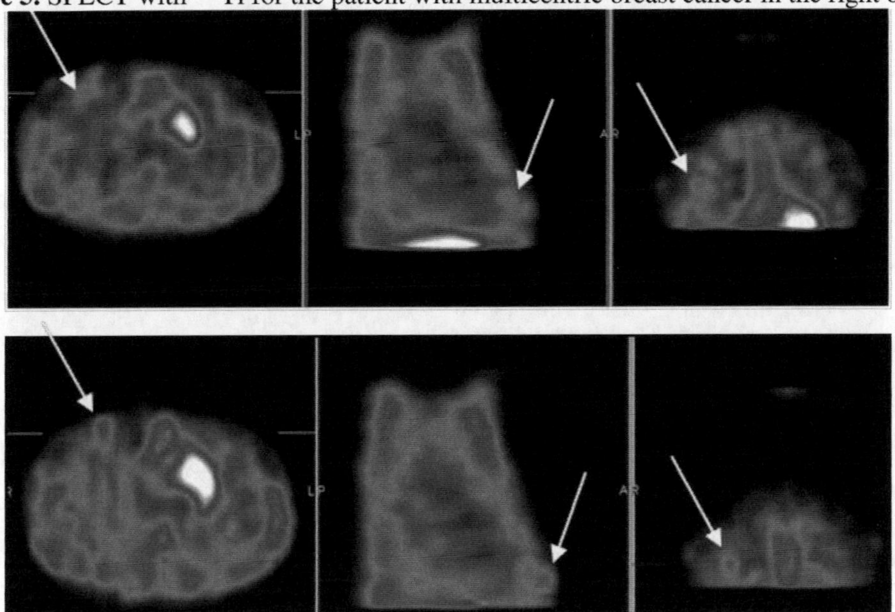

Figure 6. SPECT with ^{99m}Tc-MIBI for the patient with multicentric breast cancer in the right breast.

As regard to lymph node involvement, increased ^{199}Tl uptake in the axillary lymph nodes was observed in 15 of the 25 patients (60%) with histologically verified metastases (Figure 7). Only 2 of the 5 patients had abnormal radiotracer uptake in the pectoral and subclavicular lymph nodes (Figure 8). Thus, the ^{199}Tl SPECT sensitivity for the detection of metastatic regional lymph nodes in breast cancer patients was 60%.

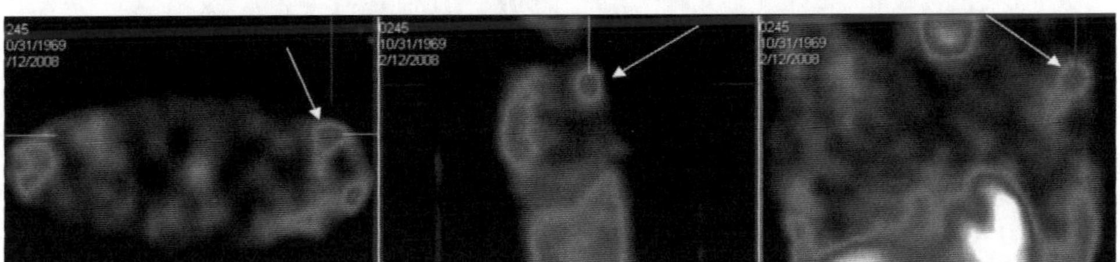

Figure 7. SPECT with ^{199}Tl in the patient with stage T2N1M0 cancer in the left breast. Axillary lymph node visualization.

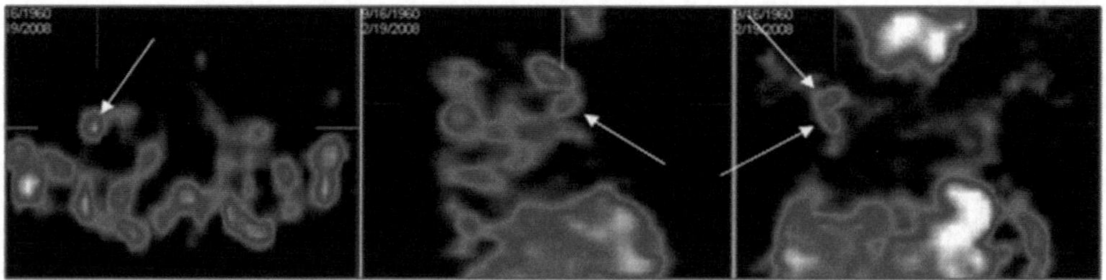

Figure 8. SPECT with ^{199}Tl in the patient with stage T2N3M0 cancer in the right breast. Visualization of pectoral and subclavicular lymph nodes.

It was possible to visualize the modified axillary lymph nodes in almost all cases during the study with ^{99m}Tc-MIBI: in 27 (93.1 %) out of 29 patients with verified metastasis pathological inclusion of the ^{99m}Tc-MIBI in axillary region was detected (Figure 9). In addition, pathological changes in the projection of pectoral lymph nodes, subclavian and subscapularis groups were detected in 5 patients. This technique allowed us to differentiate between different levels of lymph nodes (Figure 10). The sensitivity of the SPECT with ^{99m}Tc-MIBI in identifying metastatic regional lymph nodes was 93.1 %. The specificity of SPECT with ^{199}Tl was 100%, with ^{99m}Tc-MIBI – 80%.

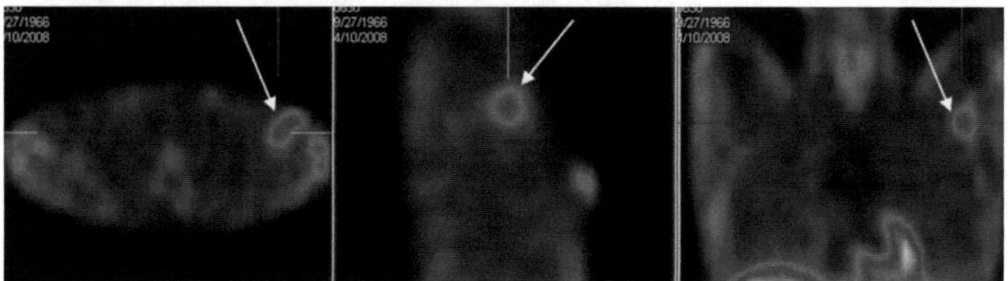

Figure 9. SPECT with ^{99m}Tc-MIBI in the patient with stage T2N1M0 cancer in the left breast. Axillary lymph node visualization.

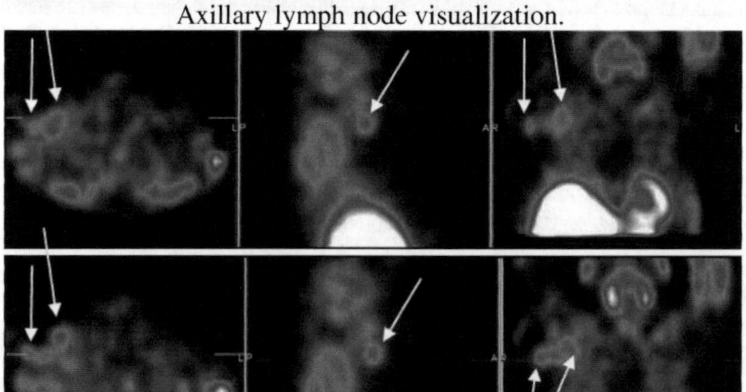

Figure 10. SPECT with ^{99m}Tc-MIBI in the patient with stage T2N3M0 cancer in the right breast. Visualization of pectoral, subclavicular lymph nodes.

In laryngeal/hypopharyngeal cancer patients, abnormal both ^{199}Tl and ^{99m}Tc-MIBI uptake was detected in 95 % patients with primary laryngeal/hypopharyngeal cancer (Figures 11, 12). It should be noted that despite limitations for SPECT in visualization of tumors less than 10 mm. The minimum lesion size was 4 mm in greatest dimension.

IOP Conf. Series: Materials Science and Engineering **135** (2016) 012008 doi:10.1088/1757-899X/135/1/012008

In 20 of the 20 patients with recurrent laryngeal/hypopharyngeal cancer was detected abnormal uptake both ^{199}Tl and ^{99m}Tc-MIBI, it was 100 % sensitivity of SPECT with ^{199}Tl and ^{99m}Tc-MIBI in the diagnosis of recurrent laryngeal/hypopharyngeal cancer.

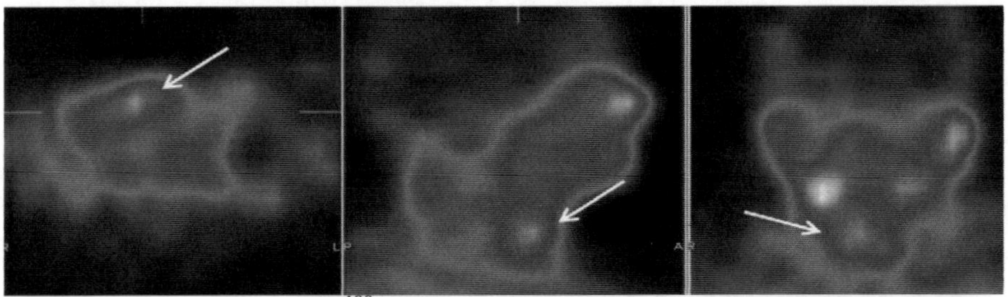

Figure 11. SPECT with ^{199}Tl for the patient with primary laryngeal cancer.

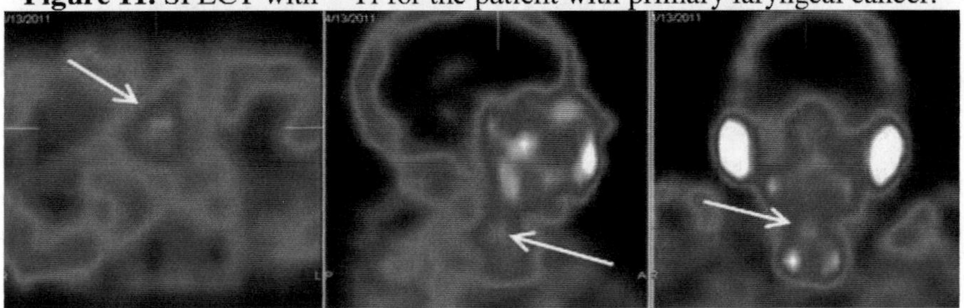

Figure 12. SPECT with ^{99m}Tc-MIBI for the patient with primary laryngeal cancer.

Histological examination of specimens obtained by ultrasound-guided fine needle aspiration biopsy showed cervical lymph node metastases in 18.3% of cases. On SPECT the lymph nodes, affected by metastases, were visualized in the form of tracer hyperfixation focus of irregular shape with rough, indistinct boundaries, expressed by a nonuniform distribution of tracer, usually of low intensity.

Single photon emission computed tomography with ^{199}Tl could detect biopsy-proven lymph node metastases in 75 % of cases (Figure 13). When performing SPECT with ^{99m}Tc-MIBI we were able to identify the accumulation of the drug in the projection of lymph nodes, affected by metastases, in 2 (17%) of 12 patients (Figure 14). We assume, that such a low information content of SPECT with ^{99m}Tc-MIBI in the detection of metastases in regional lymph nodes is primarily related to their small size and physiological, initially high accumulation of ^{99m}Tc-MIBI in the projection of the thyroid and submandibular salivary glands, which often screens the zones of the regional lymph collector (Figure 15). This effect is less pronounced when using of ^{199}Tl (Figure 16).

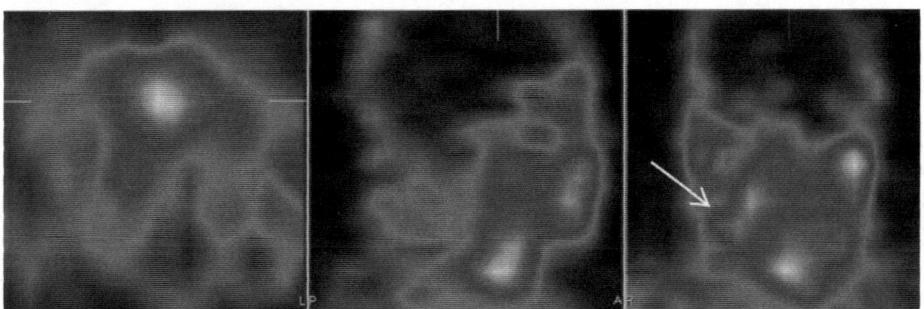

Figure 13. SPECT with ^{199}Tl for the patient with primary laryngeal cancer and cervical lymph node metastases.

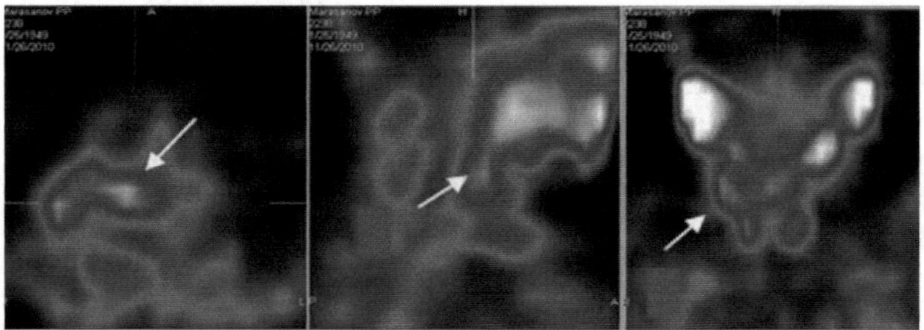

Figure 14. SPECT with ^{99m}Tc-MIBI for the patient with primary laryngeal cancer and cervical lymph node metastases.

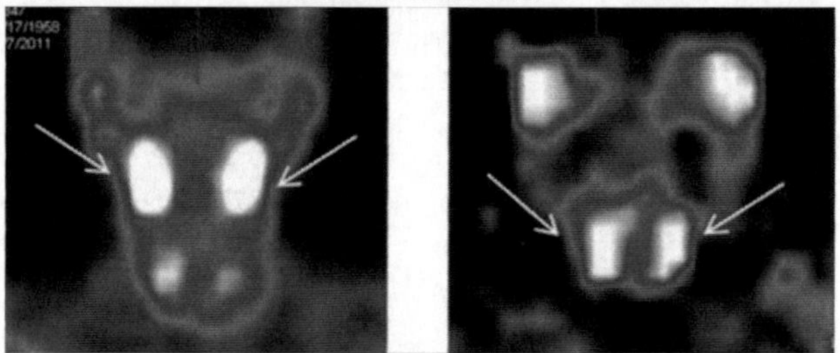

Figure 15. Screens the zones physiological accumulation of ^{99m}Tc-MIBI in the projection of the thyroid gland and submandibular salivary glands.

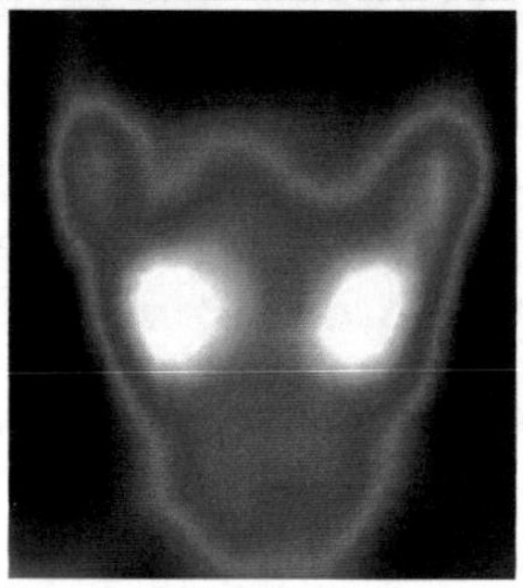

Figure 16. Physiological accumulation of ^{199}Tl.

4. Conclusion

Our first experience of using ^{199}Tl has clearly indicated that this radiopharmaceutical represents a useful diagnostic tool for the detection of breast and laryngel/hypopharyngeal cancers. Single photon emission computed tomography with ^{199}Tl showed 100% specificity for detecting breast cancer, when compared to ^{99m}Tc-MIBI with reported specificity varying from 85% to 94% [5, 13].

With regard to the assessment of regional lymph node metastasis in patients with laryngeal/hypopharyngeal cancer, ^{199}Tl SPECT demonstrated high specificity, whereas ^{99m}Tc- MIBI failed to evaluate the status of regional lymph nodes due to its physiological high uptake in the thyroid gland.

Despite the above advantages, ^{199}Tl has also disadvantages. This cyclotron-produced radiopharmaceutical has a higher cost compared to ^{99m}Tc, the most widely used radioisotope in nuclear medicine. Moreover, our findings have demonstrated a rather low sensitivity in the detection of regional lymph node metastases in breast cancer patients, thus significantly impairing the diagnostic accuracy.

Anyway, ^{199}Tl is a promising radiopharmaceutical for the detection of malignant tumors, however, further studies are required to determine its role in the cancer diagnostic algorithm.

References

[1] A.D. Caprin, V.V. Starinskay, G.V. Petrov, "Malignant neoplasm in Russia in 2014 (morbidity and mortality)"/ M .: MNIOI P.A. Herzen - 2016. - Fig. – 250p. ISBN 978-5-85502-219-3.

[2] A.M. Granov, L.A. Tyutin, M.S. Tolstanova, "The application of positron emission tomography with 18-fluorodeoxyglucose in clinical oncology", Problems in Oncology (Voprosy onkologii) 49, 5 (2003) 563-573.

[3] A. Aziz, R. Hashmi, Y. Ogawa, "Tc-99m – MIBI scintimammography; SPECT versus planar imaging". J. Cancer Bioter. Radiopharm. 14 (6) (1999) 495-500.

[4] C.K. Kuhl, "MRI of the breast cancer". J. Eur. Radiol. 10 (2000) 46-58.

[5] S.J. Kim, Y.T. Bae, J.S. Lee, I.J. Kim, Y.K. Kim, "Diagnostic performances of double-phase tc-99m MIBI scintimammography in patients with indeterminate ultrasound findings: visual and quantitative analyses". J. Ann. Nucl. Med. 21 (3) (2007) 145-150.

[6] E.L. Choynzonov, R.V. Zelchan, A.A. Titskaya, I.G. Sinilkin, S.Y. Chijevskaya, V.I. Chernov. "SPECT with ^{99M}Tc-MIBI and ^{199}TL-chloride in the diagnosis of primary and recurrent malignant tumors of the larynx and hypopharynx" A Century of progress in head & neck cancer: 5th world Congress of IFHNOS & annual meeting of the AHNS. – 2014, 26-30 july (http://ahns.jnabstracts.com/Detail.aspx?ID=56831).

[7] K. Koizumi, K. Toyama, T Araki, "Uptake of Tc-99m tetrafosmin, Tc-99m MIBI and 201Tl in tumour cell line", J. Nuc. Med. 37 (1996) 1551-1556.

[8] A. Sehweil, J.H. McKillop, R. Milroy, "^{201}Tl scintigraphy in the staging of lung cancer, breast cancer and lymphoma", Nucl. Med. Commun. 1, 4 (1990) 263-269.

[9] R.P. Spencer, "Tumour-seeking radiopharmaceuticals: nature and mechanisms". Nuclear Medicine, 1994, No 2, P. 649-662.

[10] Yu.B. Lishmanov, V.I. Chernov, N.G. Krivonogov, "Experimental perfusion scintigraphy using ^{199}Tl-chloride, Medical Radiology and Radiation Safety". Meditsinskaya radiologiya i radiatsionnaya bezopasnost' 3 (1988) 13-16.

[11] V. Chernov, S. Triss, V. Skuridin, Yu. Lishmanov, "Thallium-199: a new radiopharmaceutical for myocardial perfusion imaging". Int J Cardiac Imaging 1996; 12(2): 119-26.

[12] Yu.B. Lishmanov, V.I. Chernov, S.V. Triss, I.Yu. Mazurin, "Thallium-199 myocardial scintigraphy", Medicinskaa radiologia 4(1990) 35-38.

[13] A.A. Titskaya, V.I. Chernov, E.M. Slonimskaya, I.G. Sinilkin, R.V. Zelchan, "Mammoscintigraphy with 99mTs-MIBI in the diagnosis of breast cancer". Siberian Medical Journal, Tomsk. 2010. Vol. 25. № 4-1. p. 92-95.

A model of multicomponent extraction process in the pulsating column

Aleksandr A Denisevich. Yuriy A Chursin, Oksana P Kabrysheva, Aleksey D Terentev

[1] National Research Tomsk Polytechnic University, 30 Lenina Avenue, Tomsk, Russia

E-mail: denisevichaa@tpu.ru

Abstract. The model of multicomponent process of extraction in pulsating column is presented. Model analysis on adequacy to real apparatus was made. The received results have shown, that the presented model adequacy to the real nuclear fuel-reprocessing apparatus.

1. Introduction

In nuclear reactors there is no complete combustion of fuel, which is determined by either physical destruction of fuel (fuel elements), or premature deterioration of reactivity. Spent fuel elements contain a significant amount of fissile material, the extraction of which is economically profitable.

Nowadays spent nuclear fuel is processed using an extraction technology, which can provide separation of substances. Hardware design of this technology is based on counter current columns connected in cascade.

In order to achieve high performance and high quality products with minimum production costs, the automated control systems (ACS) are used. The implementation of a good-quality automated control system of the extraction process will enhance its safety, improve technological and economic indicators of production as well as reduce the impact on the environment, due to decrease of toxic waste. In the development of automated control systems, it is advisable to use simulation models for the selection of the output variables, for the regulator settings and for the estimation of control quality.

The aim of this work was to develop a universal model of the extraction process in a common extraction column with its further testing by ACS synthesis of extraction process in a real apparatus.

2. A mathematical model of uranium extraction in pulsating columns

The following modeling methods of extraction in column reactors are widely used: the theory of similarity, hydraulic modeling and mathematical modeling [1].

The team of Department of Electronics and Automation of Nuclear Plants and Technology of Tomsk Polytechnic University applied mathematical models for the development of column models.

The equations of convective diffusion nonsteady in time at substance transfer from one phase to another are as follows [1, 2]:

$$\begin{cases} \frac{\partial x}{\partial t} = D_x \frac{\partial^2 x}{\partial h^2} - U_x \frac{\partial x}{\partial h} + W \\ \frac{\partial y}{\partial t} = D_y \frac{\partial^2 y}{\partial h^2} - U_y \frac{\partial y}{\partial h} + W \end{cases} \quad (1)$$

where x and y - uranium concentration in the dispersed and continuous phases; Ux, Uy - the rates of phase flow; Dx, Dy - diffusion coefficients of liquids.

$$W = \beta_1 a(x^* - x) = \beta_2 a(y - y^*) \quad (2)$$

where β1 and β2 - mass transfer coefficients from one phase to another; a - the contact surface of two phases by a unit of volume; x*, y*- equilibrium values of concentration.

The diffusion coefficient of liquids D is so small that even at low rates of liquids the convective substance transport considerably dominates over molecular diffusion. In order to simplify the expression (1) the term of the equation containing D is usually neglected.

The resulting simplified equation of convective diffusion is a partial differential equation, which at a first approximation describes the processes in counter-current columns [2, 3, 4]. However, this equation assumes that the rates Ux, Uy are constant. Even in the first extraction model the flow rate changes depending on time, location and concentration of uranium. In this regard, the equation (1) becomes a nonlinear differential equation, analytical solution of which is difficult.

The analytical method for calculation by extraction stages was applied to solve this problem [1]. It takes into account the spatial distribution in the height of the column. In this method, an extraction column is divided into zones - stages. The extraction stage in packed columns is a part of the column in which equilibrium is established, i.e., the concentration of one of the considered components of the output counter flows at this stage, is characterized by a curve of the equilibrium distribution. A differential equation of material balance is used for each stage. The stages are connected in series. The coefficients are recalculated at each time point.

$$\begin{cases} \frac{d}{dt}\left(x^{out} \cdot V_x + y^{out} \cdot V_y\right) = x^{in} \cdot Q_x^{in} + y^{in} \cdot Q_y^{in} - x^{out} \cdot Q_x^{out} - y^{out} \cdot Q_y^{out} \\ y^{out} = K_p \cdot x^{out} \end{cases} \quad (3)$$

where xin and и yin - input uranium concentration in the dispersed and continuous phase; xout and yout - output uranium concentration in the dispersed and continuous phases; Kp - coefficient of equilibrium distribution of uranium; Vx,Vy - volume of the dispersed and continuous phases; Qx Qy - input and output flow rates of the dispersed and continuous phases.

The developed model is adequate under certain conditions [5], and was used to develop several ACS of uranium concentration in pulsating columns of the "Utjos" unit at the Radiochemical plant of Siberian Chemical enterprise.

However, to enhance functionality and the universalization of ACS development it was necessary to eliminate the following restrictions on the model.

- The model is single-component. In this model it is possible to calculate only uranium concentration in the organic and aqueous phases apart from influence of other components.
- In the model it is impossible to change the number of stages into which the column height is divided. Therefore, the required accuracy of the components distribution by height of the column cannot be obtained.
- Due to the constancy of the uranium distribution coefficient Kp, the model is adequate for a very narrow range of input coordinates.

The developed model does not have these drawbacks. As an example a two-component model will be considered. However, the number of components can always be increased by adding equations to the system which doesn't significantly increase complexity of the model

This mathematical model of extraction column will take the following assumptions:

- The model of ideal displacement is used for the continuous phase [1].
- At each stage equilibrium between dissolved substance content in both phases is established [1, 6].
- Extraction rate is very high (equilibrium is established instantly) [6].

- Fluid temperature in the column is constant. Temperatures of all input and output flows are the same, the heat exchange between the phases and environment is not taken into account.
- It is considered that convective counter-current transfer does not affect the extraction process and, therefore, is not modeled.

The basic equations describing the process of mass transfer in a dynamic mode of the column operation are the equations of the system (1), as well as the equations describing the extraction of nitric acid.

$$\begin{cases} \frac{\partial x}{\partial t} = D_x \frac{\partial^2 x}{\partial t^2} - U_x \cdot \frac{\partial x}{\partial h} + \beta_1 \cdot \alpha \cdot \left(x^*(y) - x\right) \\ \frac{\partial y}{\partial t} = D_y \frac{\partial^2 y}{\partial t^2} - U_y \cdot \frac{\partial y}{\partial h} + \beta_2 \cdot \alpha \cdot \left(y^*(x) - y\right) \\ \frac{\partial N_x}{\partial t} = D_x \frac{\partial^2 N_x}{\partial t^2} - U_x \cdot \frac{\partial N_x}{\partial h} + \beta_{N1} \cdot \alpha \cdot \left(N_x^*(N_y) - N_x\right) \\ \frac{\partial N_y}{\partial t} = D_x \frac{\partial^2 N_y}{\partial t^2} - U_y \cdot \frac{\partial N_y}{\partial h} + \beta_{N2} \cdot \alpha \cdot \left(N_y^*(N_x) - N_y\right) \end{cases}, \quad (4)$$

where x and Ny - nitric acid concentration in the dispersed and continuous phases; β_x, β_y , β_{N1} and β_{N2} individual mass transfer coefficients; N_x* and N_y* - equilibrium concentrations of nitric acid.

To simplify the expression (4) the term of the equation containing D is usually neglected. The system of equations describing the extraction process with regard to the material balance is shown below (5).

The model can be composed from the upper settling section (USS), the lower settling section (LSS) and the extraction section which number of stages is determined by height. Modeling the processes which are observed in the extraction column, the elements are calculated successively at each iteration: USS, the extraction section, LSS, calculation of sensor readings.

$$\begin{cases} \frac{\partial x}{\partial t} = -U_x \cdot \frac{\partial x}{\partial h} + \beta_1 \cdot \alpha \cdot \left(x^*(y) - x\right) + \frac{1}{V_x} \cdot \left(Q_x^{in} \cdot x_{in} - Q_x^{out} \cdot x\right) \\ \frac{\partial y}{\partial t} = -U_y \cdot \frac{\partial y}{\partial h} + \beta_2 \cdot \alpha \cdot \left(y^*(x) - y\right) + \frac{1}{V_y} \cdot \left(Q_y^{in} \cdot y_{in} - Q_y^{out} \cdot y\right) \\ \frac{\partial N_x}{\partial t} = -U_x \cdot \frac{\partial N_x}{\partial h} + \beta_{N1} \cdot \alpha \cdot \left(N_x^*(N_y) - N_x\right) + \frac{1}{V_x} \cdot \left(Q_x^{in} \cdot N_x^{in} - Q_x^{out} \cdot N_x\right) \\ \frac{\partial N_y}{\partial t} = -U_y \cdot \frac{\partial N_y}{\partial h} + \beta_{N2} \cdot \alpha \cdot \left(N_y^*(N_x) - N_y\right) + \frac{1}{V_y} \cdot \left(Q_y^{in} \cdot N_y^{in} - Q_y^{out} \cdot N_y\right) \end{cases}, \quad (5)$$

where x_{in}, y_{in} - input uranium concentration in the aqueous and organic phases, respectively; N_x^{in} , N_y^{in} - input concentration of acid in the aqueous and organic phases, respectively.

3. Synthesis of the automated control system of uranium concentration in the raffinate of EC1 column

The head column of Radiochemical plant of Siberian Chemical enterprise was chosen for checking the adequacy of the model to the real device and for ACS development.

The following parameters can be measured in the column:

- average density of fluid in the column U (1);
- volume of the dispersed phase in the column $L_{злф}$ (2). Measuring vessels are cut in the column. Buoy level meters are installed in the upper parts of the column. They measure weight of liquid columns and their average. In its turn, the average density in the column is proportional to the volume (fraction) of the dispersed phase, which is known as volume of the dispersed phase or dispersed phase delay (DPD) [1, 5].

IOP Conf. Series: Materials Science and Engineering **135** (2016) 012009 doi:10.1088/1757-899X/135/1/012009

- local concentration of the metal in the place of installation *c*U (3).

Model adequacy was checked using F-test and Pearson's chi-squared test. The modeling results correspond to the experimental data in a wider range of parameters than the results presented in [7].

An extraction column can be referred to multi-loop controlled objects. In such objects, each output variable depends on two or more input values. Multi-loop automated control systems with cross-connection are used for automated control of such objects [7] In this paper we study single-loop and cascaded ACS, so it is necessary to choose such an output variable, the control of which could give the best characteristics of the output product.

The analysis of the input variables in the EC1 column revealed the following:

- OP consumption cannot be a controlling variable, as it is set for the whole cascade. Its change will cause some change in an operation mode, which is unacceptable.
- AP consumption can be selected as a controlling variable. Moreover, currently it is the one.
- AP composition is prepared in advance; uranium concentration and acid in it cannot be changed promptly. Therefore, these variables are not satisfactory.
- It is fairly hard to stabilize pulsation amplitude, so this variable also cannot be controlling one.

Thus, the controlling value in the column is AF consumption.

In EC1 column it is possible to control four parameters: the level of the delay dispersion phase in the upper part of the column and the whole column (*L*зв, *L*зн), extract density (which determines the uranium concentration in the extract - cU^{EC}) and the uranium concentration in the middle of the lower extraction section (comparable with uranium concentration in the raffinate - $cU^{РФ}$). As a controlling variable *L*зв (*L*зн) or $cU^{РФ}$ is used. cU^{EC} is not considered due to the fact that this parameter was applied in similar columns, and it gave poor results.

For the development of a control system it is necessary to determine the transfer functions of the aqueous phase consumption, as this parameter is a control action. By all the output variables, TCO can be replaced with aperiodic links of the second order with different time response and delay.

The L_{3H} output parameter will not be used for the ACS development. The character of its response to the changes in AP consumption is comparable to L_{3B}, but it is less sensitive and more inertial.

In order to improve the quality of control it is necessary to turn from the single-loop automated control systems to more complex systems using additional (correction) perturbations impulses or secondary output coordinates. In this case, we used a cascade ACS. Block diagram of the cascade ACS $cU^{РФ}$ is presented in Figure 1.

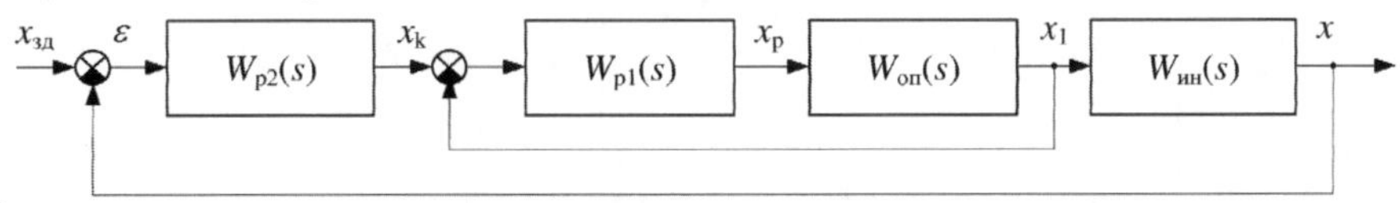

$cU^{РФ}$; $W_{p1}(s)$, $W_{p2}(s)$ - transfer functions of correcting and stabilizing regulators, respectively; $W_{оп}(s)$ - transfer function of the advanced part of the object; $W_{оп}(s)$ - transfer function of the inertial part of the object

Figure 1. Block diagram of the cascade ACS

From the approximate determination of the stabilizing regulator settings we turn to determination of correcting regulator, which circuit includes a regulator with known setting parameter [8].

Experimental study of EC-1 ACS was carried out using a computer model ACS. Perturbations were supplied to system input by all channels, and control quality was estimated.

The analysis of the experiments showed that regulator of the inner loop of cascade $cU^{РФ}$ ACS and the regulator of DPD ACS cannot be applied with the received settings. This system could damage the actuator, so the control coefficient was reduced in order to obtain a smoother signal at the regulator output.

The results of the experiments are presented in Figure 2 – 5. All perturbations were supplied at time moment $t_0 = 1000$ c.

In comparison to the single-loop ACS the cascade $cU^{РФ}$ ACS showed better quality indicators for the perturbations response of OP consumption, and the change of uranium concentration in the AP.

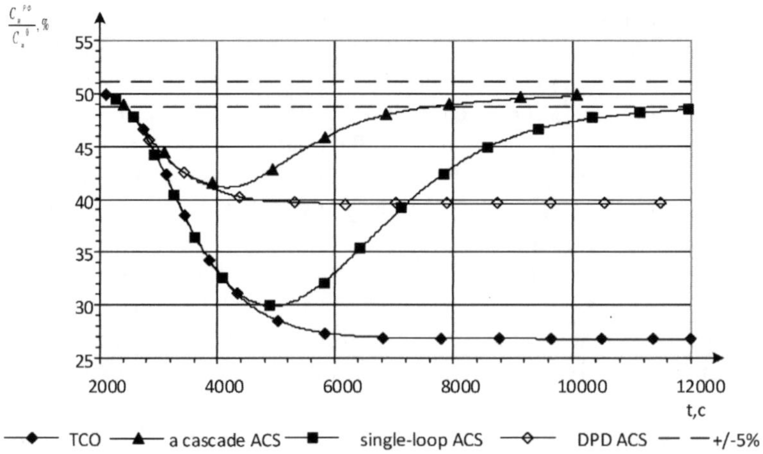

Figure 2. Response of TCO to 5 % perturbations and its influence on OP flow in the cascade, single- loop $cU^{P\Phi}$ ACS and DPD ACS

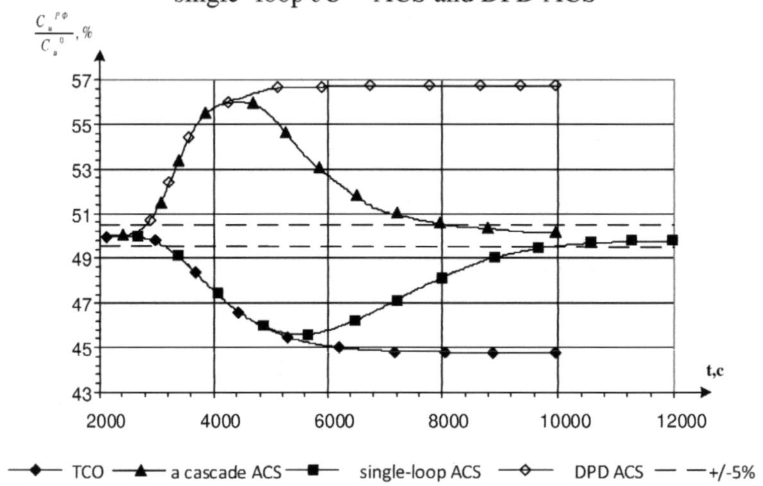

Figure 3. Response of TCO, cascade ACS, single-loop $cU^{P\Phi}$ ACS and DPD ACS to the 5 % perturbations and its influence on uranium concentration in AP

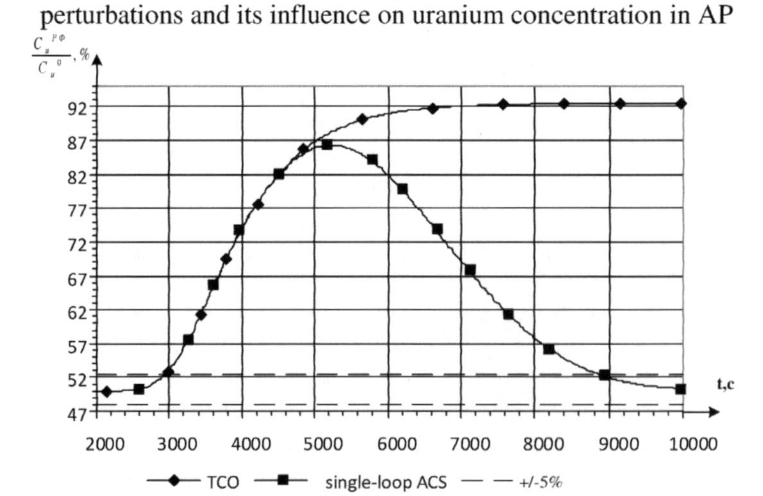

Figure 4. Response of TCO and single-loop $cU^{P\Phi}$ ACS to 7 % perturbations and its influence on AP6 flow rate

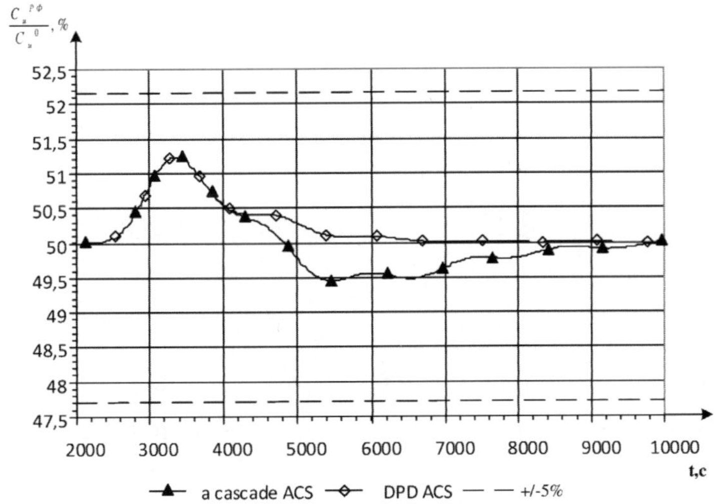

Figure 5. Response of cascade $cU^{Р\Phi}$ ACS and DPD ACS to 7 % perturbations and its influence on AP6 flow rate

4. Conclusion

A dynamic multi-component model based on the solutions of the revised convective diffusion equations under the accepted assumptions, was developed. Hydrodynamic processes in the column were calculated using the results of work.

The comparison of the model to the real apparatus proved its adequacy in a wide range of coordinate variation. On this basis, uranium concentration ACS in EC-1 raffinate was synthesized.

The developed automated control system of the extraction process will significantly improve technical and economic indicators of production by increasing the purity degree of the produced uranium and reducing the amount of uranium in the raffinate. ACS is recommended for implementation.

The work resulted in the development of a universal multi-component model which can be used for the extraction columns operating in different modes and with different types of spent nuclear fuel.

References

[1] Gafarov V V, Glebov M B 1991 *Mathematical modeling of the basic processes of chemical production* 400 p.

[2] Bagrov V G, Belov V V, Zadorozhnyy V N, Trifonov A Yu 1999 *Methods of Mathematical Physics*. Special functions. Equations of mathematical physics 771 p.

[3] Franks R 1971 *Mathematical modeling in chemical engineering* 272 p.

[4] Liventsov S N, Lysenok A A, Goryunov A G, Zelenetskaya E P, Simogaev G A 2004 *The research report "Development of control systems for the units of technological processes at a radiochemical plant."*- 56 p.

[5] Chursin Yu A, Goryunov A G, Liventsov S N 2007 *Automated control system of extraction column. Automation, Computer Science and Information technology* 149 p.

[6] Balashov A A, Lossievsky V L, Chernyshov V N, Schwab A F, Shelemin B V 1963 *Automation schemes and facilities of radiochemical production* 187 p.

[7] Puzikov E A, Silberman B J, Fedorov S, Mishin E N, Schmidt O V, Goletsky N D, Sitnic L V 2004 *Development of a mathematical model for calculation of steady-state modes of extraction cascades at NPP SNF processing using tributyl phosphate-compliant processes (Superpureks)* pp. 136-143

[8] Kulakov G T 2003 *Analysis and synthesis of automated control systems* 134 p.

Advanced ultrasonic testing of complex shaped composite structures

D Dolmatov, V Zhvyrblya, G Filippov, Y Salchak and E Sedanova

National Research Tomsk Polytechnic University, 634050 Tomsk, Russia

yasalchak@gmail.com

Abstract. Due to the wide application of composite materials it is necessary to develop unconventional quality control techniques. One of the methods that can be used for this purpose is ultrasonic tomography. In this article an application of a robotic ultrasonic system is considered. Precise positioning of the robotic scanner and path generating are defined as ones of the most important aspects. This study proposes a non-contact calibration method of a robotic ultrasonic system. Path of the scanner requires a 3D model of controlled objects which are created in accordance with the proposed algorithm. The suggested techniques are based on implementation of structured light method.

1. Introduction

Nowadays advanced technologies require development of high-quality materials. Carbon fiber composites are widely used for construction of complex shaped objects. The wide range of its properties, such as high stiffness, low weight, high tensile strength and temperature tolerance, make them very useful for different areas, such as nuclear industry as well. For instance, carbon fiber reinforced silicon carbide (SiCf/SiC) was suggested for further development of the very high temperature reactor (VHTR) due to its ability to tolerate higher operating temperature [1].

However, this complex structure of materials causes the necessity of unconventional testing techniques. One of the most promising methods is ultrasonic tomography due to the ability of visualization of the results with quite high accuracy.

Testing procedure consists of several steps and each of them effects the accuracy of the final results. This study suggests an ultrasonic testing technique applying a robotic scanner. The major challenge in this case is making setting of the scanner. Therefore, the precision of positioning of the robotic scanner and generating the right path for the scanner is very important.

Positioning requires relevant correlation between coordinate system of the scanner and given points of the controlled object. A non-contact calibration method was proposed for this purpose. The scanning path can be generated in accordance to 3-D model of the surface of the controlled object. This paper suggests creating of the model by means of 3-D reconstruction techniques based on the structured light method.

2. General aspects of testing technique

Methods of positioning of the scanner and controlled object were studied for robotic ultrasound tomography system (see Fig.1). This system consists of ultrasound electronics unit, scanner and immersion tank. Ultrasound electronics unit itself contains the unit of control of the path and PC for displaying of the results.

Ultrasound electronics unit performs mainly the functions connected with transmitting and receiving acoustic signals, multiplexing of channels, digitizing of the signals, data transfer and processing. This unit has the following technical parameters:

- 128 multiplexed channels (16 parallel channels)
- capacity of 200 MB/s
- voltage of transmitting module: from -50 to -180 V
- the minimum of input sensitivity: 100 mkV
- the maximum of input voltage: 10 V
- the resolution of AD converter: 14 bit
- sampling frequency: 20 MHz
- AD buffer: 64 000 measurements.

Figure 1. Robotic ultrasound tomography system

2.1. Non-contact calibration of a robotic scanner

As it was mentioned above, for accurate positioning it is important to estimate the relative position of the scanner and the controlled object. At present, contact calibration methods are applied for this purpose. These techniques imply that an operator determines each point of the object manually. Unfortunately, it can lead to huge uncertainties in the results and is very time-consuming.

That is why, we suggest to implement a noncontact calibration method. The proposed method applies 3D- scanner system that will make the process of determination of the coordinates automatic which means less time-consuming and more accurate.

Industrial robot is applied as an automatic scanner. It was developed by German company KUKA, model number KR10R1100 six WP. The scanner performs positioning of the transducer on the given distance from the controlled object.

IOP Conf. Series: Materials Science and Engineering **135** (2016) 012010 doi:10.1088/1757-899X/135/1/012010

The proposed calibration technique is based on structural light method. This method is commonly used for 3-D reconstruction. It is based on the projection of vertical and horizontal lines at the surface of the object in a strict order. Digital projector is used as a light source for this purpose. At the same time as projection is made the received image should be captured by photo or video camera. Further data processing of the images enables to get points values of the controlled object.

The main feature of proposed non-contact calibration technique is implementation of a special calibration disk (see Fig.2). This disk is made in a shape of a triangle and is to be placed at the surface of the robotic manipulator. The measured points within the manipulator are located in the scanning area according to three corners of the disk. The results of scanning occur in a point cloud that can be used for the correlation between coordinate system of the robotic manipulator and an object (see Fig.3).

Due to mathematical calculations it is sufficient to have not less than 21 values of measured points. That has determined the choice of the calibration disk shape. Nine coordinates corresponding to the location of the triangle corners are known and the other nine coordinates can be easily determined in accordance to zero position. These values within three coordinates of the given point at the scanner flange allow to form a matrix of values. Necessary calculations can be made in accordance with [2] or can be realized by PCL libraries.

Figure 2. Implementation of the calibration disk

The results of testing with both contact and non-contact calibration methods demonstrated almost the same values [3].

Taking into account the fact that non-contact technique is simpler, less time consuming and more convenient we propose to apply this method for effective ultrasonic testing by a robotic scanner.

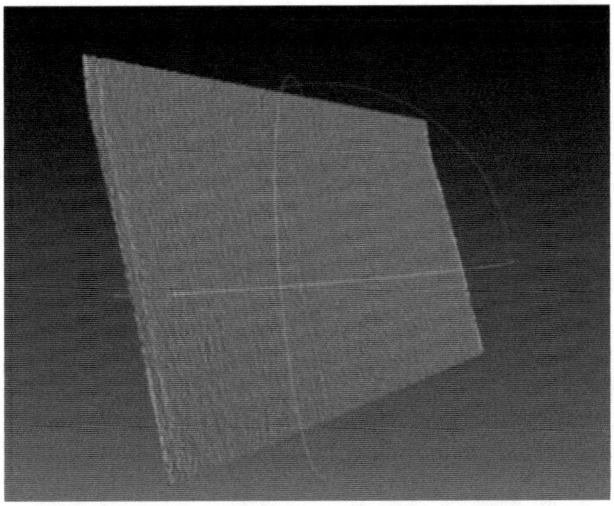

Figure 3. The point cloud of measured values

2.2. Generation of 3D CAD model

The other important aspect that effects total accuracy of testing is positioning of the scanner. It implies precise direction of the movement path. It is common to generate the path of the scanner according to the created model of an object. More accurate the model is, more useful it will be for setting the right path.

Three-dimensional reconstruction based on structured light techniques was chosen as approach for model creation. The experimental setup was described in the introduction of the second paragraph.

We propose a method of ultrasonic testing for complex shaped objects. These objects can have roughened surfaces or changed structure. Therefore, it is necessary to create a model of the object based on its condition at the time when inspection is required.

Moreover, it is evident that several defects might have different location depth. That is why it was suggested to define an object into several sectors according to depth. For example, in case there are no gates there might be the situation when signals from a defect interfere with the signal from the surface. Also, in order to improve the quality of 3-D reconstruction it was suggested to define a sector of the object by covering its surface with matte white coating.

Unlike calibration procedure the model generation requires a range set of values. The point cloud in accordance with the controlled object geometry is formed as a result of the scanning procedure.

Robotic ultrasound tomography system consists of the six-axes scanner KUKA. The special software developed by KUKA allows to set the path for the scanner if solid model of the object is applied. However, it is only possible to generate polygonal model by using a point cloud. The algorithm of scanner path setting allows implementation of T-splines or NURBS-surfaces. These splines are curves created by a complex mathematical formula that enable generation of solid models. Thus, it was suggested to transform point cloud into NURBS-surfaces.

The transformation procedure includes several steps:

1. Transformation of sparse point cloud into regular
2. Creation of the polygonal model by triangulation
3. Generation of the merged grid and its normalization
4. Division of the grid into T-splines
5. Transformation of T-splines into NURBS-surfaces

3. Experiment and results

A complex shaped composite object within 9 artificial defects is chosen as an experimental sample (See Fig. 4). There are three rows of three flat-bottom holes with diameter of 3, 6 and 9 mm. Each row is located at the different depth: 1/3; ½; 2/3 of wall thickness.

An experimental set up consists of a robotic system and single channel equipment with a frequency of 15 MHz (OLYMUS, 2-inch focus transducer).

This robotic system consists of ultrasound electronics unit, scanner and immersion tank described in paragraph 2. Ultrasound electronics was developed by I-Deal Technologies GMBH in collaboration with National Research Tomsk Polytechnic University.

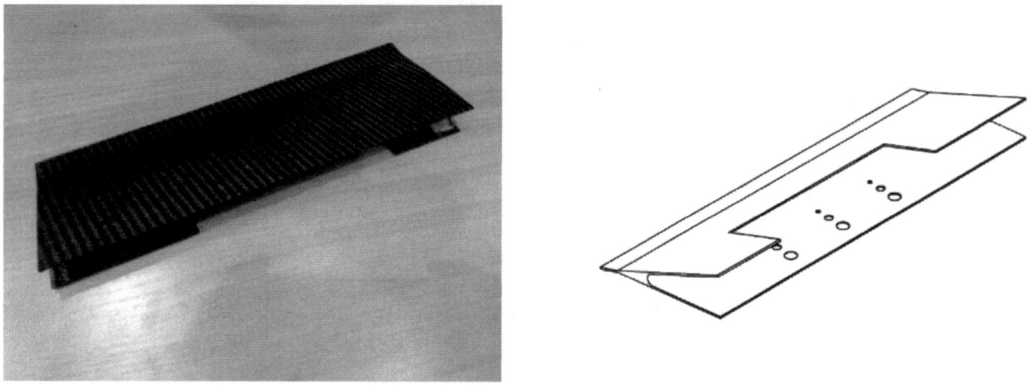

Figure 4. Experimental sample

The received results for the depth of 1/3 of the wall thickness are demonstrated in the Figure*. Three defects with different diameters are determined with high accuracy. Taking into account that the wall thickness is 1.8 mm we have estimated that the dead zone is about 0.5 mm which is rather small.

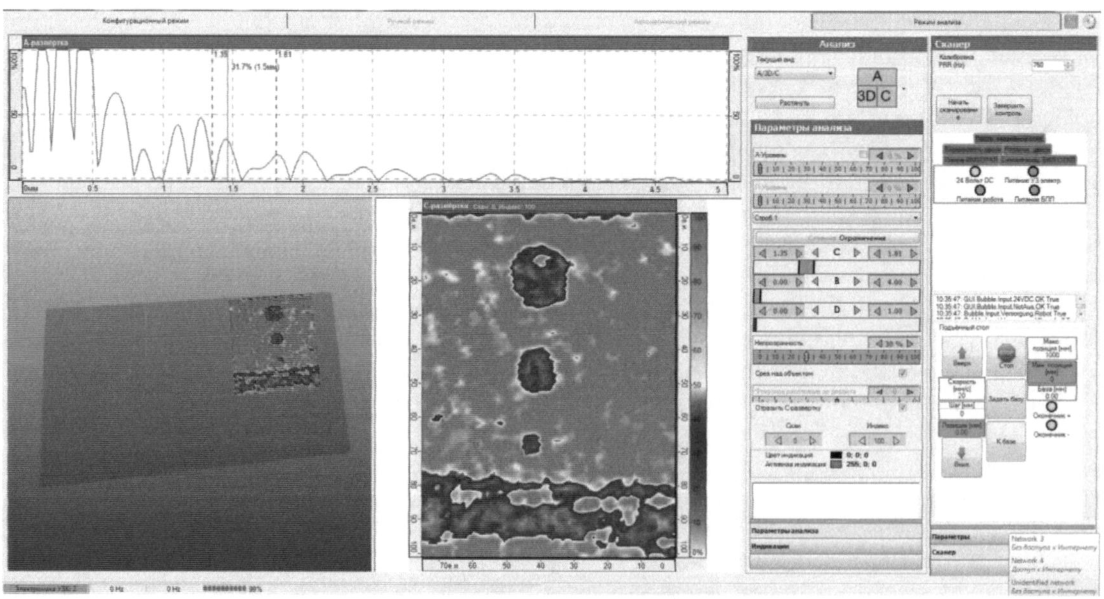

Figure 5. Experimental results

4. Conclusion

Accuracy of robotic ultrasonic testing depends not only on properties of the equipment but also on the precision of its calibration as well as on variance between the modeled path and real shape of the specimen. The path of the robotic scanner is defined according to the CAD-model of the object. The method proposed according to this study can be used for generating CAD-models of complex shaped objects. The object of any unknown shape can be controlled with implementation of the suggested optical reconstruction technique that provides accuracy about 150 μm.

Resulting from the experimental results it is evident that the proposed method provides effective nondestructive control of complex shaped carbon fiber composite materials. In fact, eight from nine defects in the experimental sample were determined unambiguously.

Furthermore, reliability of the received results also corresponds to the accuracy of calibration procedure. Non-contact calibration method proposed in this study is a rapid technique that allows to avoid human component and achieve more accurate positioning of the robotic scanner. We are currently extending this algorithm for implementation for objects of various shapes and materials.

Acknowledgements

Research has been conducted with financial support of the Government of the Russian Federation (The Ministry of Education and Science) agreement № 02.G25.31.0176.

References

[1] Ashbya M, Smidmana M 2011 *Materials for nuclear power systems* p 20
[2] Borikov V, Galtseva O,Filippov G 2016 *IOP Conf. Ser.: Mat. Sc. and Eng.* **671** 12014
[3] Filippov G, Zhvyrblya V, Sharavina S, Salchak Y 2016 *MATEC Web. Conf.* (to be published)
[4] Spies M and Jager W 2003 *Ultrasonics* **41** 125
[5] Lanman D, Crispell D and Taubin G 2007 *3DIM* **3** 107
[6] Bulavinov A, Kröning M, Pudovikov S, Oster R, Hanke R, Hegemann U, Reddy K M and Venkat R S 2008 *International Symposium on NDT in Aerospace* p 21

IOP Conf. Series: Materials Science and Engineering **135** (2016) 012011 doi:10.1088/1757-899X/135/1/012011

The first experience of using of $^{99m}Tc\text{-}Al_2O_3$ for detection of sentinel lymph nodes in breast cancer

A Doroshenko[1], V Chernov[1,2], A Medvedeva[1,2], R Zeltchan [1,2], E Slonimskaya[1], N Varlamova[2], V Skuridin[2], A Dergilev[3], I Sinilkin [1, 2]

1 Tomsk Cancer Research Institute, Kooperativny Street 5, Tomsk, Russia
2 Tomsk Polytechnic University, Lenin Avenue 30, Tomsk, Russia
3 Novosibirsk State Medical University, Krasny Prospect 52, Novosibirsk, Russia

sinilkinig@oncology.tomsk.ru

Abstract. Purpose: to study the feasibility of using the new radiopharmaceutical based on the technetium-99m-labeled gamma-alumina for identification of sentinel lymph nodes (SLNs) in breast cancer patients. Materials and methods. The study included two groups of breast cancer patients who underwent single photon emission computed tomography (SPECT) and intraoperaive gamma probe identification of sentinel lymph nodes (SLNs). To identify SLNs, the day before surgery Group I patients (n=34) were injected with radioactive 99mTc-Al_20_3, and Group II patients (n=30) received 99mTc-labeled phytate colloid. Results. A total of 37 SLNs were detected in Group I patients. The number of identified SLNs per patient ranged from 1 to 2 (the average number of identified SLNs was 1.08). Axillary lymph nodes were the most common site of SLN localization. At 18 hours after 99mTc-Al_20_3 injection, the percentage of its accumulation in the SLN was 7-11% (of the counts in the injection site) by SPECT and 17-31% by gamma probe detection. In Group II patients, SLNs were detected in 27 patients. At 18 hours after injection of the phytate colloid, the percentage of its accumulation in the SLN was 1.5-2% out of the counts in the injection site by SPECT and 4-7% by gamma probe. Conclusion. The new radiopharmaceutical based on the 99mTc – Al_20_3 demonstrates high accumulation in SLNs without redistribution through the entire lymphatic basin. Sensitivity and specificity of 99mTc – Al_20_3 were 100% for both SPECT and intraoperative gamma probe identification.

1. Introduction

Development of the sentinel node concept started at the beginning of the 1990s with evaluation and comparison of the procedures that could be used to detect the sentinel node(s).

If the SLN is free of metastatic disease, all other lymph nodes will also be free of disease [[1],[2]]. The use of colloid radiopharmaceuticals labeled with technetium-99m has appeared to be effective and safe for scintigraphic and/or intraoperative gamma probe identification of SLN [[2]]. Sentinel node biopsy may be a reasonable alternative to avoid extensive axillar lymph node dissection in breast cancer patients with negative SLNs, thereby improving the quality of life of approximately 100,000 patients undergoing lymphodissection for treating breast cancer [[1],[3],[4],[5]]. The two most widely accepted clinical scenarios for SLN biopsy are melanoma and breast cancer. Most modern medical centers worldwide now offer SLN biopsy for these malignancies, yet debate continues regarding whether SLN biopsy should be routinely used in clinical practice [[1],[2],[4],[6],[9]].

The size of radioactive particles is the most important factor in determining the choice of radiopharmaceuticals. Schauer AJ et al. [[5]] reported that a colloid with the particle size of less than 50 nm can be accumulated not only in the SLN but also in the non-SLNs. Particles of greater than 100 nm slowly migrate from the injection site. A colloid with the particle size of the order of 50-80 nm may be preferable for the SLN detection. Nanocolloids made to date are based on the compounds that form stable hydrosols, and organic substances of different structures are often chosen as starting materials for their production. Another part of such pharmaceuticals represents inorganic complexes with 99mTc-rhenium sulfide and antimony, the application of which are limited by their complex manufacturing process and high cost. Our studies have shown that stable colloidal compounds can be obtained in a simpler way - by means of adsorption on the reduced 99mTc alumina gamma [[7],[8]]. The original premise for the use of alumina as a "carrier" of 99mTc- label is its relatively low toxicity, combined with good adsorption properties, availability and low cost.

A specific feature of this compound is an organic coating of nanoparticles. During passage through the lymphatic system, nanoparticles lose their organic coating and accumulate in SLN without redistribution in the body.

Breast cancer is the most common cancer in women worldwide. Approximately 30,000 new early stages breast cancer cases are diagnosed each year in Russia [[10]]. However, in patients with early stages of breast cancer, regional lymph node metastases are found in only 10-30% of cases. Extended lymph node dissection may increase the risk of postoperative complications [[10]].

The purpose of our study was to evaluate the feasibility of using the new radiopharmaceutical, based on technetium-99m-labeled gamma-alumina for the detection of sentinel lymph nodes in breast cancer patients.

2. Materials and methods

The study included 64 patients with histologically verified breast cancer (T1-2N0-1M0). For SLN visualization, all patients were divided into two groups. Group I consisted of 34 women, who received 4 peritumoral injections of radioactive 99mTc-Al203 at a total dose of 80 MBq (20 MBq per quadrant), the day before surgery. Group II comprised 30 women, who were injected with phytate colloid at a total dose of 80 MBq (4 peritumoral injections, 20 MBq per injection), the day before surgery.

All patients underwent single photon emission computed tomography (SPECT) using the dual-head gamma camera (E.CAM 180, Siemens, energy 140 keV, window 15%) in 18 hours after the radiotracer injection. A total of 16 projections were acquired (30 seconds per projection, matrix size 64x64). The scan images were analyzed using the ESoft software (Siemens, Germany). Three-dimensional images of the sagittal, transverse and coronal sections were obtained. To obtain quantitative values, the axial sections were used. The radiotracer accumulation in the SLN with respect to the injection site was evaluated using a quantitative criterion.

For intraoperative detection of sentinel lymph nodes, the hand-held collimated Gamma Finder II® probe was used, which allowed the surgeon to precisely locate gamma radiation source and obtain accurate information about the distribution of the radionuclide in tissues and organs of the patient. The registered gamma radiation level was displayed in numerical values. The lymph node with radioactivity of at least three times more than the background counts was defined as a sentinel lymph node.

The sentinel lymph node was marked and separately sent for express cytological examination. The lymph node was sectioned transversely into 2 mm slices. Depending on the size of the node, 6–10 imprints were made from each slice by gently touching the cut surface of the node to a glass slide.

Cytological specimens were stained for 15 seconds using the kit for rapid staining of blood smears (Leucodiff 200) and examined with a Zeiss Axio Scope (Germany).

Each group of the patients was subdivided into the groups with 12 patients who underwent extended lymphodissection despite the results of histological examination. After lymphodissection, the removed lymph nodes were re-examined with gamma probe to detect SLNs missed during intraoperative examination.

3. Results and discussion

A total of 37 SLNs were identified in Group I patients. The number of identified SLNs per patient ranged from 1 to 2 (the mean number of identified SLNs was 1.08). The most frequent SLN location was the axillary region (97.2%) (Figure 1). Metastases in the excised SLNs were detected in 9 patients (26.6%), who subsequently underwent the conventional axillary lymph node dissection. Four patients (11.4%) had metastases not only in the SLN but also in lymph nodes of the level II. The presence of SLN metastasis indicates lymphatic dissemination of tumor cells and the formation of clinically undetectable regional lymph node metastases. Sentinel lymph node was free from metastasis in 25 patients (73.5%). We did not observe the phenomenon of so called "jumping" metastases in these patients.

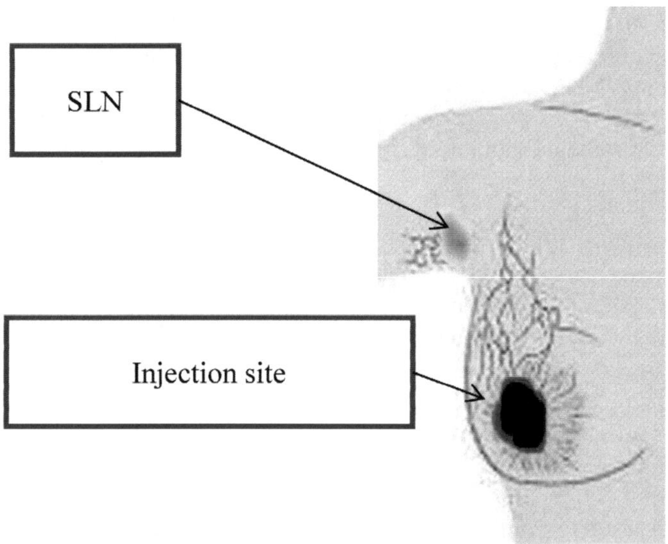

Figure 1. SPECT SLN detection in breast cancer patients.

The 99mTc-Al203 uptake in the SLN in 18 hours after injection was 7-11% (of the radioactivity in the injection site) by SPECT and 17-31% by intraoperative gamma probe detection.

Thus, the sensitivity and specificity of 99mTc-Al203 for detecting SLNs in breast cancer patients were 100%.

The intraoperative SLN detection was performed through a small skin incision. Sparing surgical procedures can significantly reduce overall surgery trauma and recovery time (Figure 2).

IOP Conf. Series: Materials Science and Engineering **135** (2016) 012011 doi:10.1088/1757-899X/135/1/012011

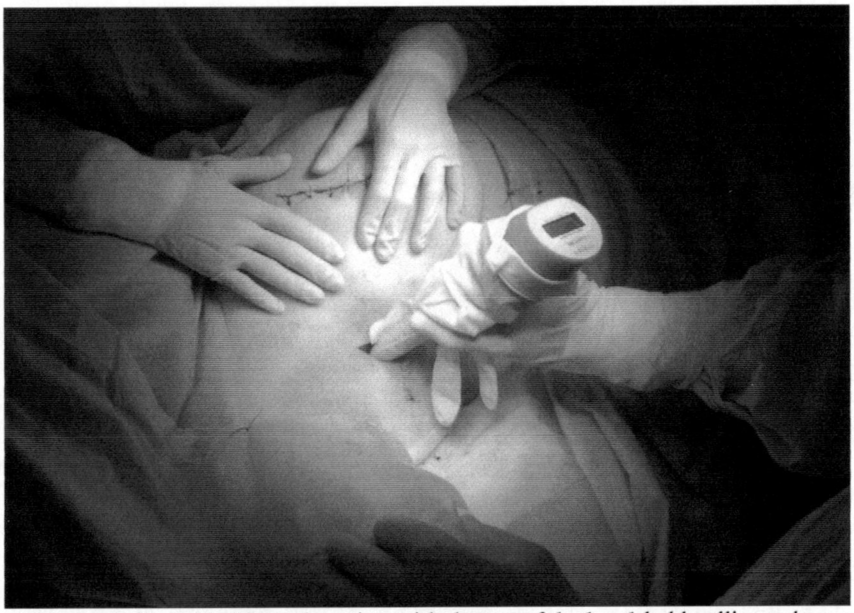

Figure 2. Intraoperative radio guided SLN detection with the use of the hand-held collimated gamma Finder II® in breast cancer patients.

In Group II, 28 SLNs were identified in 27 patients. The number of SLNs detected per patient ranged from 1 to 2 (the mean number of detected SLNs was 1.04). All detected SLNs were located in the axillary region.

Metastases in the excised SLNs were detected in 8 patients (29.6%), who subsequently underwent the conventional axillary lymph node dissection. In 6 patients (22. 2%), the removed SLN and lymph nodes of the level II had a positive histology. In 21 patients (77.7%), the SLN was negative, with no clinical evidence of lymph node metastasis during the follow-up. In this group of patients, there was also no phenomenon of so called "jumping" metastases.

The percentage of phytate colloid accumulation in the SLN in 18 hours after injection was 1.5-2% by SPECT and 4-7% by intraoperative gamma probe detection.

Thus, the use of phytate colloid for detecting SLNs in breast cancer patients demonstrated 90% sensitivity.

Various techniques of SLN identification are widely discussed in the published literature. Currently, most investigators propose that the blue dye and isotope complement each other, and the detection of SLN is more successful using combined dye- isotope technique, with sensitivity and specificity approaching 100% [[10],[12]]. In our study, the use of 99mTc-Al_2O_3, a novel molecular imaging agent has shown promising results in detecting SLNs with high sensitivity and specificity.

For histological examination of the identified sentinel lymph nodes, we used the light optical microscopy with hematoxylin-eosin. A number of authors have published immunohistochemistry and flow cytophluorometry for the detection of micrometastases in lymph nodes, allowing a three-fold increase in the detection rate to be achieved [13]. The wide use of the modern methods of histological express analysis will significantly improve the clinical significance of detecting the sentinel lymph nodes in breast cancer patients.

Thus, the use of the new radiopharmaceutical allows the precise detection of SLNs to be achieved, thereby personalizing the extent of surgery in patients with primary operable breast and cervical cancers.

4. Conclusion

The novel radiopharmaceutical based on 99mTc-Al_2O_3has high accumulation in sentinel lymph nodes, with no evidence of further redistribution along the lymphatic collector, thus significantly facilitating scintigraphic and intraoperative SLN identification.

The sensitivity and specificity of using 99mTc-Al_2O_3 were 100% for both SPECT and intraoperative identification of sentinel lymph nodes.

References

[1] Chernov VI, Sinilkin IG, Shiryaev SV. 2010 *National leadership on nuclear medicine* Ed. Lishmanov YB, Chernov VI Tomsk STT B **2** 336

[2] Paredes P, Vidal-Sicart S, Zanón G et al. 2005 *E.J.Nucl. Med.* **32** 1283

[3] Channel SV, Novikov SN, Zhukov LA, Zotov OV, Semiglazov VF, Krivorotko PV 2011 *Questions of Oncology* **57** 616

[4] Chernov VI, Sinilkin IG, Choynzonov EC, Chijevskaya SY, Titskaya AA, Zelchan RV, Bragina OD, Lyapunov AY, Skuridin VS 2015 *Eur J Nucl Med Mol Imaging* **42** (Suppl 1) 704

[5] Maza S 2003 *QJNuclMed* **30** 651

[6] Afanasyev SG, Avgustinovich AV, Chernov VI, Sinilkin IG 2009 *Siberian Journal of Oncology* **34** 27.

[7] Skuridin VS, Stasiuk ES, Varlamova VN, Rogov AS, Sadkin VL, Nesterov EA 2013 *Proceedings of TPU. Chemistry* **323** 33

[8] Skuridin VS, Stasiuk ES, Varlamova NV, Postnikov PS, Nesterov EA, Sadkin VL 2011 *The universities, Physics* **54** 332

[9] Sinilkin I, Chernov V, Zelchan R, Titskaya A, Skuridin V 2014 *Eur J Nucl Med Mol Imaging* **41** (Suppl 2) 518

[10] Malignant neoplasms in Russia in 2014 (morbidity and mortality), edited by Caprin AD, Starinskaya VV, Petrova GV Moscow 2016

[11] Kim MC, Kim HH, Jung GJ, Lee JH, Choi SR, Kang DY, Roh MS, Jeong JS 2004 *Ann Surg* **239** 383

[12] Kara PP, Ayhan A, Caner B 2008 *Ann Nucl Med* **22**(6) 487

[13] Arigami T, Natsugoe S, Uenosono Y, et al. 2006 *Ann Surg* **243** 341

IOP Conf. Series: Materials Science and Engineering **135** (2016) 012012 doi:10.1088/1757-899X/135/1/012012

PHYSICS AND CHEMISTRY OF THE HYDROGEN FLUORIDE PRODUCTION PROCESS FROM FLUORINE-CONTAINING WASTE

A N Dyachenko[1], R I Kraydenko[1], M S Lesnikova[1], L N Malyutin[1], I V Petlin[1],

[1]National Research Tomsk Polytechnic University, 30 Lenin Ave., Tomsk, 634050, Russia

E-mail: petliniv@tpu.ru

Abstract. The impact of the aluminum industry wastes on the environment is established. The resource efficient method of aluminum industry fluorine-containing wastes processing, which includes wastes oxidizing roasting to remove carbon component and the interaction of fluorine-containing particles with sulfuric acid in order to produce hydrogen fluoride, is considered. The economic and environmental effect of the proposed processing method is substantiated.

1. Introduction

Metallic aluminum world production is steadily increasing. At the end of 2015 it has increased by 15.5 % and has amounted to 56 134 tons, as compared to 48 578 tons in 2014 [1]. Russian aluminum industry is the most prosperous among national steel industries, and it is the one of the biggest world exporters of metallic aluminum. The national production capacity of primary aluminum is about 4.2 million tons per year. More than 75 % of Russian aluminum is produced by plants located in Krasnoyarsk, Irkutsk, Kemerovo, and Sverdlovsk regions.

Due to aluminum practical design characteristics and working properties, the use of metallic aluminum is increasing in all sectors of the global industry: mechanical engineering, aerospace engineering, packaging printing, shipbuilding, industrial and civil construction. Aluminum is without a rival because of its technical, technological, and economic parameters. However, the aluminum industry is characterized by significant amounts of pollutants fluxing into the environment [2].

Aluminum production loses about 18 kg of fluorine per produced metallic aluminum ton in the form of gaseous hydrogen fluoride (HF), electrofilter dust, gas purification slimes, skim flotation tailings, etc. [3].

According to the available data, the content of hydrogen fluoride in the air of a aluminum factory production building reaches nearly 0.3 mg/m^3. At the same time the maximum permissible concentration of HF in the air of the working area is 1 mg/m^3 [4].

The annual hydrogen fluoride concentration increase in cities producing metallic aluminum is as follows: Bratsk – 2-8 μg/m^3, Volgograd – 5 μg/m^3, Novokuznetsk – 5-6 μg/m^3, Pervouralsk – 6-10 μg/m^3.

It has been determined that the impact of fluorine compounds on vegetation is among the most toxic. Fluorine-containing substances pollute the atmosphere in the form of gaseous compounds (hydrogen fluoride) or the dust of fluorinated substances mainly because of the metallic aluminum production, fertilizers, glass, and fluoroorganic compounds [5].

Large-scale mapping of the Krasnoyarsk Aluminium Smelter zone (suburban area of Krasnoyarsk) has shown that the area of soils with extremely and highly hazardous contamination with fluorine of grasslands and pastures amounts to 8.1 thousand hectares or 25.7 % of the surveyed area [6].

Modern industrial production should combine the principles of resource efficiency and environmental safety, which implies the full recycled use of most reagents and the regulation of industrial emissions. Such term as "production waste" should not be present in the production vocabulary.

The so-called natural protective forest areas, mostly "dead" taiga, will extend for many hundreds of kilometers if decisive measures on the disposal of the aluminum industry waste are not taken.

The greatest harm is inflicted by solid production wastes, as the formation of liquid waste is minimal and the main portion of gaseous emissions is captured and neutralized on special gas-cleaning units comprising the steps of dry and wet cleaning of anode gases [7]. To solve this problem it is necessary to develop an effective method for processing of solid fluorine-containing wastes from the aluminum production.

The problem of waste management lies in the fact that there are no industrially implemented technologies of its processing. All wastes differ in their physical properties and chemical composition, and often require an individual approach in the selection of processing options. The main areas in the processing technology are: leaching, flotation, pyrohydrolysis, heat treatment.

2. Theoretical part

The concept of the proposed method consists in the use of solid fluorine-containing wastes of the aluminum industry as a raw material for production of HF. The schematic diagram of processing of the fluorine-containing wastes of the aluminum industry to produce hydrogen fluoride is shown in Fig. 1.

The main reactions in the production of HF are the interaction of sodium fluoroaluminates (Na_3AlF_6, $Na_5Al_3F_{14}$) with sulfuric acid:

$$2Na_3AlF_6+9H_2SO_4=12HF+6NaHSO_4+Al_2(SO_4)_3 \quad (1)$$

$$2Na_5Al_3F_{14}+19H_2SO_4=28HF+10NaHSO4+3Al_2(SO_4)_3\ Na_2SO_4+H_2SO_4+2NaHSO_4 \quad (2)$$

Sodium sulfate, present in slag as a result of capturing sulfur dioxide by soda solution in the wet gas cleaning system, also reacts with sulfuric acid to form sodium hydrosulfate:

$$Na_2SO_4+H_2SO_4=2NaHSO_4 \quad (3)$$

Aluminum and iron oxides are among main components in almost all types of wastes occurring in the aluminum production. Typically, oxides pass into a highly soluble sulfate form. The presence of silicon dioxide has a negative impact on the process of sulfuric acid decomposition of fluorine-containing wastes, because silicon dioxide interacts with the produced hydrogen fluoride and, thereby, affects the yield and the purity of HF. The pollution of production gasses by silicon tetrafluoride takes place:

$$Al_2O_3+3H_2SO_4=Al_2(SO_4)_3+3H_2O \quad (4)$$

$$Fe_2O_3+3H_2SO_4=Fe_2(SO_4)_3+3H_2O \quad (5)$$

$$SiO_2+4HF=SiF_4+2H_2O \quad (6)$$

To investigate the possibility of the process of sulfuric acid decomposition of aluminum production wastes, an equilibrium thermodynamic calculation of the equilibrium of chemical reactions of slag components has been carried out using the Temkin – Schwartzman method. The results of thermodynamic calculations are presented in Table 1.

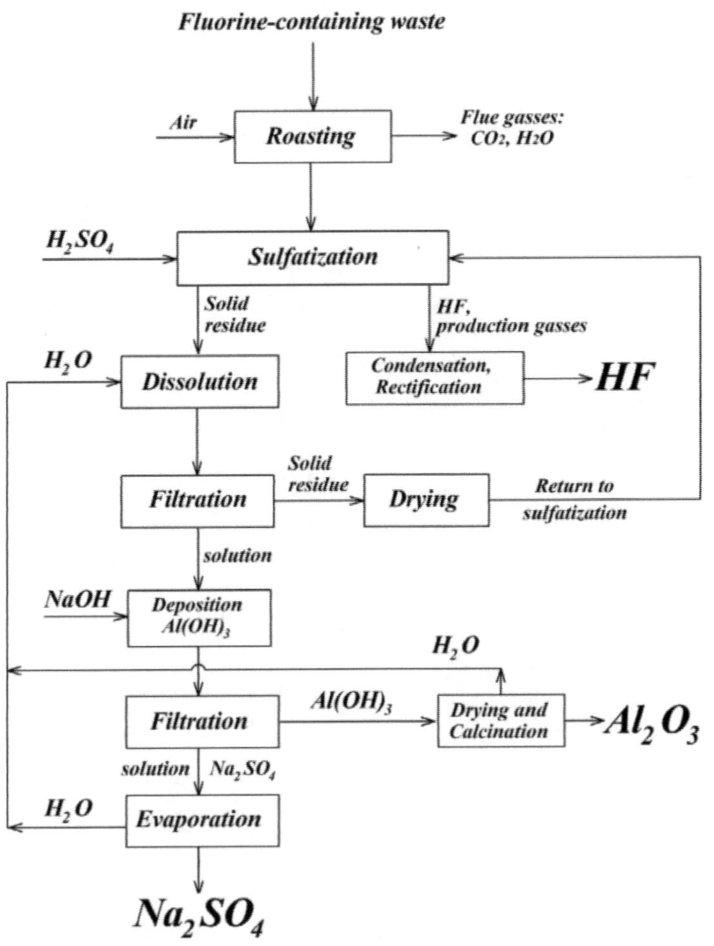

Figure 1. The schematic diagram of the fluorine-containing waste processing.

Reactions of the sulfuric acid decomposition of cryolite and hyolite are endothermic and require heat input. The probability of their occurrence towards the formation of reaction products increases with an increase in the temperature. The onset temperatures of direct reactions are 81 °C and 236 °C, respectively. The interaction reactions of aluminum oxide, sodium sulphate, and iron oxide (III) with sulfuric acid are exothermic and proceed with the evolution of heat towards the formation of reaction products even at room temperatures.

3 Practical Part

The kinetic experiment was carried out using the method of continuous weighing of the reaction mixture with an automatic mass registration at different process temperatures. The conversion degree was determined based on the weight loss which was caused by the formation of gaseous reaction products – hydrogen fluoride and water vapor. A chemically pure mixture (GOST 4204-77) consisting of 5 g of slag and 7.7 g of sulfuric acid, with regard to acid excess of 10% from the stoichiometrically required, was used in the course of the experiment. Based on the research findings, a dependency graph of the conversion degree on time was built (shown in Fig. 2).

Table 1. Gibbs energies, equilibrium constants, and thermal effects of reactions of slag component sulphatization.

T, K	298	300	400	500	600
	$2Na_3AlF_6+9H_2SO_4=12HF+Al_2(SO_4)_3+6NaHSO_4$				
ΔH, KJ/mol	420.24	420.59	443.99	477.83	521.23
ΔG, KJ/mol	66.437	64.061	-57.999	-187.21	-324.14
K_P	$2.26\cdot10^{-12}$	$7.02\cdot10^{-12}$	$3.75\cdot10^{7}$	$3.61\cdot10^{19}$	$1.65\cdot10^{28}$
	$2Na_5Al_3F_{14}+19H_2SO_4=28HF+3Al_2(SO_4)_3+10NaHSO_4$				
ΔH, KJ/mol	1366.0	1366.3	1385.6	1419.0	1466.6
ΔG, KJ/mol	524.51	519.86	278.53	23.265	-243.77
K_P	$1.16\cdot10^{-92}$	$3.06\cdot10^{-91}$	$4.25\cdot10^{-37}$	$3.71\cdot10^{-3}$	$1.67\cdot10^{21}$
	$Al_2O_3+3H_2SO_4=Al_2(SO_4)_3+3H_2O$				
ΔH, KJ/mol	-181.33	-181.35	-63.024	-76.261	-89.409
ΔG, KJ/mol	-159.71	-159.41	-153.67	-170.84	-185.94
K_P	$9.87\cdot10^{27}$	$5.69\cdot10^{27}$	$1.17\cdot10^{20}$	$7.04\cdot10^{17}$	$1.54\cdot10^{16}$
	$Na_2SO_4+H_2SO_4=2NaHSO_4$				
ΔH, KJ/mol	-68.880	-68.584	-51.894	-32.300	-10.390
ΔG, KJ/mol	-52.413	-52.303	-49.297	-50.863	-56.598
K_P	$1.54\cdot10^{9}$	$1.28\cdot10^{9}$	$2.74\cdot10^{6}$	$2.06\cdot10^{5}$	$8.46\cdot10^{4}$
	$Fe_2O_3+3H_2SO_4=Fe_2(SO_4)+3H_2O$				
ΔH, KJ/mol	-182.00	-182.05	-63.651	-73.809	-80.554
ΔG, KJ/mol	-161.75	-161.62	-163.16	-186.80	-208.70
K_P	$2.25\cdot10^{28}$	$1.38\cdot10^{28}$	$2.03\cdot10^{21}$	$3.27\cdot10^{19}$	$1.48\cdot10^{18}$

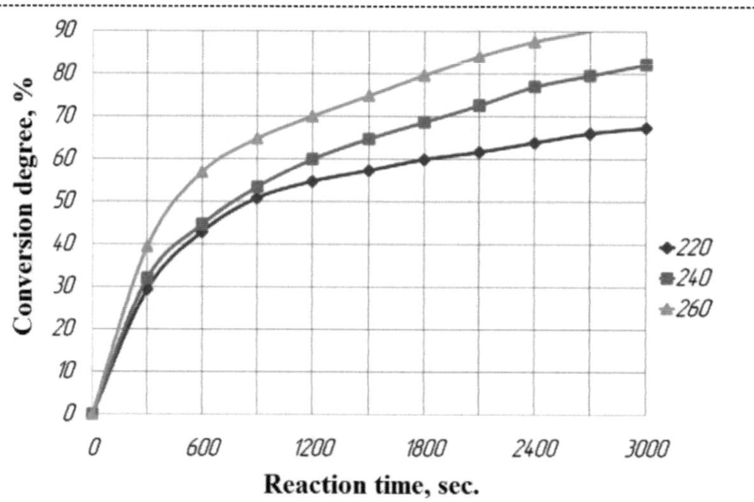

Figure 2. The dependence of the conversion degree on time.

This kind of dependence is mostly linearized under the Crank-Gistling-Braunstein equation [8]:

$$1 - {}^{2}/_{3}\alpha - (1-\alpha)^{1/3} = kt$$

The values of the rate constant (k_0 = 6.05 c^{-1}) and the value of the activation energy (E = 50.2 KJ/mol) were determined based on the dependency built in coordinates of the Arrhenius equation. Thus, the general form of dependence is:

$$1 - {}^{2}/_{3}\alpha - (1-\alpha)^{1/3} = 6.5 \cdot \exp\left(\frac{-50258}{RT}\right) \cdot t$$

In the temperature range of 220-260 °C the activation energy of the process was 50.2 kJ/mol. The process proceeds in the outer-kinetic reaction area. The limiting stage of the process is the interaction of reagents. An increase in the temperature accelerates the process.

The proposed method for processing of aluminum industry wastes has both the environmental and the economic benefit. According to preliminary calculations, the use of fluorine-containing wastes will allow reducing the cost of the produced hydrogen fluoride by 20% due to the reduction in the cost of raw materials and the sale of popular on the market by-products – aluminum oxide and sodium sulfate [9].

4 Conclusions

1. The studies have proved the possibility of the use of fluorine-containing wastes as raw materials for the production of hydrogen fluoride.
2. The use of aluminum industry wastes as raw materials for the production of hydrogen fluoride will solve the problem of its processing, and will allow manufacturers of hydrogen fluoride to give up the procurement of costly raw materials – fluorspar.
3. The possibility of the sulfuric acid processing of fluorine-containing wastes from the aluminum production has been determined in the course of thermodynamic calculations. The value 220-260 °C has been selected as the operating temperature range.
4. The activation energy of the process – 50.2 kJ/mol and the rate constant k_0= 6.05 c^{-1} have been determined as a result of the kinetic study. The dependency equation of the reaction degree on time in the temperature range of 220 – 260 °C has been established. The process takes place in the outer-kinetic reaction area. The limiting stage of the process is the interaction of reagents. An increase in the temperature accelerates the process.
5. The use of fluorine-containing wastes will allow reducing the cost of the produced hydrogen fluoride by 20 % due to the reduction in the cost of raw materials and the sale of popular on the market by-products.

References

[1] Aluminum production in 2015 [Electronic resource]. – Access mode: http://www.finmarket.ru/news/4201070 (reference date: 06/02/16)

[2] Elterman V.M. 1985 *Environmental protection in chemical and petrochemical industries.* (Moscow: Khimiya)

[3] Petlin I.V. 2014 *Processes of hydrogen fluoride production from fluorine-containing wastes of the aluminum industry: Dis. of Cand. Eng. Sciences* (Tomsk, Russia)

[4] Gromov O.B., Prokudin V.K. 2008 *On the issue of the neutralization of gas emissions from aluminum plants* (Chemical technology V. 9. No. 7)

[5] Konopelko L.A., Sidorenko A.Yu. 2013 *Development of methods of the metrological provision of control over the content of fluorine compounds in the air.* (Abstracts of the II All-Russian congress of young scientists. St. Petersburg), pp. 52-53.

[6] Belyakova T.M. 1977 *Fluoride in soils and plants in connection with endemic fluorosis* (Soil science. No. 8), pp. 55-63

[7] Istomin S.I. 2004 *Processing of the aluminum production wastes.* (Krasnoyarsk, Russia) 480 p

[8] Semiokhin I.A. 1995 *Kinetics of chemical reactions: Tutorial* (Moscow: Publishing house of MSU) 347 p.

[9] Petlin I.V. 2013 *Prospects for the use of wastes of the aluminum industry in the production of hydrogen fluoride* (Proceedings of the universities. Physics. V. 56. No. 4/2) pp. 146-150

IOP Conf. Series: Materials Science and Engineering **135** (2016) 012013 doi:10.1088/1757-899X/135/1/012013

Simulation of the aging process of insulating systems variable frequency drive

A Garganeev, A Leonov, V Merkulov, D Charkov

Tomsk Polytechnic University, 634050 Lenin str. 30 Tomsk, Russia

E-mail: leonov_ap@tpu.ru

Abstract. The paper deals with the intensity of the model in the electrical insulation variable frequency drive controlled at different temperatures and electric fields. It is shown that aging of insulation mechanism associated with the development of corona discharges caused by transients when the frequency adjustment. Laws aging of insulation can be described from the point of view of the theory of thermal destruction of dielectrics vibrations.

1. Introduction

The use of frequency converters (FC), in the structure using semiconductor switches, led to a new problem - sharp decrease in the life and the insulation of supply cables of the windings of electric machines. This is due to insulation failure caused by surge [1-2].

FC using a technology known as "Pulse width modulation" (PWM) is generated at the output of the necessary basic voltage and frequency. The inverter circuit "switching" transistors operating at high speed, producing a carrier frequency over which the useful basic voltage and frequency.

The output voltage of the autonomous voltage inverter (VSI) with PWM is a high-frequency sequence of rectangular pulses having different polarities and duration with the same amplitude DC input on the VSI. The high rate of rise of the voltage pulses is determined by the speed switching VSI power switches and by using different semiconductor devices is about 0.05 ... 0.1 ms.

Such a high speed has a negative impact on the flow of transients in the circuit VSI - power cable - squirrel-cage induction motor. Passage of pulse signal at a high speed causes a voltage rise in the cable wave processes, which give rise to surges in the cable line to the motor terminals [3].

In addition, the effect of the reflected mode caused by high-rise speed of tension (dU/dt) and length of the feeding cable, which works as the transferring line, affects. For example, for the class of inverters 3.3 kW speed voltage changes (dU / dt) can exceed 7500 (B / μsek). Due to the mismatch the wave resistance at different ends on the cable (the cable - the cable and the inverter - the motor) high-frequency part of the wave that reaches the motor windings and reflected back the source. As reflected peaks superimposed on the peaks of the waves coming, their values added, causing voltage surges. In operation, an increasing number of cable lines peaks "stacked" on each other, at the same time, resulting in a greater amount pulse voltage per unit time [4-5]. Having said that, there is the problem of determining actual values of voltages and taking into account their influence on the processes aging and deterioration the electrical insulation.

2. Experimental part

The problem of estimating the values of electric voltages solved in a graphical environment simulation MATLAB Simulink.

IOP Conf. Series: Materials Science and Engineering **135** (2016) 012013 doi:10.1088/1757-899X/135/1/012013

A general view a simulation model of the variable frequency drive, made in MATLAB Simulink and including FC based on a PWB (1), the cable line (2) and squirrel-cage induction motor (3), Fig. 1. When building simulation models take into account the wave parameters of cable lines: linear capacity C_x, inductance L_x and wave resistance Z_x.

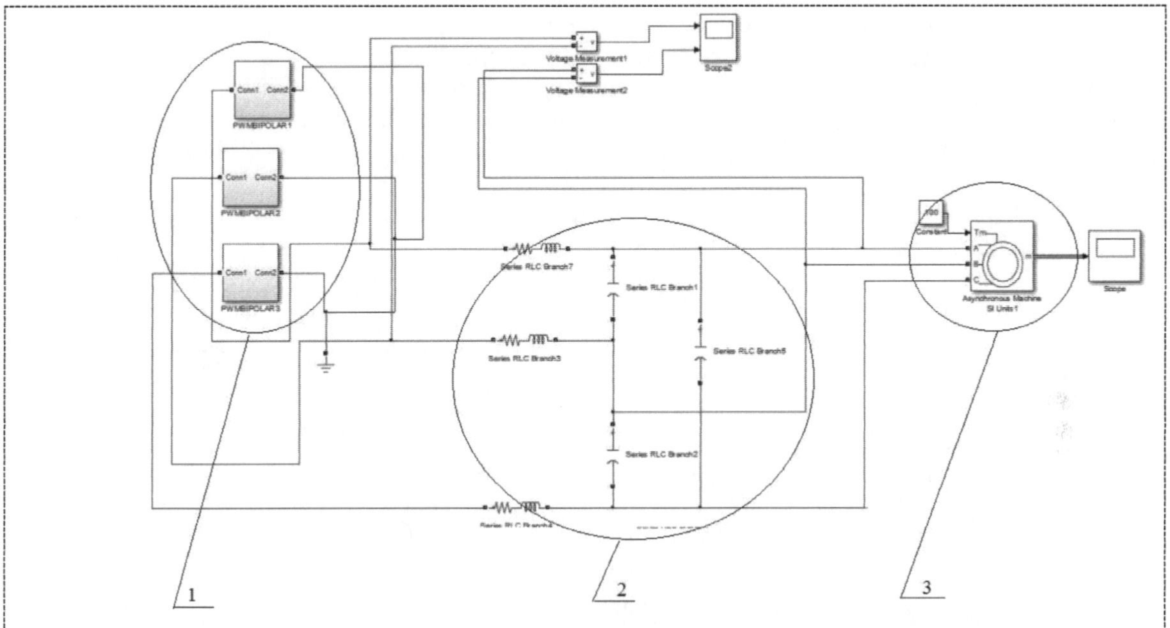

Figure 1. The simulation model of the variable frequency drive (VFD)
1 - generator PWM pulse signals; 2 - cable line VVG 1x2,5; 3 - squirrel-cage induction motor series RA1004L

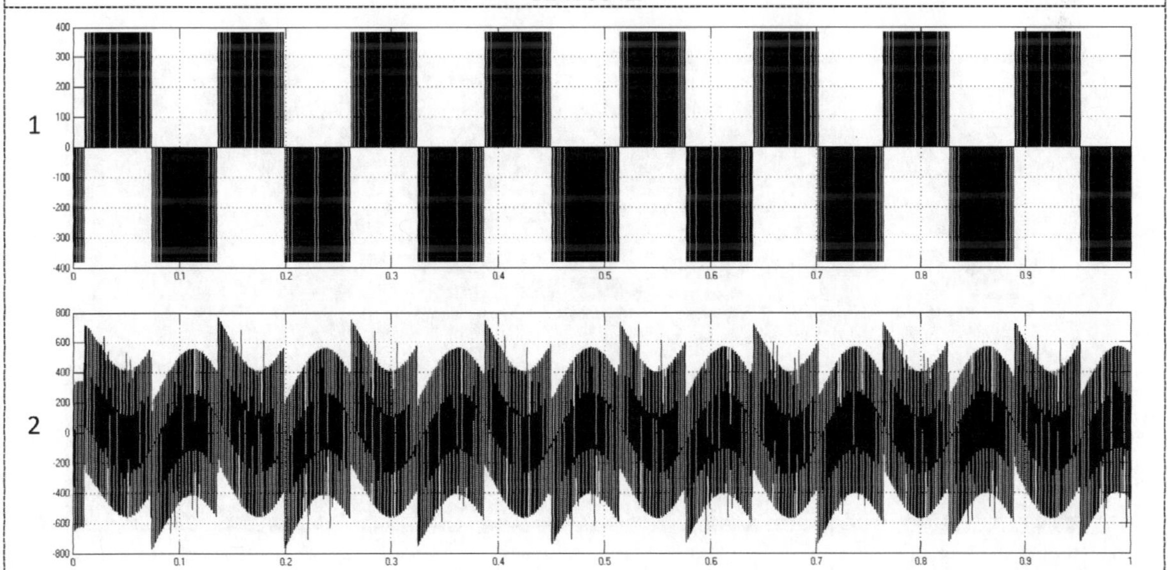

Figure 2. The nature of the voltage stress in the VFD 1 - effective value of line voltage generator after the pulse-pulse signals, 2 - the current value of the linear voltage after the cable line VVG 1x2,5

The simulation results allow to note:

1. In the absence of a sinusoidal filter is a sharp increase in the voltage amplitude on the cable line and motor terminals. This high level of stress in combination with high-frequency components of the PWM voltage pulses should definitely accelerate the electrothermal aging of insulation.

2. The supply voltage curve comprises high-frequency components that distort the voltage waveform (Fig. 6) and the resulting PWM operation, the effect of "reflected wave" and the high-frequency harmonics, belonging to the cable line from FC.

Definition of mean time to breakdown of the cable insulation

Subject to certain electric magnitudes of the stresses acting on the VFD element insulation, comparative tests were carried out cable samples under commercial frequency voltage and high-frequency modulated signal. The basis adopted by the procedure described in [6-8]. The results shown in Table 1.

Table 1. Temporal dependence before breakthrough on type of insulation and temperature

Type	Wire insulation	The mean time to breakdown (τ_{cp}), s.		
	Temperature, °C	180`°C	190°C	200°C
1	layer (inner) – polyester varnish , layer (outer) – polyamide imide varnish	3082,6	2642,2	2429,4
2	1 layer (inner) – 3-hydroxiethylcyanurate varnish, comprising at least 0,1 volume percent of silicon nanoparticles,2 layer (outer) – polyamide imide varnish	9692	5326	3441,4

The presence of corona discharges, as well as appropriate forms of electrical loads supplied signal characteristic for VFD confirmed glowing corona surface twists and waveform values, characteristic currents and voltages, removed from the sample (Fig. 3 a, b)

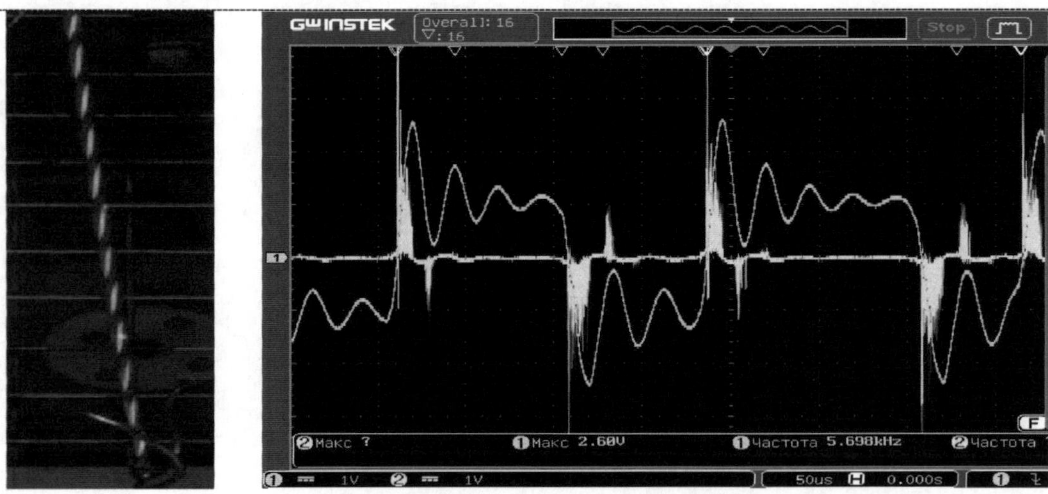

a) b)

Figure 3. General view of the twist test and electrical load applied to the sample: a - glow corona discharge on the sample; b - waveform of current and voltage applied to the sample

Table 1 shows that the greatest time to have a breakdown of the samples the wires with insulation type 1 with corona resistance enamel.

Explanation of the results can be discuss on the thermofluctuational destruction theory proposed Zhurkov. Based on the presentation of this theory in [9] it has shown that the basis of presentation breaks the chemical bonds of insulation is the failure mechanism when subjected to applied loads, which include temperature, electric field strength and frequency. In this model, the time until insulation breakdown can be determined:

$$\tau = \tau_0 e^{\frac{D\cdot\varphi(x)}{2KT}}, \quad (1)$$

where: τ_0 – time constant, s; D – tear energy of chemical bonds, J; $\varphi(x)$ – function of acting loads (x), causing a decrease of potential energy barrier:

$$\varphi(x) = \sqrt{1-2x} - x\,ln\left[\frac{1}{x} + \frac{1}{x}\sqrt{1-2x} - 1\right], \quad (2)$$

$$x = \frac{1}{D}\sqrt{\left(Ae^{-bT}\beta\eta E\right)^2 + (\gamma\sigma)^2}. \quad (3)$$

A,γ - determined experimentally equation parameters (1); η - coefficient taking into account the increase of electric field strength by electrode shape; β - coefficient taking into account the increase of electric field strength by homogeneity of dielectric material structure; σ - mechanical load, [H/m]; E – strength of applied electric field, [V/m]; b - coefficient taking into account the change of material modulus of elasticity with temperature, 1/K.

To determine the parameters in the equation and the time to breakdown technique can be use as described in [10].

Results were obtain on basis of the calculations shown in Figure 4

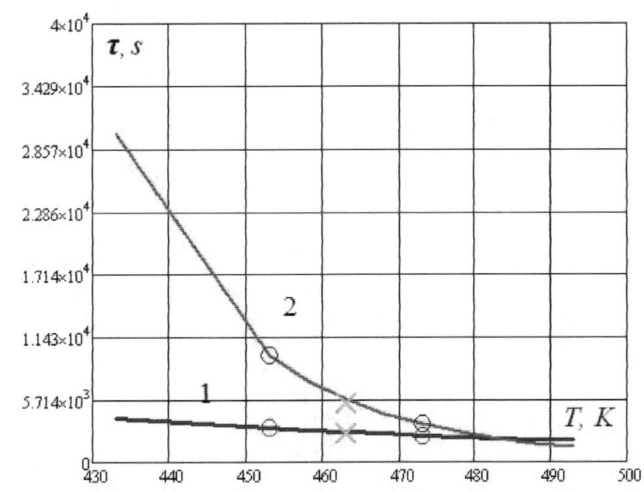

Figure 4. Dependence of the mean time to breakdown temperature: type 1; type 2

As presented from Fig. 4, calculating the data according thermofluctuational theory are in good agreement with the experimental results shown in the graph in the form of points.

Conclusions

1. It shown that a mathematical model based on the thermofluctuational theory of destruction can used for explaining the mechanism of breakdown of polymer dielectrics.
2. The method of determining the equation parameters based on the preliminary insulation tests on breakdown at two temperatures of electrical aging are proposed and approved.
3. It is found that the calculated dependences of the time to breakdown based on thermofluctuational theory are good accordance with the experimental results and allow predict the time to breakdown of dielectrics taking into account various factors.
4. At calculating the time to breakdown, it is necessary to focus on the average values of equation parameters by the experimental data considering the inhomogeneity of electrical insulation and the statistical nature of impact loads.

References

[1] Kojkov S.N., Cikin A.N 1968 *Jelektricheskoe starenie tverdyh dijelektrikov i nadezhnost' dijelektricheskih detalej* (M.-L.: Jenergija) p 287
[2] M. G. Minnick. The effect of winding stresses on the pulse endurance of corona resistant magnet wire. Electrical Insulation, 2004. Conference Record of the 2004 IEEE International Symposium on. 19-22 Sept. 2004, pp. 169-173. doi: 10.1109/ELINSL.2004.1380503
[3] Bartenev G.M. *Prochnost' i mehanizm razrushenija polimerov*. – M.: Himija, 1984. – 280 s. L. A. Saunders, G. L. Skibinski, S. T. Evon, D. L. Kempkes. Riding the reflected wave-IGBT drive technology demands new motor and cable considerations. Petroleum and Chemical Industry Conference, 1996, Record of Conference Papers. The Institute of Electrical and Electronics Engineers Incorporated Industry Applications Society 43rd Annual. 23-25 Sep 1996, pp. 75-84. doi: 10.1109/PCICON.1996.564866
[4] Bogh, D., Coffee, J., Stone, G., Custodio, J. Partial discharge inception testing on low voltage motors // Industry Applications, IEEE Transactions on, Jan.-Feb. 2006, pages (148 - 154)
[5] B. Basavaraja, D. V. S. S. S. Sarma. Application problem of PWM AC drives due to long cable length and high dv/dt. Power Electronics, Drives and Energy Systems, 2006. PEDES '06. International Conference on. 12-15 Dec. 2006, pp. 1-6. doi: 10.1109/PEDES.2006.344380
[6] Leonov Andrey Petrovich. Estimation of winding insulation resistance to the corona discharges [Electronic resource] / A. P. Leonov, V. V. Redko, E. Yu. Soldatenko // IOP Conference Series: Materials Science and Engineering. — 2014. — Vol. 66: 20th International Conference for Students and Young Scientists: Modern Techniques and Technologies (MTT'2014), Tomsk, Russia, 14-18 April 2014
[7] Determination of enamel insulation corona resistance by high-frequency modulated pulses / A. P. Leonov [et al.] // IOP Conference Series: Materials Science and Engineering. — 2015. — Vol. 81 : Radiation-Thermal Effects and Processes in Inorganic Materials. — [012094, 7 p.]
[8] Pulse width modulation simulator for testing insulating materials. US Patent No. 6051980, Issued on April 18, 2000
[9] Merkulov V.I. Evaluation of time to breakdown for polymeric insulation / V. I. Merkulov A. P. Leonov, V. A. Bolgova// Electromechanical energy converters: Proceedings of the VII International Scientific and Technical Conference. — Tomsk, Russia, 14-16 October 2015 pp. 267 – 270
[10] Merkulov V.I. Determining the parameters for the curve of the equation of life, based on the theory thermofluctuational / V. I. Merkulov A. P. Leonov, K. P. Arefiev // Physical and technical problems in science, industry and medicine: Book of Abstracts of the VII International Scientific and Practical Conference. — Tomsk, Russia, 6 June 2015 pp. 114 – 115

Simulation of the effectiveness evaluation process of security systems

A V Godovykh[1], B P Stepanov[1], A A Sheveleva[1], K R Sharafieva[2]

[1] Tomsk Polytechnic University, 634050 Russia, Tomsk, Lenin Avenue, 30
[2] Tomsk State University, 634050 Russia, Tomsk, Lenin Avenue, 36

e-mail: aas@yandex.ru, ruf_84@mail.ru

Abstract. The paper is devoted to issues of creation of cross-functional analytical complex for simulation of the process of operation of the security system elements. Basic objectives, a design concept and an interrelation of main elements of the complex are described. The proposed conception of the analytical complex provides an opportunity to simulate processes for evaluating the effectiveness of physical protection system of a nuclear facility. The complex uses models, that take into account features of the object, parameters of technical means and tactics of adversaries. Recommendations were made for applying of this conception for training specialists in the field of physical protection of nuclear materials.

1. Introduction

Smooth functioning of the security system is one of the essential parts of the operation of a nuclear facility. The effective security system combines an activity of staff, administrative measures and correctly functioning set of engineering-technical means [1]. Management of the security system is carried out by an automated system that presupposes the presence of people. In this case, competent and prompt actions of the security staff define the system capabilities to resist to threats with regard to an object of defense.

Training of the security staff is an effective way to increase the protection level of an object. However, training of security staff on management of the technical equipment and response forces is impossible on an active security system. Depending on area of activity of the staff, different approaches to training can be implemented. It can be consideration of technical and legal documentation, usage of specialized stands with technical equipment of security systems. This methodological approach allows to simulate the operation of the security system to the maximum extent possible. Nowadays are widely used integrated electronic training devices that based on modern modelling technologies. These training simulators are usually implemented as specialized computer workstations. Its features are only limited by assigned tasks.

2. Design of the analytical complex

The cross-functional analytical complex is considered as an automated workstation that has a wide range of features to simulate an interaction between adversary and security system of the object. The automated workplace provides training opportunities to perform specified functions of security systems.

Main goals of the analytical complex are:

- performance of the analytical work on the effectiveness evaluation of the security system of an object;
- simulation of the interaction "adversary - security system";
- skills training of the operators of control center;
- formation of capabilities to sensing and analyzing of the information by the security system staff;
- learning the decision-making methods caused by unauthorized activity.

The main element of the developing analytical complex is dedicated software. It stores and archives a data, executes algorithms, organizes a connection between units and fundamental mathematical model, visualizes the processes.

The software configuration is divided to target blocks - units, which are allocated according to set tasks [2]. The units, in turn, operate data, that represented by embedded or customer databases. In general, databases are considered as systematized and categorized data. Its purpose is information support of the organization of the simulation process.

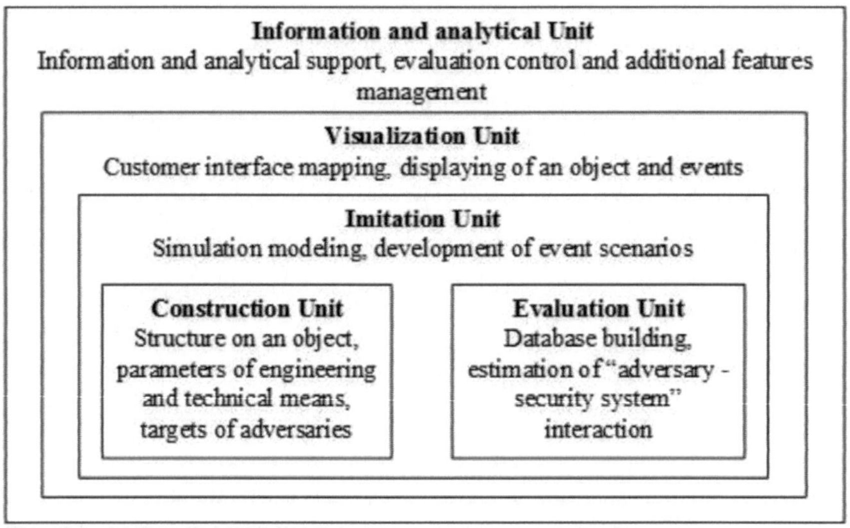

Figure 1. Main units of analytical complex

Constructive, evaluation and interface units are the basis of the analytical complex. The first two is responsible for the imitation of the first level model – "static" model. At this level, basic characteristics and infrastructure of the object are described: features of surrounding area, properties of engineering and technical means of physical protection system (PPS) etc. In addition, general presumptions are defined: number of response forces and adversaries, its "tactic" etc. Sufficiency and correctness of source data are verified for further PPS effectiveness evaluation based on incorporated mathematical model.

The imitation unit is responsible for the second level of the simulation. This level is determined by dynamic parameters of the model. It forms model's extended functions. At this stage, interaction algorithms of elements of the first level model are defined. Examples are the implementation of different "tactics" and versions of scenarios for response forces and adversaries; influence of intruder; simulation modeling within emergency; PPS effectiveness calculation algorithm etc.

The visual unit presents the third level of the simulation. Its function is to display the prototype object and events in a form that is convenient and realistic for user. Another purpose is to visualize a graphical user interface. Along with the interface unit, the visual unit maintains connection between user and resources of analytical complex.

The information-analytical unit is responsible for an informational support of the analytical complex. It is reference-methodical maintenance and legislative and regulatory framework related to physical protection of nuclear materials. Extra tools for work represent analytical functions with evaluation unit. This unit also allows to choose the modes of operation of the analytical complex depending on the demands in training of several types of the staff of the security systems.

The interface unit (it is not pictured in the figure) administrates software interaction between databases and units. It is also responsible for linking organization of hardware, which is combined into automatized workplace.

Table 1. Primary data sources

Name and short description	Supported units	Content
"Attacker" – characteristics of outside adversary	Construction Imitation Evaluation Information and analysis	- number; - type; - probability of conspiracy; - qualification; - equipment etc.
"Guard" – characteristics of response forces		- number; - type; - qualification; - equipment etc.
"Barriers" – parameters of engineering means of physical protection		- type; - time of delay; - location; - way to bypass; - strength and durability etc.
"Sensors" – parameters of technical means of physical protection		- type; - location; - triggering probability; - false triggering probability; - strength and durability etc.
"Library" – legislative and regulatory framework, supporting assistance	Imitation Information and analysis Visualization	- information support; - fragmentary and full text documents; - visual material etc.
"Facility" – description and features of secure facility for simulation purposes	Imitation Information and analysis Visualization	- characteristics of standard objects; - targets (object of defense); - threats; - operation mode; - site area parameters; - others features: climate, political and geographical parameters etc.
"Media" – video, sound data, graphical primitives	Construction Information and analysis Visualization	- primitives and elements of graphical maintenance; - video elements; - graphical objects; - acoustic elements; - dynamic elements etc.

The databases are a resource base for simulation of the object. They are represented as numerical characteristics, independent or combined parameters and as graphical and other elements of visualization. One or few units can use the elements of the databases.

All the databases used by the analytical complex are shown in Table 1.

The minimal configuration of the analytical complex can be presented as a personal device. It is enough for training of the operator of security systems. The automatized workstation is organized with simultaneous operation of the server and client parts of the complex. But this case only locally pledged situations can be processed through running of specialized software complex. Also parallel analysis of operator actions in a simulated situation is significantly reduced.

An expert mode allows a synchronous organization up to four roles. Active Members are "operator of the security system", "intruder" and "response forces". Passive participants are "observer experts". The number of participants of the "training" determines the maximum number of personal devices.

The local configuration of each of the personal devices includes specialized software, the subunits of the security staff and the adversary, the data on engineering and technical means. Thus, each workstation is a separate system and can be used independently from other workstations. The interaction between the users and subsystems of the instructor and operator is implemented through the appropriate interfaces. Considered configuration suggests the possibility of a separate and independent work in training.

3. Effectiveness evaluation

The effectiveness evaluation is one of the core tasks when creating of security system. Estimation of this criterion is executed within the assigned target, method of threat response and selected criteria of efficiency (probability to interrupt of adversary). Depending on the criteria, efficiency of the security system can be estimated based on different methods [3]. It can be experimental evaluation or analytical estimation in terms of specialized software. In addition, security system effectiveness can be estimated during different stages of its development.

There are psychophysical problems that occur during operation of the security staff. Among these are a monotony of the work, an operator's fatigue, a necessity to analyze a lot of displayed data. However, efficiency of the security system depends much on operator's quality of work. Nevertheless, efficiency of the security system depends much on operator's quality of work, his quick reaction capability, adaptability and flexibility to changing of situation. Identification of alarm at the appropriate time, detection of malfunctions of sensors and defined response – are the confirmations of operator's preparedness to control and management of security system.

Methods of security system effectiveness evaluation were considered within the proposed conception of analytical complex. It was highlighted, that generalized criterion of the efficiency of security system is a decision-making time by an operator of the system. In these terms, the response time divides into length of the main stages.

In this case the main stages of security system staff activity are the detection of an intruder, the decision-making time (which includes time of alarm communications and time of an assessment), the time of preparation of the response forces and their further time of interruption of adversary. The travel time of intruder divides into parts that correspond to the real adversary's tasks. The whole action sequence of adversary includes overcoming of barriers, travel on territory of an object, access to building and travel within it, room access, and access to the defense object. All stages of actions of adversary and response forces are pictured on the figure 2.

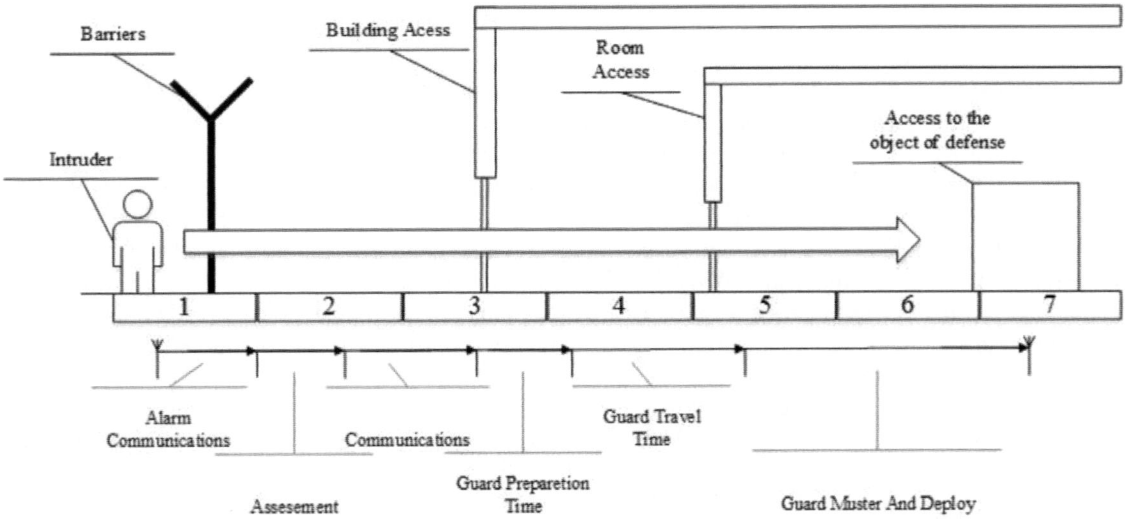

Figure 2. Stages of actions of adversary and response forces within the protected object

The total time of the intruders' action on each stage is formed based on parameters of used engineering and technical means as well as characteristics of adversary and response forces. As a result of interaction a diagrammatic model is constructed, on which the time of adversary and response forces are recorded. The classic methods based on timeline represents a quantitative approach to the effectiveness evaluation [3,4]. The result of these methods is a determination the probability of interruption the adversary's actions [5]. This probability depends on the strategies of intruders and response forces, level of an equipment of the object by engineering and technical means of the security system.

Based on the obtained data, the time of operator actions as well as probability of penetration interruption of the offender is evaluated. These results characterize the accuracy and speed of the executed operator's procedures. In setting the initial data for modeling the interaction in the system "intruder - security" for the effective working of the operator takes time and the probability of successful performance of the task for an adjustable time interval [6].

In terms of presented conception of analytical complex, the main criterion of security system effectiveness is the effectiveness of operator's work.

Within the performance of analytical complex, proposed model allows to evaluate the effectiveness of a whole security system of an object as well as effectiveness of the security system's staff work.

The effectiveness of the operator's work can be determined as the probability of the successful performance of the task for a specified time interval.

Operator error is defined as fail in performance of the task (or performance of forbidden actions), which can cause disruption of the security system [7]. There are also several factors that can be regarded as the operator errors. The operator can be committed to achieving the false target. The target cannot be reached because of the wrong actions of the operator. The operator is inactive at the moment when it is necessary to participate.

Thereby, the efficiency of the operator's work is characterized by response speed and reliability. The criterion of the response speed is the decision-making time; in other words, it is time from the response to the incoming signal until the end of control actions.

Special attention was paid to the factors that are associated with the organization of the operator's workstation. For better performance of the training functions, the imitation must correspond to the real conditions of the work.

Based on the described above suppositions and features of design of the analytical complex, the modelling capabilities of the object are represented at appropriate level according to the related goals. Basic requirements regarding the effectiveness evaluation of the operator's work are:

- Conformance of developed display mode to the real elements of the control system used by the operators;
- Sufficient authenticity and approximation of mathematical interpretation of interaction between adversary and security system;
- Appropriate level of abstraction in description of technological processes.

A number of standard actions is identified for the security staff. They must to be made when the alarm signal occurs or in case of false alarm as well as in case of disruption of a technical mean. When these situations are being simulated within the analytical complex, the response speed, the quality of conformance of actions to the standard, operator's actions are being assessed.

It is worth noting that since the adversary in this interaction is also a person, the details and specifics of the situations will be different due to the nature of each individual's personality.

4. Conclusion

As a result of a work the conception of the analytical complex to assess the effectiveness of security systems is developed. The proposed set of the core units enables to model an object security system, and then on this basis, to evaluate the effectiveness of used engineering and technical means. In addition, the complex has enough functionality to assess the effectiveness of operator's actions. This provides a comprehensive approach to the effectiveness assessment of security systems, which takes into account the individual characteristics of the personality.

This analytical complex can be used within security system operator's automatized workstation for training of "person - security system" interaction.

One of the distinctive features is function that provides maximum variability when specifying the object and editing the existing interaction mechanisms and methods of evaluation. Besides, functional capability on realization of experts' analytical work on effectiveness of security systems' assessment is provided.

References

[1] Garcia M L 2005 *Vulnerability Assessment of Physical Protection Systems* 1st ed (Butterworth-Heinemann) p 400
[2] Godovykh A, Stepanov B 2015 Development and Creation of Software and Information Environment for Simulation of Nuclear Facility *Advanced Materials Research* **1084** 652-654
[3] Lovecek T, Ristvej J, Simal L 2010 Critical Infrastructure Protection Systems Effectiveness Evaluation *J. of Homeland Security and Emergency Management* 7 p
[4] Fennelly L 2012 *Effective Physical Security* 4th ed (Butterworth-Heinemann) p 384
[5] Garcia M L 2007 *The Design and Evaluation of Physical Protection Systems* 2nd ed, (Butterworth-Heinemann) p 370
[6] Bukovetskiy A V, Stepanov B P, Tatarnikov D A 2016 Initial Data Forming for Process Simulation in System "Intruder – Physical Protection System" *Key Engineering Materials* **685** 148-152
[7] Fischer R J, Hailibozek E P, Walters D C *Introduction to Security* 9th ed (Elsevier) p 544

Synthesis and properties of the materials obtained by SHS mode for radiation protection

D S Isachenko, O Yu Dolmatov, A O Semenov, S S Chursin and M S Kuznetsov
National Research Tomsk Polytechnic University, Tomsk, Russia

E-mail: kms@tpu.ru

Abstract. The article shows the process of protective composite materials manufacturing. Also, the analysis of experimental results concerning the composite materials protective properties is given. The advantages of SHS method are considered in comparison with traditional materials. The uniqueness of SHS obtained products based on combination of nuclear-physical properties and parameters is presented.

1. Introduction

Nuclear facilities for various purposes are powerful sources of integrated ionizing radiation which are based on the fission of heavy isotopes stimulated by neutrons. Inside such devices and their surrounding structures, a lot of neutron-stimulated reactions take place causing damage to various structural materials.

A range of research works [1-3] indicated that the creation of a fine-grained structure in the sintering process leads to the increased resistance characteristics of ceramics, which supports the idea that such ceramic materials including metal-ceramics are very promising for use in nuclear and energy engineering in terms of their radiation resistance characteristics.

Now, materials obtained in the mode of self-propagating high temperature synthesis (SHS) are finding increasing application in various fields of science and technology. The advantages of the SHS technology include simplicity of hardware implementation, short time periods of the synthesis course, relatively low energy consumption [3]. In addition, the synthesis is characterized by a unique feature of the existence of a high-temperature medium during the time of reaction, which allows different types of additional external influences, which make it possible to regulate the structure and properties of final products [4]. From this perspective, it can be described as a directed synthesis, whose preparation and implementation of the regimes deliver/provide with the material with the desired combination of properties.

For the protection from the integrated flow of ionizing radiation it is advisable to use materials which include various elements [5, 6]. In case of high-energy neutron radiation flow, it is advisable to use a mixture of light and heavy elements. Heavy protection component must, on the one hand, have a high cross section for inelastic scattering of neutrons of fission spectrum and the low threshold of the inelastic neutron scattering (tungsten, niobium), on the other – to have a high slowed-down neutron capture cross section while emitting low-energy capture gamma-ray emission (indium, hafnium, tungsten). Lightweight component should effectively slow down the fast neutrons in the process of elastic scattering (beryllium, carbon, boron) and have a high capture cross section of low-energy neutrons (boron). Consequently, for protection against ionizing radiation flows of nuclear power plants it is possible to use materials that contain materials such as boron, tungsten, indium, and others.

Borides of metals of a high atomic number with high concentrations of a heavy nuclear component, for example, can be attributed to such materials.

Thus, it can be assumed that a mixture of tungsten and light elements can be considered as promising materials for the protection from the flow of fission neutrons and gamma rays.

This paper presents the results of theoretical and experimental studies on the production and use of new materials used as a protection from the fields of ionizing radiation.

2. Experiment

The synthesis of materials for the following reactions were carried out:

$$W + B \rightarrow WB, \quad (1)$$

$$4B + C \rightarrow B_4C. \quad (2)$$

In the experiments were used powders of boron (chemical purity - 93%), tungsten (99%), carbon (98%), nickel (99%), aluminum (99%), which were mixed in a planetary ball mill AGO-2c for 10 minutes at a frequency of 50 Hz. The pressing of initial blend is performed by a hydraulic press model P338 (pressure is 3.5 MPa). Metal molds are made from tool steel. The inner diameter of the matrix is 30 mm. Duration of compression was 30 minutes. The resulting samples were placed in the furnace, where the synthesis process was carried out the air atmosphere, as shown in Figure 1. Control of combustion temperature was performed by spaced tungsten-rhenium thermocouples (TBP A-2) with an output to a memory storage device for temperature measurement and temperature control.

Different times of synthesis of the sample

0 s 2 s 4 s 6 s

Figure 1. Propagation of the combustion wave at the time of the synthesis process of the sample.

The analysis of qualitative and quantitative composition of the synthesized samples was carried out on X-ray diffractometer Shimadzu XRD6000.

Experimental study on the protective properties of boron-containing materials from neutron flows was carried out with the setup which is shown in Figure 2.

The graphite prism was composed out of 18 blocks of graphite each sized 200x200x600 mm with through-holes of a diameter of 65 mm. In the openings of all blocks (except the one where the source IBN-10 type (fast neutrons flow $1.02 \cdot 10^7$ 1/s), the protective sample and detector (BDKN-03) were located), there were cylindrical graphite rods of the same length in the sleeves.

Experiments were carried out with variation of the thickness of protective shields made of boron-containing materials B_4C and WB. Safety properties of the synthesized materials depending on fast neutron flow were studied and compared to the analogous properties of the traditional material of nuclear technology - graphite.

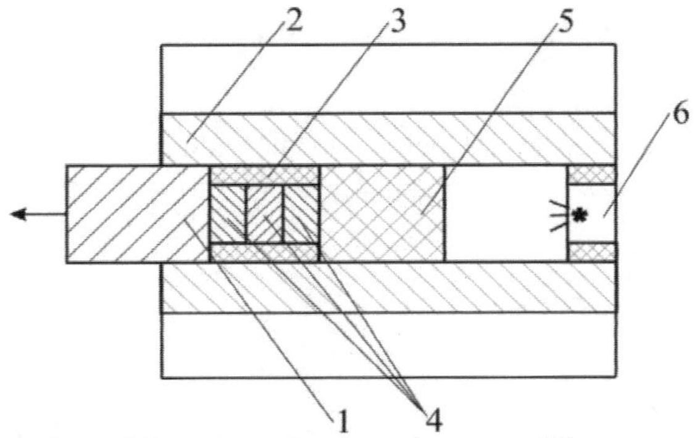

Figure 2. Experimental graphite prism: 1 – face detector of fast neutrons (BDKN-03); 2 – block of graphite prism; 3 – composing graphite sleeve; 4 – inlaid shield of ceramic samples; 5 – a removable protective shield of graphite; 6 – Pu-Be source of fast neutrons.

3. Results and discussion

Figure 3 shows the temperature distribution in the initial (a) and final (b) the time for the W-B system, obtained with the implementation of experimentation results.

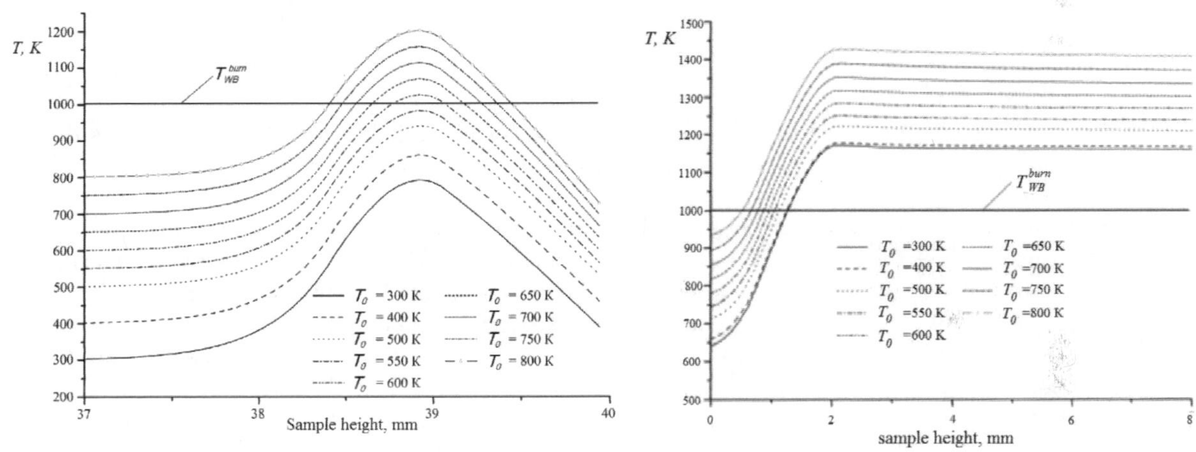

a) b)

Figure 3. Distribution of temperature in the initial (a) and final (b) the time for the W-B system after the initiation of the fusion reaction.

From the distributions obtained it can be seen that with the values of the initial temperature up to 600 K, the maximum temperature of the process in the initial period is lower than the initiation temperature of the tungsten-boron system, then it can be assumed that the propagation of the combustion wave would be unstable, or the synthesis process would not be observed. When the initial combustion temperature is 700 K and above, the minimum phase formation is observed in this sample. Similarly, the temperature distribution was obtained for the system B-C (Figure 3).

There is a weak dependence of the process temperature on initial pre-heating temperature. Thus, at values of from room temperature to 700 K the temperature at the initial moment of time has a value lower than the B-C system combustion temperature value. It can be assumed that the distribution of the combustion wave will be unstable.

X-ray analysis of obtained materials in the course of synthesis was carried out in order to determine their phase composition. Investigations of the phase composition revealed that the structure of the obtained materials, beyond those required, also contains a large number of other phase components. This is due to the staging of the process of combustion in these SHS systems where each step corresponds to a certain temperature. Because of the rapidity of the SHS process and performed during burning inhomogeneity of the temperature field over the volume of the sample synthesized, in the structure of the sample local areas were formed, where the formation of the final product stopped at one stage or another.

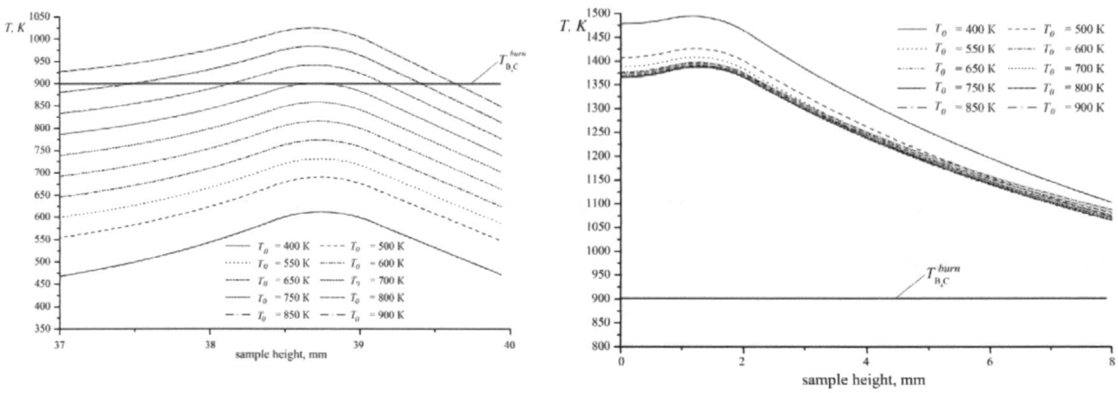

a) b)

Figure 4. The distribution of temperature in the initial (a) and final (b) time for the B-C system after the initiation of the fusion reaction

Obtaining necessary compounds required for the synthesis of temperature regimes throughout the sample volume can be achieved by increasing the thermal effect of the combination of reactions occurring in the system. For this purpose, the initial blend of reagents has been extended by introducing additional components entering into an exothermic reaction, thereby increasing the total heat effect. Reactive mixture of nickel and aluminum are used as these additives. With their introduction to the process temperature is increased from 1500-1700 K to 2500 K.

Figure 5 shows the fragments of an X-ray picture of a material based on W-B compounds, characterizing its phase structure without any additional reagents (Figure 5a) and in the presence of these additives (Fig. 5b). It is evident that the introduction of reactive Ni-Al mixtures, leading to increased exothermicity of the SHS systems, allowed to avoid multiphase. In this case the material was a nearly two-phase system of the required phase components. A downward shift of α-WB phase quantity was observed, an upward shift of tungsten diboride phase, which makes them competitive on the content. Phase of tungsten oxide WO_3 and intermetallic compounds of nickel are present at the level of background.

In the experiments shown in Fig. 6 the dependence of the attenuation of fast neutrons on the geometric and mass thickness of the boron-containing materials and graphite was determined. Approximation was performed by the method of least squares (precision 7%).

Experiments have shown that the use of boron-containing SHS-ceramics for the protection from fast neutrons provides gains with regard to dimension indicators. This is illustrated by the dependences shown in Fig. 4. For example, the weakening of the multiplicity of neutrons is equal to 1.3 is provided by a graphite screen media 45.6, ceramic WB - 7 Ceramic B_4C - 2 g/cm^2 thick. The optimal dimensions and weight of the material for protection against fast neutrons is B_4C, WB and then, finally, C.

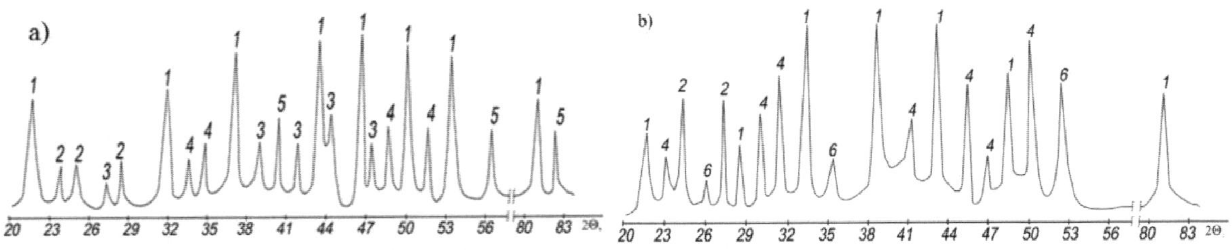

Figure 5. Fragments of an X-ray picture of a material based on tungsten boride (a) and boride of tungsten with a nickel-aluminum modifier (b), characterizing multiphase SHS product: 1 – tungsten boride α-WB phase; 2 – tungsten oxide WO_3 phase; 3 – phase connection W_2B_4; 4 – tungsten diboride phase; 5 – tungsten boride β-WB phase; 6 – nickel aluminide Ni_3Al phase.

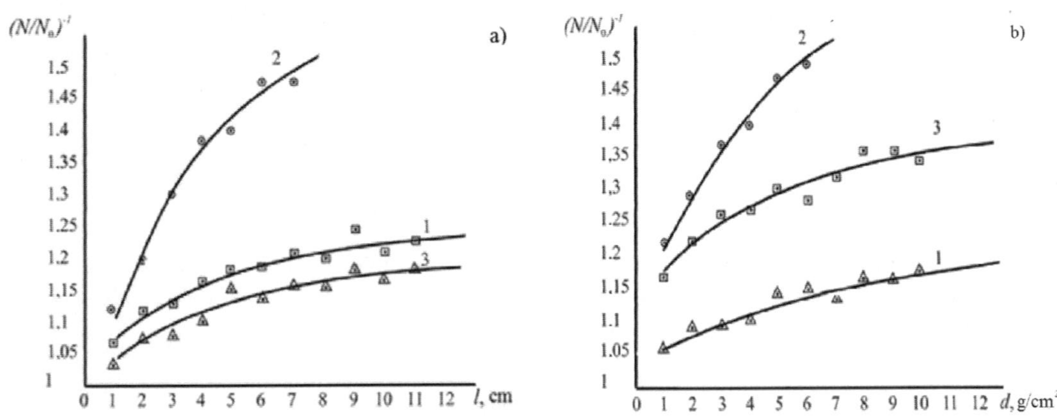

Figure 6. Dependence of the attenuation coefficient $(N/N_0)^{-1}$ on the geometric (a) and the mass thickness (b) of the shielding material: 1 – graphite, 2 – boron carbide, 3 – tungsten boride.

4. Conclusion

Composite materials intended for the manufacturing of protective elements from the neutron flows were synthesized in the process of combustion. Analysis of the results of instrument experiments on protective properties of composite materials, resulting from the combustion process, allows to formulate the following conclusions:

- products obtained in the SHS mode, provides a unique combination of nuclear-physical properties and parameters of weight and size, which allows them to use for effective protection against fast-neutron fluxes;
- SHS method allows to obtain materials that provide advantages in mass and size parameters for the protection of the fluxes of fast neutrons compared with traditional materials.

References

[1] Cao J, Lu J, Jiang L, Wang Z 2016 *Ceram Int* **42** 10079-10084
[2] Kothiyal G, Ananthanarayanan A, Dey G 2012 *Adv Funct Mater* **9** 323–386
[3] Xu X, Ren S, Geng H, Du H, Liu J 2016 *J. Eur. Ceram. Soc* **36** 1791-1796
[4] Huang L, Wang H, Qui F, Jiang Q 2006 *J. Mater. Sci.***422** 309-315
[5] Tashlykov O, Shcheklein S, Lukyanenko V, Mikhaylova A 2016 *Nucl. Energ. Technol.***2** 42-44
[6] Ahmed S 2014 Physics and Engineering of Radiation Detection (Elsevier) 784p

IOP Conf. Series: Materials Science and Engineering **135** (2016) 012016 doi:10.1088/1757-899X/135/1/012016

Low resistance Cu_3Ge compounds formation by the low temperature treatment of Cu/Ge system in atomic hydrogen

A I Kazimirov[1], **E V Erofeev**[1], **, I V Fedin**[1], **V A Kagadei**[1] **and Y N Yurjev**[2]
[1] Research Institute of Electrical Communication Systems of Tomsk State University of Control Systems and Radioelectronics, 47 Vershinina street, Tomsk, 634034 Russian Federation
[2] Institute of Physics and Technology, National Research Tomsk Polytechnic University, 30 Lenina av., Tomsk, 634050, Russian Federation

E-mail: erofeev@micran.ru

Abstract. The research deals with the regularities for Cu_3Ge compound formation under the low temperature treatment of a double-layer Cu/Ge system deposited on *i*-GaAs substrate in atomic hydrogen flow. The treatment of a Cu/Ge/*i*-GaAs system with layer thicknesses, respectively, of 122 and 78 nm, in atomic hydrogen with a flow rate of 10^{15} at.$\cdot$cm$^{-2}\cdot$s^{-1} for a duration of 2.5–10 min at room temperature, leads to an interdiffusion of Cu and Ge and formation of a polycrystalline film containing stoichiometric phase Cu_3Ge. The film consists of vertically oriented grains of dimensions 100–150 nm and has a minimum specific resistance of 4.5 μΩ cm. Variation in the treatment duration of Cu/Ge/*i*-GaAs samples in atomic hydrogen affects Cu and Ge distribution profiles, the phase composition of films formed, and the specific resistance of the latter. As observed, Cu_3Ge compound synthesis at room temperature demonstrates the stimulative effects characteristic of atomic hydrogen treatment for both Cu and Ge diffusion and for the chemical reaction of Cu_3Ge compound generation. Activation of these processes can be conditioned by the energy released during recombination of hydrogen atoms adsorbed on the surface of a Cu/Ge/*i*-GaAs sample.

1. Introduction

The effects of treatment in hydrogen plasma or atomic hydrogen (hydrogenation) on the properties of near-surface layers of various semiconductor materials and solids have been intensively studied since the early 1990's. A number of effects have been discovered, including passivation of structural defects, deep and minor impurity centers, and also altering of electrophysical properties of materials [1–7]. The mechanisms of these effects are conditioned by atomic hydrogen penetration into the near-surface layers of semiconductor materials, followed by the formation of passive complexes with various types of crystal lattice defects.

In other studies, it is shown that the recombination of hydrogen atoms adsorbed on the surface of solids is accompanied by an energy release equaling ≅4.5 eV per single recombination event. Under certain conditions, this energy can efficiently stimulate diffusion processes on the surface both in the surface and near-surface layers of semiconductor crystals and likewise, the desorption of residual impurities from the surfaces [8].

In paper [9], it is shown that under the action of atomic hydrogen generated in the gaseous discharge plasma, low temperature crystallization of thin polycrystalline Ni films occurs.

In paper [10], the research focused on low temperature diffusion of In into the near-surface layers of Ge stimulated by treatment of the samples in hydrogen plasma. The diffusion of In atoms from the thin film deposit on the surface of a Ge sample was observed in the so-called "active zone" (≈10 nm), where, according to the authors, the processes of defect formation take place and multiphonon dissipation mechanisms are brought into effect in the energy release, during recombination of hydrogen atoms. Localization of In atoms was observed both in the vacancies and in the interstitial sites of Ge crystal lattice. The factor of In diffusion into Ge, calculated for the hydrogen plasma treatment mode used in the study, amounted to $\sim 1.4 \cdot 10^{-19}$ m^2/s, which corresponds to In thermal diffusion into Ge at $T = 610$ °C. The actual temperature of samples in these experiments did not exceed 30 °C.

The authors of paper [11] carried out research into low temperature diffusion of Au films into Ge substrate, stimulated by treatment in the hydrogen plasma of a gaseous discharge. It has been shown that treatment of this kind stimulates diffusion of Au atoms into Ge at temperatures close to room temperature. It has been shown that diffusion of Au within the crystal occurs by the interstitial mechanism, and a considerable part of Au atoms are localized on Ge lattice sites.

We have found earlier that treatment of semi-insulating GaAs samples, with thin Cu and Ge films deposited on their surface, in atomic hydrogen flow at room temperature, leads to interdiffusion of Cu and Ge and formation of polycrystalline film with vertically grains of dimension 100–150 nm [12].

This paper presents the results of further studies of phenomena occurring under the low temperature treatment in atomic hydrogen flow of a double-layer Cu/Ge system deposited on an *i*-GaAs substrate and, in particular, the regularities of Cu and Ge interdiffusion accompanied by formation of a polycrystalline film containing a considerable portion of the Cu_3Ge stoichiometric compound.

2. Experimental details

The semi-insulating *i*-GaAs (100) substrates were used in experiments. All substrates were cleaned from native oxides and contaminants in HCl:H_2O (1:10) solution, followed by rinsing in deionized water and drying in purified nitrogen flow.

The first series of experiments focused on a study of the regularities of profile changes in Cu and Ge distribution variant to the depth of samples, and of the changes of thin film phase composition variant to the duration of Cu/Ge/*i*-GaAs samples treatment in atomic hydrogen flow at room temperature. For that purpose, thin films of Ge and Cu, with thickness of 78 and 122 nm, respectively, were deposited on the surface of GaAs substrates by electron-beam evaporation in a vacuum, with residual pressure $p = 5 \cdot 10^{-7}$ mbar. The thickness of the films was chosen to complete mixing of the layers occurs, the Ge content in the mixture would amount to $N_{Ge} = 25$ at.%. This was supposed to ensure generation of the Cu_3Ge stoichiometric compound. On withdrawal of samples from the evaporator chamber, they were loaded into an experimental unit to treatment in atomic hydrogen flow with flow density $j = 10^{15}$ at.·cm^{-2}·s^{-1} for $t = 1$, 2.5, 5, and 10 minutes at room temperature.

A reference sample of Cu/Ge/*i*-GaAs was subjected to heat treatment in a vacuum with residual pressure $p = 5 \cdot 10^{-6}$ mbar at temperature $T = 200$ °C for $t = 30$ min. According to the data [13], this mode of heat treatment leads to complete interdiffusion of Ge and Cu thin films and to generation of Cu_3Ge stoichiometric compound.

The second series of experiments dealt with a study of the dependence of the specific resistance of Cu/Ge thin film with a total thickness of 200 nm and with varying thicknesses of Ge film ($d = 65$–102 nm) on the duration of Cu/Ge/*i*-GaAs samples treatment in atomic hydrogen flow at room temperature. After cleaning the surface of substrates from native oxides by electron-beam evaporation in a vacuum, deposition of Ge and Cu thin films would be performed, in which case Ge content amounted to $N_{Ge} = 20$, 25, 30, 35 at.%. Next, each sample would be divided into three parts. Each sample thus obtained would then be treated in atomic hydrogen flow with flow density $j = 10^{15}$ at.·cm^{-2}·s^{-1} at room temperature; $t = 5$, 15, and 30 minutes, respectively.

The experiments were carried out using the atomic hydrogen source described in paper [14].

A cross section of the samples was examined using scanning electron microscopy (SEM), and the distribution of Cu and Ge elements by depth was examined using Auger electron spectroscopy (AES), with layer-by-layer etching. To determine the qualitative composition of phases of the obtained compounds, X-ray Diffraction (XRD) was applied. The specific resistance of Cu/Ge thin films was studied using a four-point probe method, with measurement error margin not exceeding 10 %.

3. Results and discussion

The profiles of Cu and Ge elements distribution by depth and phase composition of near-surface layers of the samples from the first series of experiments are presented in Figs. 1–5.

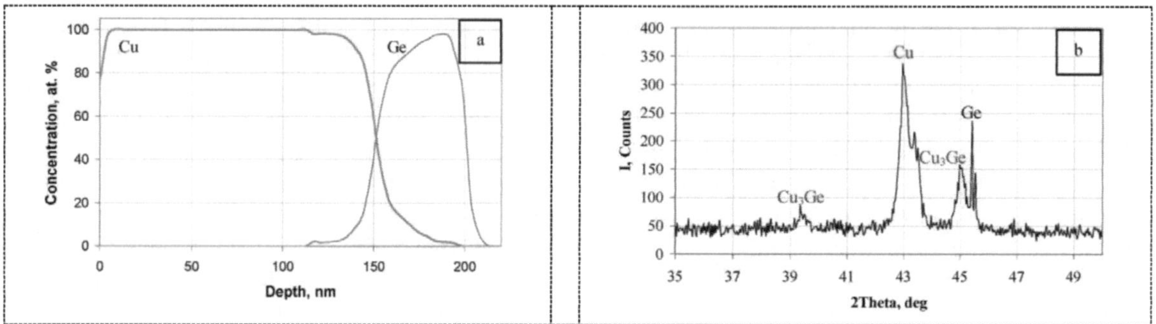

Figure 1. Profiles of Ge and Cu elements distribution by depth (*a*) and composition of XRD phases (*b*) of a Cu/Ge/*i*-GaAs sample treated in atomic hydrogen flow for t = 1 min.

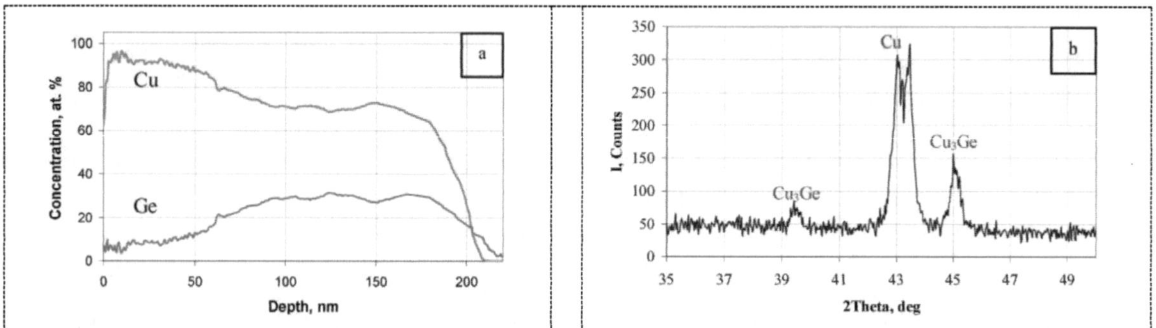

Figure 2. Profiles of Ge and Cu elements distribution by depth (*a*) and composition of XRD phases (*b*) of a Cu/Ge/*i*-GaAs sample treated in atomic hydrogen flow for t = 2.5 min.

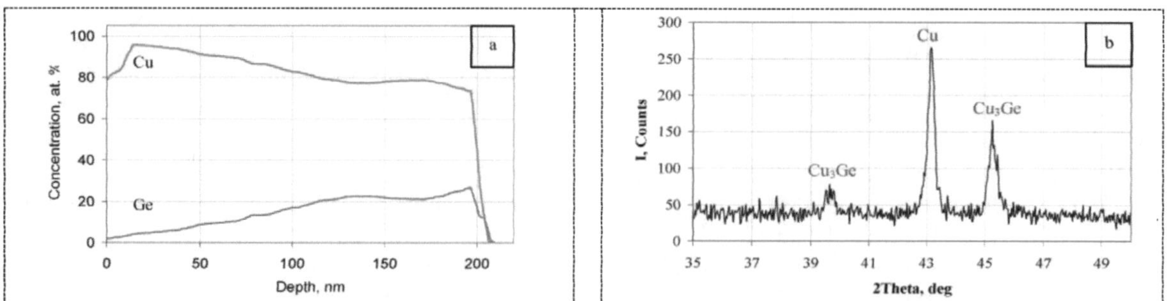

Figure 3. Profiles of Ge and Cu elements distribution by depth (*a*) and composition of XRD phases (*b*) of a Cu/Ge/*i*-GaAs sample treated in atomic hydrogen flow for t = 5 min.

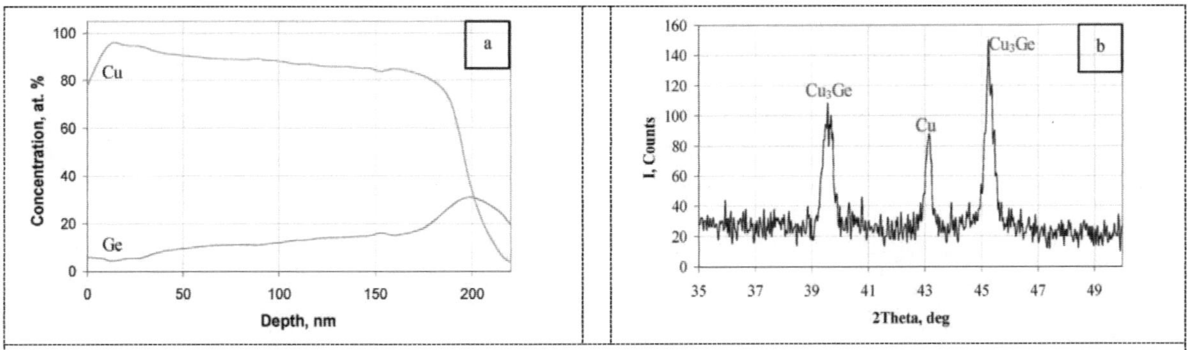

Figure 4. Profiles of Ge and Cu elements distribution by depth (*a*) and composition of XRD phases (*b*) of a Cu/Ge/*i*-GaAs sample treated in atomic hydrogen flow for t = 10 min.

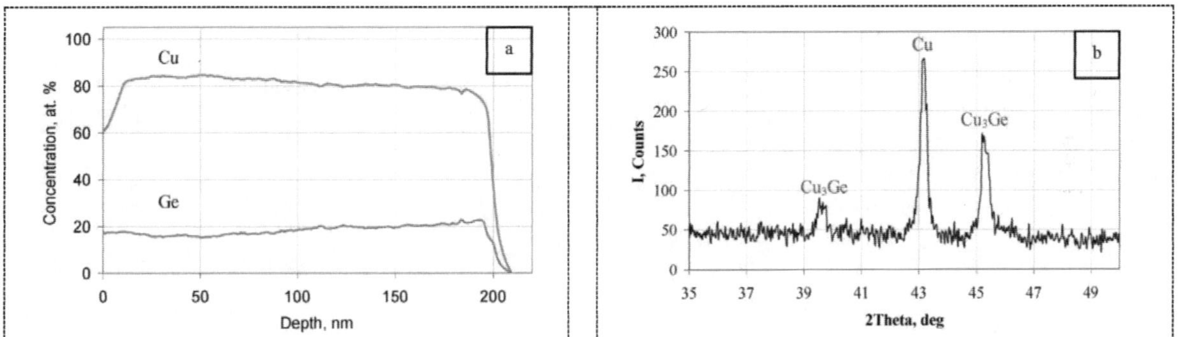

Figure 5. Profiles of Ge and Cu elements distribution by depth (*a*) and composition of XRD phases (*b*) of Cu/Ge/*i*-GaAs samples after high-temperature vacuum annealing at temperature T = 200 °C for t = 30 min.

The data in Fig. 1, *a* demonstrates the fact that treatment of a two-component Cu/Ge/*i*-GaAs system in atomic hydrogen flow at room temperature for t = 1 min does not result in any noticeable diffusion of Cu and Ge by sample depth. The profiles of elements distribution are similar to the profiles characteristic of the sample before its treatment in atomic hydrogen flow (not shown in the figure).

With the time of sample treatment in atomic hydrogen flow being t = 2.5 min (Fig. 2, *a*), the profiles change considerably, forming a pattern typical of a complete mixing of Ge and Cu films. With the treatment time increased to t = 5 and 10 min (Figs. 3 and 4, *a*), no further considerable change in the distribution profiles pattern occurs. The concentration of Ge in the near-surface layers remains at the level of 4–8 at.%, gradually increasing to 25–30 at.% at a depth of 200 nm.

A reference sample obtained through vacuum annealing at temperature T = 200 °C for t = 30 minutes (Fig. 5, *a*) also demonstrates complete mixing of Ge and Cu films, but with more uniform elements distribution by depth. The concentration of Ge is at a level of 20 at.% through the entire thickness of the film examined.

The X-ray pattern of a sample treated in atomic hydrogen flow for t = 1 min (Fig. 1, *b*) demonstrates that with treatment durations being small, considerable amounts of Cu and Ge are in a chemically free state, which correlates with data obtained using the AES method. However, even in the absence of any substantial Cu and Ge interdiffusion, the X-ray pattern features a peak corresponding to Cu_3Ge phase.

An increase in the duration of Cu/Ge/*i*-GaAs samples treatment in atomic hydrogen flow (t = 2.5–10 min, Fig. 2–4, *b*) results in a decrease of the intensity of the peaks which correspond to elementary Ge and Cu and in an increase of the intensity of peaks of the Cu_3Ge stoichiometric compound phase. It goes to show that even though a more uniform distribution of Ge and Cu elements by film thickness is not achieved with treatment time $t = 5$ and 10 min, chemical reaction of compound formation proceeds successfully, leading to an increase of Cu_3Ge phase content in the samples.

A reference sample obtained through vacuum annealing (Fig. 5, *b*) has an X-ray pattern similar to the one which is typical for a sample obtained by treatment in atomic hydrogen flow for $t = 5$ min (Fig. 3, *b*). This demonstrates the fact that, although film elements in the reference sample have a more uniform distribution, the maximum content of the Cu_3Ge phase is not achieved in the film, whereas the content is observable in a sample treated in atomic hydrogen flow for $t = 10$ min.

It should be noted that changes in the sequence of Ge and Cu films deposition on *i*-GaAs substrates during treatment of samples in atomic hydrogen flow has never resulted in any considerable changes in the behavior regularities of the processes of formation of a continuous polycrystalline film containing the Cu_3Ge phase.

To further elaborate the pattern of Cu and Ge films interaction, microscopic examination was carried out for the cross section of the initial Cu/Ge/*i*-GaAs samples, as it was for samples having undergone a control vacuum annealing and for those treated in atomic hydrogen for $t = 5$ min (Fig. 6).

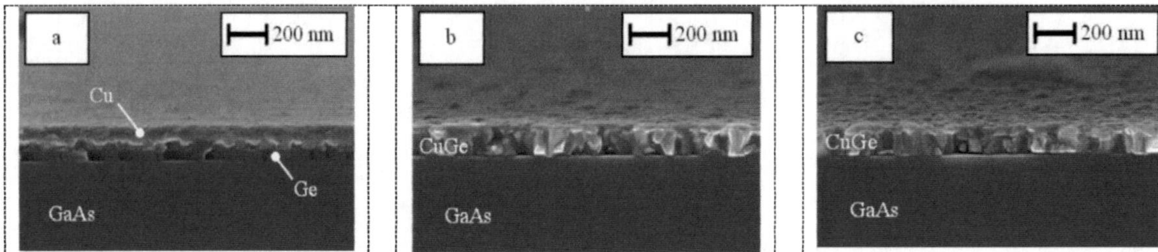

Figure 6. Cross section of Cu/Ge/*i*-GaAs samples before treatment (*a*), after treatment in atomic hydrogen flow for $t = 5$ min (*b*), and after heat treatment in a vacuum at temperature $T = 200$ °C for $t = 30$ min (*c*).

For the initial Cu/Ge/*i*-GaAs sample (Fig. 6, *a*), the boundary between the two films of Cu and Ge is clearly visible. After heat treatment in a vacuum ($T = 200$ °C, $t = 30$ min), interdiffusion of films occurs, leading to formation of a polycrystalline film with vertically oriented grains (Fig. 6, *b*), similar to those obtained in study [15]. Treatment of a Cu/Ge/*i*-GaAs sample in atomic hydrogen at room temperature for $t = 5$ min leads to a result similar to that obtained in a reference sample with vacuum annealing. Formation of a polycrystalline film with vertically oriented grains of 100–150 nm average size (Fig. 6, *c*) is observed.

The results shown in Figs. 1–6 correlate well with each other and testify to the fact that during treatment of Cu/Ge/*i*-GaAs samples in atomic hydrogen flow at room temperatures, interdiffusion of Cu and Ge occurs, accompanied by generation of a Cu_3Ge compound having polycrystalline structure.

Fig. 7 shows the dependencies of specific resistance, ρ of Cu/Ge thin film on the duration of the treatment of Cu/Ge/*i*-GaAs samples from the second series of experiments in atomic hydrogen flow at room temperature. The content of Ge, N_{Ge} in the samples varied within a range of 20 to 35 at.%.

The minimum value of Cu/Ge film specific resistance was $\rho \approx 4.5$ μΩ·cm, this being observed for samples with $N_{Ge} = 20$–25 at.% and $t = 5$ min. Increase of samples treatment time from $t = 5$ to 30 min results in a gradual increase of thin film specific resistance value, irrespective of N_{Ge} value. At the same time, for all treatment durations, the minimum value of ρ is observed for samples with Ge content at the level of 20–25 at.%, which is close to a stoichiometric composition characteristic in a Cu_3Ge compound.

The value of film specific resistance of a sample having a similar N_{Ge} magnitude but formed through vacuum annealing at temperature $T = 200$ °C for $t = 30$ min amounted to $\rho = 7.0$ μΩ·cm. The obtained data concurs with the results of study [13], in which the value of specific electrical resistance of the film of Cu_3Ge compound, with Ge content in the system amounting to about 27–30 at.%, after heat treatment in a vacuum at $T = 400$ °C for $t = 30$ min resulted in $\rho \approx 5$–6 μΩ·cm.

Thus, proceeding from the results obtained from the experiments a conclusion can be drawn concerning the simulative character of the effect of treatment in atomic hydrogen flow on both the interdiffusion of Cu and Ge and the chemical reaction of Cu_3Ge compound generation in Cu/Ge/*i*-GaAs samples. Activation of these processes, according to the data of studies [8–11], can be conditioned by the energy released during recombination of hydrogen atoms on the surface of Cu/Ge/*i*-GaAs samples.

4. Conclusion

It has been shown that treatment of Cu/Ge/*i*-GaAs samples in atomic hydrogen flow with density $j = 10^{15}$ at.·cm^{-2}·s^{-1} for $t \geq 2.5$ min at room temperature leads to interdiffusion of Cu and Ge atoms and formation of a polycrystalline film with a Cu_3Ge compound stoichiometric phase. The film consists of vertically oriented grains of 100–150 nm average size. The minimum value of polycrystalline film specific resistance, with Ge content at the level of 20–25 at.%, is $\rho \approx 4.5$ μΩ·cm. Activation of diffusion processes, as well as the process of Cu_3Ge chemical compound generation, can be conditioned by the energy released during recombination of hydrogen atoms on the surface of Cu/Ge/*i*-GaAs samples.

Acknowledgments

The authors of this study express their appreciation to the staff of the “Nanotechnology” Education & Research Center, Tomsk State University of Control Systems and Radioelectronics, and of the “Microelectronics” Research & Production Department, “Micran” Research & Production Company, for their involvement and assistance in the work, and also to Ivan A. Shulepov and Artyom A. Kachayev, staff members of the National Research Tomsk Polytechnic University, for carrying out measurements using the AES and XRD methods.

This work was supported by the Ministry of Education and Science of the Russian Federation under the Contract № 14.577.21.0204 from 27.10.15.

References

[1] Tyurin Y I 2000 *The storage properties of hydrogen in solids* (Moscow: Energoatomizdat)
[2] Alefeld E 1978 *Hydrogen in metals. Basic Properties* (New York: Springer-Verlag)
[3] Golban A N 1976 *The interaction of atomic particles with solids* (Kharkov)
[4] Pearton S J 2010 *Chem. Vap. Deposition* **16** 266
[5] Seager C H, Myers S M, Wright A F 2002 *J. Appl. Phys.* **92** 7246
[6] Seager C H, Mayers S M, Wright A F 2000 *J. Appl. Phys.* **88** 4676
[7] Polyakov A Y, Smirnov N B, Govorkov A V 2003 *J. Appl. Phys.* **94** 3960
[8] Lavrenko V A 1973 *Hydrogen atoms recombination on solid surfaces* (Kiev: Naychnaya misl)
[9] Zhavzharov E L, Byalik G A, Matyushin V M 2007 *Technical Physics Letters* 33 571
[10] Matyushin V M 2001 *Semiconductors* **35** 287
[12] Matyushin V M 1999 *Zh. Tekh. Fiz.* **69** 73
[13] Kazimirov A I, Erofeev E V, Kagadei V A 2013 *Proc. of SPIE* **8700** 870005
[14] Kagadei V A, Proskurovsky D I 2004 US Patent 6765216 B2
[15] Aboelfotoh M O, Oktyabrsky S, Narayan J 1997 *J. Mater. Res.* **12** 2325

IOP Conf. Series: Materials Science and Engineering **135** (2016) 012017 doi:10.1088/1757-899X/135/1/012017

Analysis of highly charged ion states in a simplified HFD approximation

V A Kilin

Tomsk Polytechnic University, Lenin 30 av., 634050 Tomsk, Russian Federation

vak@tpu.ru

Abstract. The analysis of calculating the total and single electron energies of a sequence of highly charged Li - and Be - like ions in Hartree-Fock-Dirac (HFD) approximation is presented. It is ascertained an unexpected order of location of some energy levels at Z>50 for both the ground and the excited configurations. Possible reasons of such deviations are discussed from theoretical and computational points of view.

1. Introduction

The data on energy levels and wave functions of heavy highly charged ions is very important in different physical theoretical models and experimental applications. As an example, one could point out the processes of channeling such ions through a crystal lattice [1-3]. Due to strong Coulomb interaction with the crystal lattice periodic potential field the ion electrons could be transferred to some excited states. Post channeling, high-energy photons should be radiated due to transitions to the lower-lying states. The analysis of registered photon spectra may be helpful in revealing the essence of both the channeling process physical mechanisms and the details of multi-dipole crystalline field potential exciting ionic electrons.

2. Elements of theory

Here we do not describe the basic of well-known HFD approximation (see, e.g. [4]) but present some specific features which may be useful in the below discussion of the obtained results.

In atomic calculations, the total wave function is the determinant of single-electron wave functions. They could be found by solving the system of simultaneous integro-differential equations under the condition of minimum of total energy.

The single-electron wave function $\varphi_i(\vec{r},\sigma)$ of a bound electron i has a form of Dirac bispinor

$$\chi_i(\vec{r},\sigma)=\frac{1}{r}\begin{pmatrix} F_{n\kappa}(r)\,\Omega_{jlj_z}(\vec{n}) \\ i\alpha_0 G_{n\kappa}(r)\,\Omega_{jl'j_z}(\vec{n}) \end{pmatrix}. \quad (1)$$

Here $F_{n\kappa}(r)$ is the large component and $G_{n\kappa}(r)$ is the small component of the wave function radial part, $\Omega_{jlj_z}(\vec{n})$ and $\Omega_{jl'j_z}(\vec{n})=-(\vec{\sigma},\vec{n})\Omega_{jlj_z}(\vec{n})$ present the spin-angular parts, correspondently, $\alpha_0=1/c$ is the fine structure constant, n is the electron principal quantum number, j and l are the total and orbital electron momenta, κ is the "effective" orbital momentum defined as follows:

$$\kappa = \begin{cases} -(j+1/2) = -(l+1) \ \ if \ \ l = j - 1/2 \\ j + 1/2 \ \ if \ \ l = j + 1/2 \end{cases} . \tag{2}$$

By variation procedure, the following simultaneous equations could be derived [5] (atomic units are used, $\hbar = m = e = 1$, energy in Ry) in the form

$$F_i'(r) + \frac{\kappa_i}{r} F_i(r) - [2 + \alpha_0^2(\hat{\varepsilon}_i - U_i(r))]G_i(r) = -\frac{\alpha_0^2}{r} \sum_{k,j \neq i} D(i,j,k)\hat{Y}_{ij}^k G_j(r) \tag{3}$$

$$G_i'(r) - \frac{\kappa_i}{r} G_i(r) + (\hat{\varepsilon}_i - U_i(r))F_i(r) = \frac{1}{r} \sum_{k,j \neq i} D(i,j,k)\hat{Y}_{ij}^k F_j(r) . \tag{4}$$

Here $i = n\kappa$, $j = n'\kappa'$ run over all atomic subshells, $U_i(r)$ is the self-consisted potential in which the electrons of a subshell i move, $\hat{\varepsilon}_i = \varepsilon_{ii}$ is the single electron energy. The terms with non-diagonal parameters ε_{ij}, responsible for the wave functions orthogonality, are equal to zero for closed shells and rather small for the ground states of most atoms [5,6]. Therefore they are omitted in the above equations (3) and (4). Additional simplification done in (3) and realized in the computer codes ATOM [7] used in our calculation is the neglect of the right hand side term in (3). This term stands for correcting the single electron energies and wave functions because of so-called "exchange in small components". Estimations show that this neglect results in inaccuracy of 0.1% for the most inter shells and of 0.01% for outer shells [8].

3. Main results and discussion

Here we present some results on the total and single-electron energies for Li -, Be - like ions in the nuclear charge range Z up to 100 calculated in HFD approximation simplified as mentioned above. Computer codes ATOM [7] have been used.

Since the single-electron wave functions of a highly charged ion are expected to be strongly localized nearby a nuclear, the radial variable r variation is restricted by a value of R=20 a.u. The wave functions are calculated at 600 points of r. LSJ approximation is used in calculating spin-angular coefficients in equations (1) and (2). The total and single electron energies as functions of nuclear charge Z, (increment ΔZ=10) are listed in tables below.

The below presented analysis of the obtained results is intended to reveal some unexpected features, along with standard ones, which may be a subject of the discussion on simplified theoretical approximations used in the research of extra-ordinary objects such as highly charged ions.

3.1. Be-like ion configurations $1s^2\ 2s,^2\ 1s^2\ 2s\ 2p$ and $1s^2 2p^2$

The analysis of results presented in tables 1-3 for Be-like ions of ground $1s^2\ 2s^2$, excited $1s^2\ 2s^1\ 2p^1$ and doubly excited $1s^2\ 2p^2$ configurations reveals the following:

(a) the relationship between the total energies of mentioned above configurations in the nuclear charge Z range up to 100 corresponds to the expected usual sequence, i.e., $E_{tot}(1s^2\ 2s^2) < E_{tot}(1s^2\ 2s\ 2p) < E_{tot}(1s^2 2p^2)$;

(b) the single electron energies ε_{1s} of the most inner sub-shell $1s$ also follows the above relationship, except for Z=4;

(c) the single electron energies ε_{2s} of $1s^2\ 2s^2$ and $1s^2\ 2s_{1/2}\ 2p_{1/2}$ configurations also follows the above relationship up to Z=50 but then changes it to the inverse one;

(d) the single electron energies ε_{2p} of $1s^2\ 2s_{1/2}\ 2p_{1/2}$ and $1s^2\ 2p^2$ configurations follows the relationship (a) for all values of Z;

(e) for configurations $1s^2\ 2s^2$, $1s^2\ 2s_{1/2}\ 2p_{1/2}$ and $1s^2\ 2p^2$ the single electron energies show an expected usual relationship $\varepsilon_{2s} < \varepsilon_{2p}$ for all values of Z.

Table 1. The total E_{tot} and single electron energies ε of Be $1s^2\,2s^2$ (in Ry).

Z	E_{tot}	ε_{1s}	ε_{2s}
4	-29.152	-9.467	-0.619
10	-220.509	-81.177	-15.002
20	-944.020	-362.452	-79.490
30	-2186.007	-850.315	-195.826
40	-3973.058	-1554.007	-366.575
50	-6348.905	-2488.496	-595.730
60	-9380.383	-3676.653	-889.147
70	-13166.178	-5152.791	-1255.275
80	-17850.796	-6968.754	-1706.445
90	-23649.073	-9205.204	-2261.262
93	-25651.055	-9975.930	-2451.193
100	-30894.050	-11994.819	-2949.469

Table 2. The total E_{tot} and single electron energies ε of Be $1s^2\,2s\,2p$ (in Ry).

Z	E_{tot}	ε_{1s}	ε_{2s}	ε_{2p}
4	-29.029	-9.489	-.789	-.483
10	-219.725	-80.740	-15.394	-14.213
20	-942.128	-361.278	-80.234	-77.587
30	-2182.934	-848.372	-196.917	-192.720
40	-3968.680	-1551.228	-368.006	-362.113
50	-6343.036	-2484.782	-597.490	-.589.680
60	-9372.764	-3671.857	-891.218	-881.167
70	-13156.451	-5146.708	-1257.633	-1244.865
80	-17838.462	-6961.088	-1709.053	-1692.860
90	-23633.424	-9195.521	-2264.061	-2243.343
93	-25634.227	-9965.527	-2454.772	-2432.389
100	-30874.022	-11982.455	-2952.365	-2925.305

Table 3. The total E_{tot} and single electron energies ε of Be $1s^2\,2p^2$ (in Ry).

Z	E_{tot}	ε_{1s}	ε_{2p}
4	-28.591	-9.676	-.389
10	-217.436	-80.321	-13.337
20	-936.705	-359.784	-75.376
30	-2174.269	-845.760	-189.142
40	-3956.589	-1547.422	-.357.115

50	-6327.244	-2479.664	-583.184
60	-9352.865	-3665.259	-873.062
70	-13131.850	-5138.387	-1234.997
80	-17808.261	-6950.681	-1681.011
90	-23596.199	-9182.477	-2229.188
93	-25594.490	-9951.536	-2417.454
100	-30827.370	-11965.859	-2908.327

3.2. *Li-like ion configurations* $1s^2\ 2s\ 1s^2\ 2p_{1/2}$

The analysis of results presented in tables 4 and 5 for the Li-like ground $1s^2$ 2s and excited $1s^2\ 2p_{1/2}$ configurations show the following:

(a) the relationship between the total energies of Li-like ions in the nuclear charge range Z up to 50 corresponds to the expected usual sequence, i.e., $E_{tot}(1s^2\ 2s)<E_{tot}(1s^2\ 2p_{1/2})$, but unexpectedly changes the relationship to the inverse one for Z>50;

(b) the single electron energies ε_{1s} of the most inner sub-shell 1*s* follows the relationship (b) for Be, except for Z=3;

(c) starting from Z>50 the single electron energies of configurations $1s^2$ 2s and $1s^2\ 2p_{1/2}$ reveal unexpected relationship $\varepsilon_{2s} > \varepsilon_{2p}$.

The latter statement (c) probably serves as one of the reasons responsible for the relationship formulated in (a).

Table 4. The total E_{tot} and single electron energies ε of Li $1s^2$ 2s (in Ry).

Z	E_{tot}	ε_{1s}	ε_{2p}
3	-14.867	-4.956	-.393
10	-205.531	-84.460	-17.565
20	-864.562	-369.787	-85.101
30	-1990.169	-861.810	-204.523
40	-3606.146	-1569.856	-378.342
50	-5751.687	-2508.995	-610.346
60	-8486.766	-3702.238	-905.911
70	-11900.010	-5184.111	-1272.538
80	-16121.245	-7006.824	-1720.790
90	-21343.487	-9251.739	-2266.009
92	-22531.726	-9761.819	-2388.553
100	-27865.795	-12053.041	-2931,715

Table 5. The total E_{tot} and single electron energies ε of Li $1s^2\ 2p_{1/2}$ (in Ry).

Z	E_{tot}	ε_{1s}	ε_{2p}
3	-14.748	-5.014	-.274
10	-204.838	-84.138	-16.881

20	-863.105	-369.072	-83.648
30	-1987.987	-860.673	-202.272
40	-3603.610	-1568.233	-375.357
50	-5750.060	-2506.778	-607.029
60	-8489.173	-3699.245	-.903.473
70	-11912.993	-5180.015	-1273.900
80	-16157.267	-7000.978	-1732.334
90	-21425.097	-9242.708	-2301.129
92	-22626.508	-9751.837	-2431.077
100	-28033.001	-12037.293	-3018.681

4. Conclusion

Firstly, it should be mentioned that calculated in the same approximation energies of exited Li-like $1s^2\ 2p_{3/2}$ ion states, which are not presented here, show a usual behavior of the excited state energies with respect to corresponding ground configuration $1s^2\ 2s$ for all values of Z.

Secondly, it could be assumed that the mentioned unusual energy positioning of Li-like ion states for Z>50 does not have any real physical reasons but mainly arises from the simplifications of HFD approximation realized in the used computer codes [7]. To obtain the highly charged ion energy levels with an accuracy closed to spectroscopic one, approximations that are more exact should be used.

5. Acknowledgements

The author thanks to Prof. Yu. Pivovarov and E. Sinitsin, Tomsk Polytechnic University, who stimulated this investigation for using the results in studying the channeling of ions through a crystal.

References

[1] Babaev A, Pivovarov Yu L. 2012 *Computer Physics Communications* **183** 705
[2] Nakano Y, Sokolik A A., Stysin A V, Nakai Y, Komaki K, Takada E, Murakami T and Azuma T 2015 *J. Phys. B: At. Mol. Opt. Phys.* **48** 144026
[3] Sinitsyn E A, Kilin V A , Pivovarov Y L 2015 *Book of Abstracts of XI International RREPS-15 Symposium, Saint Petersburg*
[4] Grant I P. 1970 *Adv. Phys* **19** 747
[5] Lindgren I, Rosen A 1974 *Case studies in Atomic Physics* **4** № 3 105
[6] Hartree D 1960 *Calculation of atomic structures* (Moscow: Atomizdat, in Russian)
[7] Chernysheva L V and Amusia M Ya 1983 *Automation System for Atomic Structure Researches* (Leningrad: Nauka, in Russian)
[8] Fischer F 1970 *Computer Physics Communications* **1** 151

Hard X–ray Laue monochromator

V.R. Kocharyan[1,2], A.S. Gogolev[1], A.A. Kiziridi[1], A.V. Batranin[1], T.R. Muradyan[2]

[1] National Research Tomsk Polytechnic University, Tomsk, Russia
[2] Institute of Applied Problems of Physics of NAS RA,Yerevan, Armenia

E-mail: Vahan2@yandex.ru

Abstract. Experimental studies of X-ray diffraction from reflecting atomic planes $(10\bar{1}1)$ of X-cut quartz single crystal in Laue geometry influenced by the temperature gradient were carried out. It is shown that by using the temperature gradient it is possible to reflect a hard X-ray beam with photon energy near the 100 keV with high efficiency. It has been experimentally proved that the intensity of the reflected beam can be increased by more than order depending on the value of the temperature gradient.

1. Introduction

The obtaining of intensive sources and base elements of short wavelength hard X-ray "optics" with controllable working parameters (energy, intensity, and focus length) gives an opportunity to improve technology in medicine and broaden the scope of application of X-ray diffractometry. In this regard, the issues of diffraction under external influence have special and actual place. For example, in order to optimize the required dose of exposure of patients 30 keV to 250 keV energy needs to be emitted from the white spectrum of the X-ray with the necessary bandwith, luminosity and radiation density. For the purpose of obtaining such beams the questions of diffraction under external influence occupy their special and actual place.

The authors of works [1, 2] were the first to detect the complete transfer of X-rays from the direction of passing to that of reflection in single crystals of quartz in Laue geometry when exposed to temperature gradient or ultrasonic vibrations. In [3] it was theoretically and experimentally proved that, using a temperature gradient and acoustic field, one can control the position of the focus of reflected radiation in space and time and can also convert a spherical wave into a plane one. These works do not experimentally investigate the transfer and control of the spectrum of reflected hard X-rays (over 25 keV).

2. Experiment

In order to obtain an intensive hard X–rays (over 25 keV) with controllable parameters, we have considered the X-ray diffraction on quartz monochromator in Laue geometry influenced by the temperature gradient. A rectangular plate of single crystal quartz (30x30 mm^2) with a thickness of 9 mm was used as the sample under investigation. The spectrums of reflected and passing X-ray beams from the reflecting atomic plane $(10\bar{1}1)$ of quartz single crystal were experimentally studied with various values of the temperature gradient. The experimental scheme is shown on figure 1. The experiments were carried out with the energy dispersive silicon spectrometer [4]. The temperature gradient in the crystal was created with the help of a heater 3 on figure 1. The heated part of the plate was parallel to the reflecting atomic planes ($10\bar{1}1$), i.e. the temperatureo gradient was perpendicular to the reflecting atomic planes. The temperature gradient vector and the diffraction vector were antiparallel. A white spectrum of X-ray was used in the test which was generated by the tungsten X-ray tube RAP-150MN [5, 6] under up to 120 kV voltage and 95 μA current. The beam was collimated with two slits width of 1 mm before falling on the sample. The spectrum of the reflected beam was observed at the distance of 500 mm and at the angle near of 4.5°.

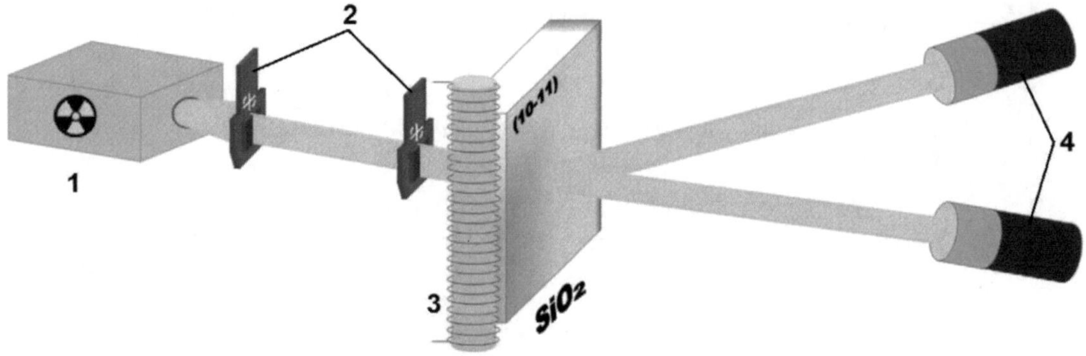

Figure 1. Experimental scheme: 1 – RAP-150MN; 2 – slit collimator; 3 – heater; 4 – detector.

In order to make sure of the effect of “transfer” the spectrums of passing and reflecting X-ray radiations were measured. For the chosen orientation of the crystal the diffraction first order matches 47 keV and second order 94 keV. Exactly for these energies in the solid spectrum of the passing beam the flops were observed, the depth of which is determined by the value of the temperature gradient and the energy resolution of the detector. The heating of the crystal side was made from 23 to 400°C, the opposite side was cooled with the help of convection. An increase in the intensity of reflecting beam in Laue geometry was observed in 45 times compared to the uniform temperature regime of the crystal quartz with 9 mm thickness. On figure 2 shown the filtered with 1.6 mm Cu radiation spectrum of “transfer effect” depending on the temperature gradient for the crystal quartz of 9 mm thickness. Energy of the first and the second order were determined from the spectrum and it values are equal to 46.86±0.05 and 93,33±0.05 keV, FWHM is 2.2±0.1 and 3.4±0.3 keV, correspondently.

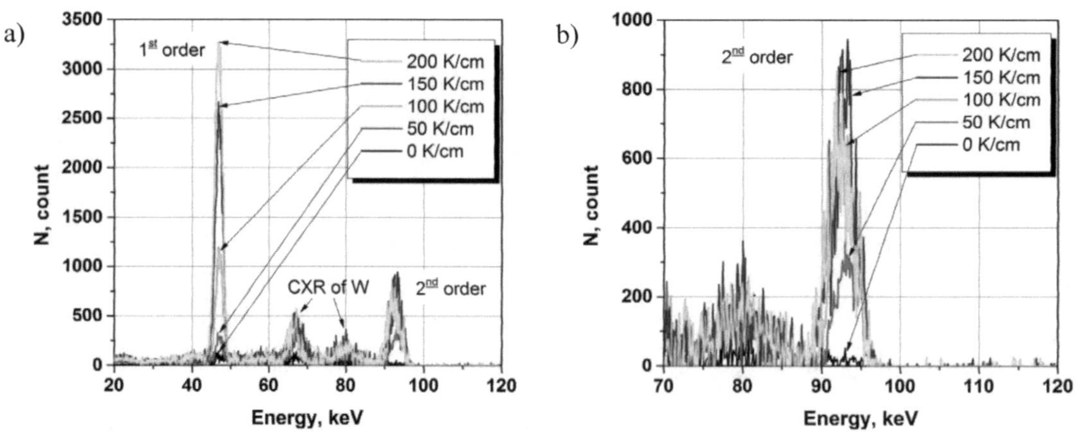

Figure 2. The spectrum of the reflected beam (a) and zoomed part (b) for the different value of the temperature gradient ΔT/Δx applied to the quartz single crystal with the thickness 9 mm.

The spectral measurements confirm the effect of transfer from passing beam of X-ray radiation to reflecting beam and its dependency from the temperature gradient that is created in the crystal. The multiple increases in the intensity is caused by the phenomenon of the full pumping of the X-ray from the passing direction to the reflecting direction with a big angular width which is much bigger than the angular width of the Darwin table and depends on the thickness of the observed mono-crystal. The saturation of the intensity parallel to the increase of the temperature gradients due to the fact that in big deformations the extinction length becomes much bigger than the effective thickness of the diffraction of each monochromatic X-ray waves.

3. Theory

Usually, an X-ray beam incident on the crystal has divergence and spectral width and in each direction spreads a totality of plane waves with different wavelengths.

In the calculations of diffraction the incident X-ray wave is decomposed into totality of plane monochromatic waves, each of which is scattered in accordance to the dynamic theory in the plane-wave approximation of the deformed crystal. Then the scattering wave integrated over the spectral and angular variable in the detector plane. In the calculations for each transmitted and reflected planar and monochromatic wave were obtained the solutions used in [7, 8].

In the XOZ-plane scattering, based on the experimental results [9, 10], in the presence of temperature gradient perpendicular to the reflecting atomic planes at a certain distance from the heating verge of the crystal the $U_x$displacement function can be written as:

$$U_x = \frac{t^2-(t-2z)^2}{8R}, \tag{1}$$

where t is the thickness of the crystal, R is the curvature radius of reflecting atomic planes.

Let, in the Laue geometry, the monochromatic plane wave of X-rays is incident on the crystal. In this case, considering (1) for the two-wave approximation the Takagi's equations take the following view:

$$\begin{cases} \frac{\partial D_h}{\partial z} - i\frac{C\chi_h K}{2\cos(\theta)}e^{-i\vec{h}\vec{u}}D_0 - i\frac{(\chi_0 - b)K}{2\cos(\theta)}D_h = 0 \\ \frac{\partial D_0}{\partial z} - i\frac{\chi_0 K}{2\cos(\theta)}D_0 - i\frac{C\chi_{\bar{h}}K}{2\cos(\theta)}e^{i\vec{h}\vec{u}}D_h = 0 \end{cases}, \tag{2}$$

where D_0 and D_h are the amplitudes of the reflected and transmitted field, the parameter $b = -2\sin(\theta)\Delta\theta$ characterizes the deviation from the exact Bragg angle, χ_0 and χ_h are the Fourier components of the dielectric permittivity, K is the wave number, θ is the angle of incidence, $\vec{h}$ is the diffraction vector, $\vec{u}$ is the displacement vector, and C is the polarization multiplier.

The boundary conditions for amplitudes of each monochromatic plane waves of X-rays D_0 and D_h on the input surface of the crystal $z = 0$ take the following form: $D_0 = D_{ins}$ and $D_h = 0$. Introducing the designations:

$$\eta = \frac{bK}{4\cos\theta}, \; \alpha = \frac{1}{2}\frac{\partial \vec{h}\vec{u}}{\partial z}, \; C_h = \frac{C\chi_h K}{2\cos\theta}, \; C_{\bar{h}} = \frac{C\chi_{\bar{h}}K}{2\cos\theta},$$
$$D_h = \tilde{D}_h e^{i(\frac{\chi_0 K}{2\cos(\theta)}z - \frac{bK}{2\cos(\theta)}z - \frac{\vec{h}\vec{u}}{2})}, \; D_0 = \tilde{D}_0 e^{i(\frac{\chi_0 K}{2\cos(\theta)}z - \frac{bK}{2\cos(\theta)}z + \frac{\vec{h}\vec{u}}{2})}. \tag{3}$$

Based on the boundary conditions and the designations (3), from equation (2) for amplitudes of each monochromatic plane wave of X-rays D_0 and D_h we get finally:

$$D_h(z) = \frac{C_h D_{ins}}{2\sqrt{W(0)W(z)}}\left(\exp\left(\frac{2A - C_i}{C_r} f(z) - \frac{K\chi_{0i}}{2\cos(\theta)} z \right) \exp\left(i[\varphi(z) + \psi_1(z) + f(z)]\right) \right.$$
$$\left. - \exp\left(\frac{C_i - 2A}{C_r} f(z) - \frac{K\chi_{0i}}{2\cos(\theta)} z \right) \exp\left(-i[\varphi(z) + \psi_2(z) + f(z)]\right) \right), \tag{4}$$

$$D_0 = \frac{D_{ins}}{2\sqrt{W(0)W(z)}}\left(\left(i(W(z)-\eta-\frac{\partial\alpha}{\partial z})-\frac{1}{2W(z)}\frac{\partial W(z)}{\partial z}\right)\exp\left(\frac{2A-C_i}{C_r}f(z)-\frac{K\chi_{0i}}{2\cos(\theta)}z\right)\exp\left(i[\varphi(z)+\psi_1(z)+f(z)]\right)\right. ,\quad (5)$$

$$\left.+\left(i(W(z)+\eta+\frac{\partial\alpha}{\partial z})+\frac{1}{2W(z)}\frac{\partial W(z)}{\partial z}\right)\exp\left(\frac{C_i-2A}{C_r}f(z)-\frac{K\chi_{0i}}{2\cos(\theta)}z\right)\exp\left(-i[\varphi(z)+\psi_2(z)+f(z)]\right)\right)$$

where

$$f(z)=\frac{2Az+\eta-At+\sqrt{(2Az+\eta-At)^2+C_r}}{\eta-At+\sqrt{(\eta-At)^2+C_r}},$$

$$\varphi(z)=\frac{(2Az+\eta-At)\sqrt{(2Az+\eta-At)^2+C_r}-(\eta-At)\sqrt{(\eta-At)^2+C_r}}{4A},$$

$$\psi_{1,2}(z)=\pm\left(\frac{K\,\mathrm{Re}\,\chi_0}{2\cos\theta}+2\eta\right)z\pm A\left(zt-z^2\right)$$

$$C_r=\mathrm{Re}(C_hC_{\bar{h}}),\qquad C_i=\mathrm{Im}(C_hC_{\bar{h}}).$$

A is the deformation parameter, which is equal to $A=\frac{h}{2R}$ in the presence of temperature gradient.

For the purpose of qualitative comparison of reflecting and passing radiation with experimental results calculations are made with the help of this method when an X-ray beam with white spectrum falls on the crystal quartz. In the calculations the diffraction in Laue geometry is viewed for reflecting atomic planes $(10\bar{1}1)$ of single crystal of quartz with 9 mm thickness when the crystal is in Bregg condition for 94 keV photon energy.

The results of the calculations are brought in the figure 3. As we can see, the calculations also confirm the existence of multiple increase of intensity which is caused by the phenomenon of full pumping of the X-ray from passing direction (figure 3b) to the reflecting direction (figure 3a) with big spectrum width. In the experimental results in the spectrum of the passing beam near the Bregg angle the flop does not reach 0 value which is due to the resolution of the spectrometer.

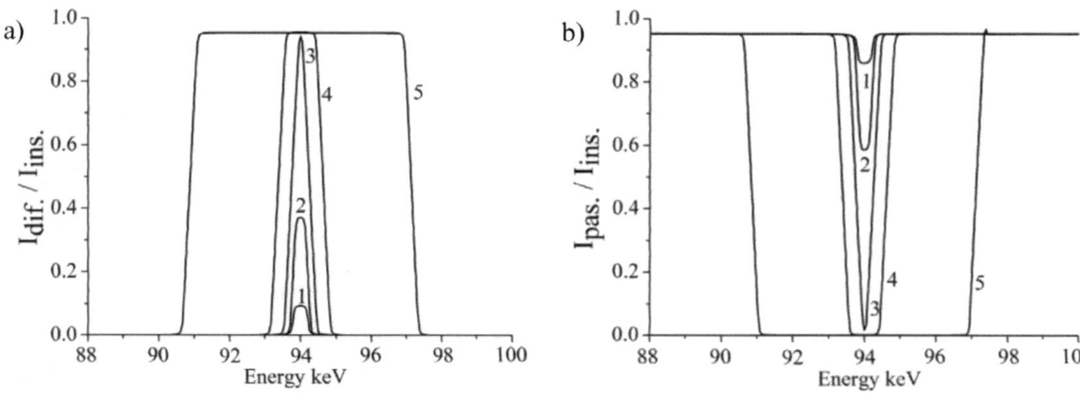

Figure 3. The spectrums of the reflected (a) and passed (b) beams for the different values of the parameter deformation of the reflecting atomic planes of the quartz single crystall with the thickness of 9 mm: 1) $A = 0$; 2) $A = 250$; 3) $A = 800$; 4) $A = 2000$; 5) A=10000.

On figure 4 shows the spectrum of reflecting beam from reflecting atomic planes of a single crystal of quartz to different thicknesses at a certain value of the parameter A. The calculations are made for the X-rays with an energy of 94 keV at a radius of curvature of the reflecting atomic planes ~ 15 m (parameter A = 10000). From figures it is clear that at a certain value of the parameter A, than greater the thickness of the single crystal, the wider the spectrum of the reflecting X-ray beam i.e., the wider the spectrum of full pumping beam. For example, a single crystal of quartz with a thickness of 1 cm, the atomic planes are curved with the curvature of 15 m fully reflect the X-rays with a spectral width of $\Delta E \approx 6 \div 7$ keV, a crystal with a thickness of 5 cm fully reflects $\Delta E \approx 25$ keV.

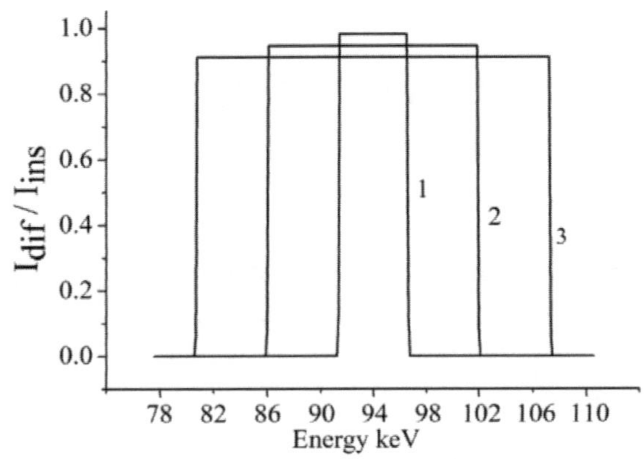

Figure 4. The spectral distribution of the reflected X-rays with energy of 94 keV from the single crystal of quartz for value of parameter A = 10000: 1) t = 1 cm, 2) t = 3 cm, 3) t = 5 cm.

Conclusion

Thus, we have obtained multiple times increase in the intensity of the reflecting hard X-rays, full pumping of X-ray from passing direction to the reflecting direction with big spectral width, its dependency from the thickness of the researched single crystal and value of the applied temperature gradient. It is shown that by using quartz crystals in Laue geometry with external temperature gradient it is possible to create a monochromators with higher of luminosity and linear filters in the region of hard X-ray radiation with controllable parameters.

Acknowledgement

This work was supported by The Ministry of education and science of the Russian Federation within "Nauka" Program project # 3761. We also acknowledge the Russian-Chinese Laboratory for Radiation Control and Inspection for the equipment.

References

[1] Mkrtchyan A.R., Navasardyan M.A., Mirzoyan V.K. 1982 *Pis'ma v ZhTF.* **8** 677
[2] Mkrtchyan A.R., Navasardyan M.A., Gabrelyan R.G. *at al* 1983 *Pis'ma v ZhTF.* **9** 1181
[3] Mkrtchyan A.R., Navasardian M.A., Gabrielyan R.G. *at al* 1986 *Solid State Communications.* **59** 147
[4] A.S. Gogolev, D. Hampai, A.Kh. Khusainov *et al* 2015 *NIM B* **355** 268
[5] Tungsten X-ray tube RAP-150MN http://www.tsnk.ru/content/view/5/12/
[6] A.V. Batranin, S.V. Chakhlov, D.V. Grinev, *et al* 2013 *Applied Mechanics and Materials* **379** 3
[7] A.R. Mkrtchyan, A.S. Bagdasaryan, S.H. Khlopuzyan *et al* 2015 *Nelineinyi mir (Nonlinear World)* **13** 47
[8] V.R. Kocharyan, R.Sh. Aleksanyan, K.G. Truni 2010 *Journal of Contemporary Physics (Armenian Academy of Sciences)* **45** 190
[9] S.N. Noreyan, V.Gh. Mirzoyan, V.R. Kocharyan 2004 *Journal of Surface Investigation. X-ray, Synchrotron and Neutron Techniques (Poverkhnost'. Rentgenovskie, Sinkhrotronnye i Neitronnye Issledovaniya)* **1** 18
[10] V. R. Kocharyan, A. S. Gogolev, A. E. Movsisyan *et al* 2015 *J. Appl. Cryst.* **48** 853

Experiments on rehabilitation of radioactive metallic waste (RMW) of reactor stainless steels of Siberian chemical plant

G N Kolpakov, V V Zakusilov, N V Demyanenko and A S Mishin

National Research Tomsk Polytechnic University, Tomsk, Russia

E-mail: kgn@tpu.ru

Abstract. Stainless steel pipes, used to cool a reactor plant, have a high cost, and after taking a reactor out of service they must be buried together with other radioactive waste. Therefore, the relevant problem is the rinse of pipes from contamination, followed by returning to operation.

1. Introduction

The acuteness of the problem was how to deal with the pipes of the first cooling circuit of a reactor, taken out of service. Pipes made of expensive stainless steel (X18H9T) had a contamination of both external and internal surfaces. They could be buried as radioactive waste, or one could try to "rinse" them to acceptable values of β- and γ-activity, take them off the books and return them to the production cycle. The latter was more preferred.

2. Experiment

These were the so-called weakly fixed deposits. Whereas external contamination of pipes was formed by deposition of radioactive aerosols, the internal surface of the same pipes had a film with thickness of about 0.5 mm, formed due to the deposition of water impurities at temperatures less than 100 °C. Research of film composition has shown that it is composed of oxides of elements: 48-63% ;Al_2O_3; 17-21% Fe_2O_3; 6-17% MgO and CaO 14-21% SiO_2; less than 0.39% CrO_2 and NiO_2 and contaminated by radioactive isotopes ^{60}Co, ^{137}Cs, ^{152}Eu. Under certain conditions, a film of pipes of "cold" leg was easily removed and thus was weakly fixed. This could mean that it has no strong chemical or epitaxial substrate. In appearance, it had pale brown color and looked like a scale. By speculating about scale and how it is removed in steam boilers, we came to the conclusion that alkaline solutions NaOH (potash) can be used.

The advisor on this issue, Prof. N.A. Kolpakova proposed not to use ready alkali NaOH, but to obtain it in the process of electrolysis of NaCl solution. In this case, chemical effect of NaOH, formed at the cathode (of the rinsed surface) will be intensified by the flowing of current.

The next step was to select the material of anode. There were two possibilities: either stainless steel or graphite. The benefits of graphite were quickly revealed. The specific resistance of graphite is about 10 times higher than that of steel, so graphite caused higher voltage on the cathode surface and decomposition of NaCl. At the same time, when current densities are suitable for rinse, (~ 1000 A/m^2) a steel electrode, made of stainless steel, is exposed to partial electrodissolution of its components, and especially chromium, having a lower potential of electrodissolution than other stainless steel components. Chromium is deposited on the rinsed surface in the form of crystals of deep black color, preventing the rinse.

Further, optimizing the process according to current density, concentrations of NaCl and time of electrolysis, the results, shown in Table 1 and the figure, were obtained.

Table 1. The results of electrolytic rinse when the concentration of salt is 40 g/l

Sample number	Electrolysis conditions		Contamination *before*/*after*		Note
	Current, A	Time, min	Beta, min^{-1}	Gamma, μR/hr	
1	30	20	1200/10	170/42	pH = 6.5
2	30	25	580/10	724/34	pH = 6.0
3	30	20	220/20	247/40	pH = 6.0
4	30	40	1000/30	159/45	pH = 6.0

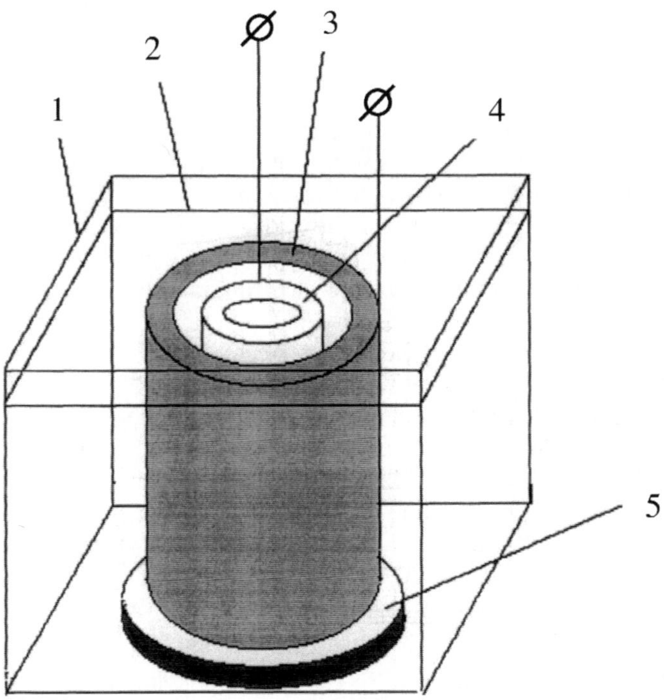

Figure 1. Installation diagram for deactivation: 1 – vessel with capacity of 9 liters; 2 – electrolyte level; 3 – rinsed pipe – first electrode; 4 – second electrode; 5 – rack-insulator for electrodes.

The analysis of obtained results led to the important conclusion that the porosity of a rinsed film has crucial importance for the process of rinse, if it is well wetted, due to good porosity, the process goes successfully. In this case, long-term "soaking" of samples can help. The driving force of further process of electrolysis is the formation of H_2 bubbles at the film-substrate interface, whereby the oxide film is peeled off almost without dissolving, that reduces LRW.Experiments have also shown that the external contamination of pipes is removed when there is a simultaneous impact on internal contamination.

3. Results and discussion

Research [1, 2] has shown, that the composition of such a film is a mixture of oxides, formed as a result of both corrosion of the shells of fuel elements and circuit walls, and water impurities, Table 2.

Table 2. Composition of deposits on the internal surface of pipe samples of "hot" branch of cooling, wt. %

Substance	Data of authors [1]		Data of authors [2]
	Scale of light brown color	Scale of red-brown color	
Al_2O_3	60.4	38.6	71.0-17.0
Fe_2O_3	2.86	9.8	15.2-1.2
MgO, CaO	2.0	7.9	1.0-0.0
CuO	0.14	-	0.10-0.01
Cr_2O	0.2	0.16	0.6-0.0
NiO	0.06	0.22	2.19-0.03
SiO_2	10.0	8.2	21.8-1.0
So_4^{+2}	2.35	0.86	3.9-0.9
P_2O_5	-	4.5	19.1-2.1

Such polyoxide deposition is firmly fixed, since it was formed at increased temperatures close to 200 °C. This was pointed out in [3, 4], in which, inter alia, it is stated that under conditions of operation of nuclear power plant with RBMK, a film is formed on steel OH18N10T. This film consists of maghemite and mixed spinels of iron, chromium and nickel of common composition $[Fe, Ni]O[Fe, Cr]_2O_3$. It is assumed that the film is formed and is growing both due to adsorption of oxygen on metal surface, and due to the solid-state diffusion mechanism. In this case, thin films are formed, with thickness of tens of angstroms, dense with low porosity, protecting metal.

Attempts to remove an oxide film in a single step electrolytically in solution of sodium chloride were unsuccessful. The result was not affected by the changes in external conditions: electrolyte concentration, its temperature, handling time, current density and change of electrode polarity. However, there was a partial reduction of radioactivity according to beta particles by 5-7 times, but it still remained unacceptably high: up to 50 or more particles ($cm^2 \cdot min$).

From the previous experience it was concluded that the efficiency of the electrolytic method of deactivation in our case depends on the permeability of the film, i.e. on its porosity. This is indicated by the fact that pre-exposure of the samples in solution significantly improves the result of the electrolytic deactivation and reduces its time. The solution wets an oxide film, and the effect of the current destroys it. If the pre-exposure (i.e. chemical handling) is not performed, the oxide film is not removed, deactivation factor is negligible.

Special experiments were carried out to determine the porosity of polyoxide film. When carrying out experiments on the electrolytic removal of firmly fixed deposits, we used:

- many hours of "soaking" of samples in the background electrolytes;
- as background electrolyte we used acids, affecting the components of oxide film (NaCl, H_3PO_4, HNO_3, HCl), and the mixture of acids leading to formation of soluble complexes (30 g/l H_3PO_4 + 50 g/l HNO_3 + 20 g/l $H_2C_2O_4$).

The most successful experiences on rehabilitation were the experiments with sample №5.

Sample №5 was subjected to the rinse in anodic mode in solution of common salt, handling time is 30 minutes, current density is100 mA/cm^2, exposure in nitric acid solution (60 g/l) is 96 hours, handling in anodic mode in solution of nitric acid (60 g/l). The anodic handling was performed in the acid solution under the following conditions: handling time is 1 minute, current density is 3000 mA/cm^2, the solution is boiling [5, 6].

Table 3. Deactivation results

Sample number		Radioactive contamination / dose rate of ionizing radiation $\frac{frequency}{\min \cdot cm^2} / \frac{\mu Sv}{hr}$			
5	internal	2600/4.60	490/3.2	40/0.31	natural background
	external	96/2.07	75/2.5	14/0.12	natural background
Note		Baseline characteristics	Handling in NaCl	Soaking in HNO_3	Electrolysis HNO_3

4. Conclusion

1. A combined method to rinse stainless steel scrap from firmly fixed and weakly fixed polyoxide deposits saturated with radionuclides has been proposed.
2. The essence of the method lies in the combination of chemical and electrolytic methods of deactivation. The first step included a many-hours exposure of contaminated sample in alkali or acid solution (electrolyte), and the second step included short electrolytic handling in the same solution in the anodic mode.
3. The holding time in the first step depends on the porosity of deposit. The worse the porosity, the longer the exposure, which is determined experimentally.
4. The current density of electrolysis is low, 100-600 mA/cm^2, electrolysis time is up to 10 minutes, which provides the surface quality.
5. Unlike the method of chemical deactivation, the concentrations of solutions (electrolytes) refer to “soft” ones, not more than 2-3%.
6. The number of steps and the type of electrolyte depend on the number and type of oxide components, as well as their partial concentration (usually 15-17% or more).
7. The sequence of steps is the following: first, the alkaline electrolytes are applied and oxides are exposed, reacting in alkali, and then acid solutions are applied. At that, in the final step, a solution of the acid is used, which can react with metal and can remove its surface layer containing radioactive contamination.

References

[1] Report of IPhCh RAS № 12105-2113 1/2066 dated 29.12.92.
[2] Experience in operating reactors at SChP. Report. Reg. Number 30-21/03 dated 02.25.91.
[3] Patent № 2328050 dated 10.12.2002.
[4] Brust K.N., Krutikov P.G., Osminin V.S., Chekmarev A.M. Corrosion products in the circuits of nuclear power plants. - M.: Energoatomisdat, 1988. – 336 p.
[5] Patent № 241746 dated 27.04.2011.
[6] Patent № 154091 dated 19.12.2014.

Capability assessment for application of clay mixture as barrier material for irradiated zirconium alloy structure elements long-term processing for storage during decommissioning of uranium-graphite nuclear reactors

S G Kotlyarevskiy[1], A O Pavliuk[1,2], E V Zakharova[3] and A G Volkova[3]
[1] Pilot & Demonstration Center for Decommissioning of Uranium-Graphite Nuclear Reactors, Seversk, Tomsk region, Russia
[2] Tomsk polytechnic university, Tomsk, Russia
[3] Frumkin IPCE RAS, Moscow, Russia

E-mail: seversknet@rambler.ru

Abstract. The radionuclide composition and the activity level of the irradiated zirconium alloy E110, the radionuclide immobilization strength and the retention properties of the mixed clay barrier material with respect to the radionuclides identified in the alloy were investigated to perform the safety assessment of handling structural units of zirconium alloy used for the technological channels in uranium-graphite reactors. The irradiated zirconium alloy waste contained the following activation products: ^{93m}Nb and the long-lived ^{94}Nb, ^{93}Zr radionuclides. Radionuclides of ^{60}Co, ^{137}Cs, ^{90}Sr, and actinides were also present in the alloy. In the course of the runs no leaching of niobium and zirconium isotopes from the E110 alloy was detected. Leach rates were observed merely for ^{60}Co and ^{137}Cs present in the deposits formed on the internal surface of technological channels. The radionuclides present were effectively adsorbed by the barrier material. To ensure the localization of radionuclides in case of the radionuclide migration from the irradiated zirconium alloy into the barrier material, the sorption properties were determined of the barrier material used for creating the long-term storage point for the graphite stack from uranium-graphite reactors.

1. Introduction

The decommissioning of uranium-graphite reactors (UGR) presents a wide range of issues that demand the solutions both in the preparation phase and directly during the process of decommissioning. Among them, the problem of handling structural elements of zirconium alloys is defined by the combination of properties and characteristics inherent to the initial material and acquired by it in the process of the irradiation in the reactor core.

The alloys of zirconium with up to 2.5% by wt. of niobium are used for manufacturing pipes for technological channels (TC) in the uranium-graphite channel type RBMK reactors currently in operation. Besides, the TC pipes of zirconium alloy were used in the final stages of operation of some production uranium-graphite reactors (PUGR); the pipes were installed in certain zones of the graphite stack to maintain its integrity.

The primary issue of handling irradiated parts of zirconium alloy is associated with the long-lived ^{94}Nb ($T_{1/2}$=2.03×10^4 years) and ^{93}Zr ($T_{1/2}$=1.5×10^6 years) radionuclides contained in the material. The decay of ^{94}Nb is accompanied by the emission of high energy gamma quanta (E_γ=703 keV, 100% yield and E_γ= 871 keV, 100% yield). So far as the activity of ^{94}Nb can reach values substantial to induce high gamma dose rate levels, serious restrictions are imposed on the access of personnel to the irradiated zirconium parts handling.

According to the attribution criteria stated in the Russian Government Regulation No.1069 of 10.19.2012, various options are considered for handling UGR radioactive waste (RW) of zirconium. If the zirconium waste is attributed to the disposable RW category, it is to be retrieved and conditioned; if it is attributed to the special RW category, it is not retrieved from the reactor (the burial option).

The aim of this study is to determine the nuclide composition and activity of the irradiated zirconium alloy, radionuclide leach rates, sorption properties of the clayey barrier material with respect to the radionuiclides in question. The study is performed to justify the economically feasible technique for zirconium RW handling.

2. Determination of the radionuclide composition and activity of irradiated zirconium samples

Samples of the material for the investigation were taken from four irradiated TC pipes of the E110 zirconium alloy. The operation period of the pipes in the reactor ranged from 2/5 to 10 years; the cooling time was ~20 years. Sampling was performed in several points along the height of the TC pipes as follows: at the half-height (C), at the distance of ~0.5-1 m from the top (T), and bottom (B) end.

Prior to the radiometric measurements the irradiated zirconium alloy sample was dissolved in the 10 g/L nitric acid solution in the presence of 5 g/L ammonium chloride. The content of γ-emitting radionuclides in the solution, including ^{93m}Nb and ^{94}Nb, was determined by the digital gamma spectrometric complex with the coaxial GEM30 and planar GLP-36360 detectors of ORTEC Company. The prepared solution was neutralized by ammonia with pH of 5-6 until a precipitate was formed. The precipitate was allowed to settle and filtered.

The precipitate of zirconium hydroxide was dissolved in 4 mole/L HCl solution. One part the solution was used to determine actinides, the other was analyzed for ^{93}Zr. Actinides were separated by the extraction-chromatographic technique on TRU resin. The actinides were successively eluted by solutions of different composition followed by the electrolytic deposition of an actinide on the metal support plate. The measurements were performed on the low-background semiconductor alpha-spectrometer equipped with BDEC-01 detector unit.

Zirconium was separated from the solution by the liquid extraction technique [1]. The samples were examined on the LC TRI-CARB 3170 TR/SL spectrometer by PerkinElmer. Recorded spectra were processed using the RadSpectraDec application program complex.

The radionuclides identified in the irradiated zirconium samples as a result of the spectrometric studies were the following: ^{93m}Nb, ^{94}Nb, ^{93}Zr, ^{60}Co, ^{137}Cs, ^{90}Sr, and certain actinides.

It was determined that the shape of ^{93m}Nb, ^{94}Nb, and ^{93}Zr activation product distribution is close to the distribution of the neutron flux during the reactor operation.

To date after ~20 year cooling the primary dose forming radionuclide (the activity of up to 1×10^6 Bq/kg) in the E110 zirconium alloy was the long-lived ^{94}Nb radionuclide with the half-life of 2.03×10^4 years. Despite the ^{93m}Nb radionuclide had the higher activity of up to 3×10^6 Bq/kg, its decay was accompanied by the low energy gamma emission, which for the practical purposes did not contribute into the external irradiation of the personnel. The long-lived ^{93}Zr radionuclide with the half-life of 1.5×10^6 years and the specific activity of ~3×10^4 Bq/kg was also present in the irradiated zirconium, but its decay was accompanied merely by the low energy beta-particles.

The contact with graphite and leaks of the cooling agent resulted in the contamination of the external TC pipe surface. Deposits formed on the internal walls contained ^{60}Co (up to 3×10^4 Bq/kg), ^{137}Cs (up to 2×10^3 Bq/kg), ^{90}Sr (up to 3×10^1 Bq/kg), and some actinides transferred by the coolant from the surface of the primary loop pipework.

The distribution of ^{90}Sr, ^{137}Cs radionuclides and actinides did not depend on the neutron flux distribution and was probably associated with the corrosion deposits on internal walls of the channels and with consequences of the local accidents in the course of the reactor operation.

The distribution of ^{60}Co along the channel height was defined by different sources of its formation. It came both from the activated cobalt impurities in the alloy itself and from cobalt incorporated in the corrosion deposits formed on the channel walls.

Thus, to date the primary dose forming radionuclide in the zirconium alloy was the ^{94}Nb radionuclide; its activity did not exceed 1×10^6 Bq/kg. Taking into account the distribution of the activity along the channel height, matching the shape of the neutron flux distribution, the average specific activity of ^{94}Nb was approximately 2 times as less, and amounted 5×10^5 Bq/kg.

Counting by the basic ^{94}Nb, ^{60}Co, and ^{93}Zr radionuclides, the maximum specific beta-activity of TC pipes was defined by ^{93}Zr and did not exceed 1×10^6 Bq/kg. Similarly to ^{94}Nb, the average specific activity of ^{93}Zr was 2 times as less of 5×10^5 Bq/kg.

The content of actinides did not exceed 1×10^2 Bq/kg. No isotopes of neptunium, americium, and curium were detected. The gross activity of irradiated zirconium channels in the graphite stack of PUGR at JSC "Pilot & Demonstration Center for Decommissioning of Uranium-Graphite Nuclear Reactors" (JSC "DNR Center") did not exceed 1% of the gross activity of the whole stack.

3. Determination of radionuclide leach rates from the irradiated zirconium samples

To determine a possible option for handling irradiated zirconium, the strength of radionuclide localization in the alloy was evaluated by determining leach rates of the radionuclides in question. Samples of the irradiated E110 zirconium alloy taken from two TC pipes were immersed into distilled water in glass beakers. The volume of distilled water in each beaker was 50 cm^3. Water in the beakers was replaced according to the technique defined in GOST P52126-2003. The samples were analyzed according to the technique described above.

The leach rate for an individual radionuclide R^i_n g/(cm^2·day) was determined according to the formula (1):

$$R^i_n = \frac{a^i_n}{A^i_o S t_n}, \qquad (1)$$

where

a^i_n – denoted the activity of an individual nuclide leached for the given period of time, Bq;

A^i_0 – the specific activity of the nuclide in the initial sample, Bq/g;

S – the square area of the open geometric surface of the sample, cm^2;

t_n – the duration of the nth leaching period, days.

Thus, the R was the ratio of the activity of a radionuclide transferred into the solution from 1 cm^2 of the zirconium alloy surface within 24 hours to the specific activity of the given radionuclide in the sample.

Results of the experimental determination of leach rates for the primary dose forming radionuclides from the irradiated zirconium were presented in Fig.1.

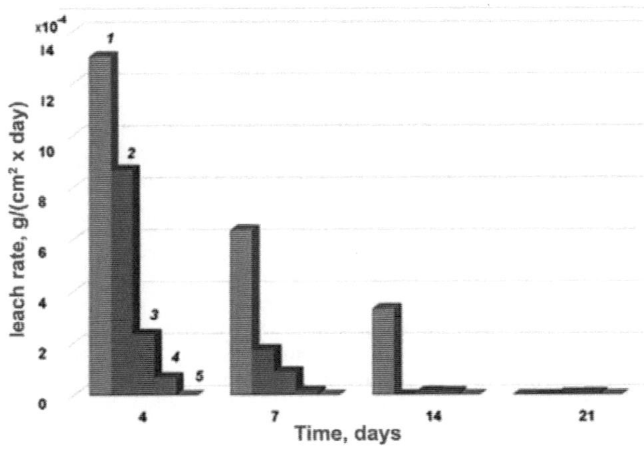

Figure 1. Radionuclide leach rates from the zirconium alloy: 1 – ^{90}Sr; 2 – actinides; 3 – ^{137}Cs; 4 – ^{60}Co; 5 – ^{94}Nb.

As it can be seen from Fig.1, ^{60}Co, ^{90}Sr, ^{137}Cs, and actinides were present in the liquid after leaching. The leach rate depended on the contamination of the deposits, radionuclide localization strength, and radionuclide speciation in the deposits.

It should be noted that the leach rates for all radionuclides were decreased with time. For the 35 day duration of the runs the statistically significant values were obtained for cobalt and cesium. The decrease in leach rates being detected merely in the final stage of monitoring. Leaching of strontium and actinides was discontinued much faster. Those radionuclides were not leached from all samples.

Not more than 2% of the activity of strontium, cesium, cobalt, and actinide radionuclides localized in the surface deposits was leached during the runs. By the end of the run the leach rates of all radionuclides were decreased by several orders of magnitude. As the contact time was increased, no further leaching was observed.

Leaching of ^{94}Nb was observed merely for the 1st day of the sample contact with water. This phenomenon can be explained either by the micro particles of the alloy remained on the sample surface during the sampling process, or by the presence of those isotopes in the irradiated graphite that was in contact with the zirconium pipe channels during the reactor operation, or both.

Results of the zirconium and niobium leaching studies were in good agreement with the data of the corrosion tests of the non-irradiated zirconium [2, 3, 4]. They proved the high corrosion resistance of the technological channels of the irradiated zirconium alloy. Obviously, the extended radiation and thermal impact on the zirconium alloy during the reactor operation did not result in any significant change in the corrosion stability of the material. It should be taken into account during the consideration of different options for handling parts of the irradiated zirconium alloy, particularly in long-term storage or burial.

4. Sorption properties of the clayey barrier material with respect to radionuclides contained in the irradiated zirconium alloys

The investigation of sorption properties of the barrier material with respect to the radionuclides present in the irradiated zirconium alloy was directly associated with the demand to justify the environmental and radiation safety of its long-term storage, disposal, or burial. The clay currently used for the engineered safety barriers was the effective sorption material, affordable by its low-cost and available resources. At the JSC "DNR Center" similar barrier was installed in the PUGR EI-2 uranium-graphite reactor vault and adjacent premises to create the long-term storage point for graphite stack and construction steelwork [5]. Results of the predictive modeling showed that the installed barrier confined the distribution of radionuclides in the geological medium and maintained their concentrations below the intervention level outside the boundaries of the created object.

Sorption experiments were conducted with the sample of the barrier material obtained by LLC "PIK", the City of Krasnoyarsk. Protective properties of the material were increased using a mix of several clays prepared by the preliminary grinding and mixing of components. To increase the sorption and capacity characteristics of the material, the poly-mineral mixes contained different minerals, namely, montmorillonite, kaolinite, chlorite, vermiculite, micas and hydromicas, etc.

The water simulant composition was chosen close to the natural water at the location area of the PUGR of the JSC "JSC "DNR Center"". The simulated water contained, mg/L: ^{2+}Ca – 40.7; ^{2+}Mg – 4.5; Na^+ - 6.9; $\sum CO_3^{2-} + HCO_3^-$ - 20.6; SO_4^{2-} - 14.3, and Cl^- - 72.3; pH 7.3. The S/L ratio was 1/20; the phases were separated by the centrifugation. Solutions for the sorption runs contained, mole/L (Bq/L): ^{90}Sr – 2.2×10^{-10} (1.0×10^5); ^{137}Cs – 1.1×10^{-9} (5.0×10^5); ^{233}U – 2.0×10^{-7} (1.66×10^4); ^{238}Pu – 5.0×10^{-10} (7.5×10^4); ^{60}Co – 4.0×10^{-11} (1.0×10^5); ^{93}Zr – 5.0×10^{-6} (4.4×10^4), and ^{94}Nb – 5.0×10^{-8} (3.25×10^4).

Sorption processes with the different mechanisms were characterized by the cumulative index, namely, the interphase distribution coefficient. It was equal to the ratio of the amount of the radionuclide adsorbed in the solid phase to its equilibrium content in the liquid (2):

$$K_d = \frac{N}{C_f} = \frac{C_i - C_f}{C_f} \cdot \frac{V_l}{m_s}, \quad (2)$$

where:
N - denoted the activity of the rock in the sample, Bq/kg;
C_i and C_f - the initial and final activity of the solution, Bq/L, mole/L;
V_l -volume of the solution, cm^3;
m_s -weight of the sample, g.
K_d - values were used in the radionuclide migration calculations. Retention properties of the sample were characterized by the effective porosity P_e and the retention factor R according to the formulas (3, 4):

$$P_e = \frac{N \cdot \rho + C_0 \cdot P_0}{C_0} = K_{id} + P_0, \quad (3)$$

$$R = \frac{P_e}{P_0} = 1 + \frac{K_p}{P_0} = 1 + \frac{K_d \rho}{P_0}, \quad (4)$$

where:
P_0 denoted the active or opened porosity of the rock;
$K_p = K_d \cdot \rho$ - the interphase distribution coefficient;
ρ - density of the sample, g/cm^3.

The retention factor R was one of the parameters defining the radionuclide behavior in the process of the interaction with rocks. It also characterized the ratio of the adsorbed radionuclide component velocity to the distribution velocity of the water flow.

For the irradiated zirconium alloy the most interest represented the behavior of long-lived zirconium and niobium isotopes during their interaction with the barrier material. In aqueous solutions zirconium and niobium were subjected to the hydrolysis and polymerization processes. Depending on the metal concentration, solution acidity, and temperature different zirconium species were formed, such as monomers $Zr(OH)_4^{(4-n)+}$, where n=0,1, 2, 3, and 4, or trimers and tetramers. The formation of species started in the solution with the concentration over 10^{-4} – 10^{-3} g-atom/L. The zirconyl ZrO^{+2} ion could also exist in the solution. The poly-nuclear nature of zirconium hydroxide remained even in high dilutions. It existed in the ionic form just below the pH of 2. Therefore in natural systems, in neutral and weak alkaline media as well as at the low zirconium and niobium concentrations those radionuclides would exist as the hydrolyzed ion species. Their behavior and speciation were similar to plutonium.

When the simulated water containing ^{93}Zr and ^{94}Nb was brought into contact with the barrier clayey material, the sorption equilibrium was attained rather quickly, in 1-2 days (see Fig.2).

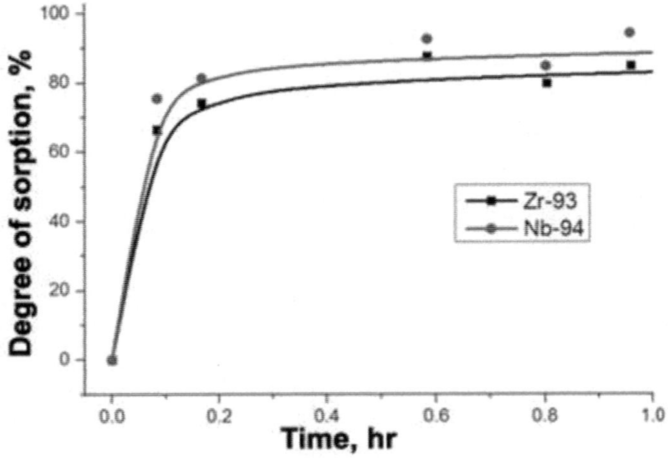

Figure 2. The sorption kinetics of ^{93}Zr and ^{94}Nb radionuclides on the barrier material.

After the equilibrium in the system was attained, the distribution coefficient K_d and retention factor R values were determined. The K_d and R values for the other radionuclides leached from the irradiated zirconium after 35 days of the interaction were presented in Table 1.

Table 1. The sorption characteristics (K_d, cm^3/g, R) of the LLC "PIK" clay sample with respect to radionuclides.

Parameter	^{93}Zr	^{94}Nb	^{90}Sr	^{137}Cs	^{233}U	^{239}Pu
K_d	0.4×10^3	0.8×10^3	1.3×10^2	0.6×10^3	2.3×10^2	2.0×10^3
R	1.3×10^3	2.6×10^3	2.6×10^2	2.1×10^3	5.4×10^2	0.6×10^4

The R value was determined with the sample of the bulk density of 0.8-1.0 g/cm^3 corresponding to the minimum packing of the rock and its porosity of 0.5. This supposed a minor packing of the barrier material during the construction stage. It was the barrier material which was used at the initial phase of backfilling the barrier for the EI-2 PUGR.

It should be noted that the distribution coefficient characterized not only the most common mechanism of the sorption process (the ion exchange) but the other mechanisms, too, Among the specific sorption mechanisms were: the chemisorption – the formation of a chemical compound on the phase boundary; the physical adsorption – the capture of organic molecules by the surface of the solid body; the molecular adsorption – the formation of ion pairs and complex associations on ion exchange materials [6]. Therefore, the technique used to determine the distribution coefficient was true when the separation of a radionuclide from the liquid phase occurred not only by ion exchange, but by means of the other specific processes, too.

The sorption capacity of clay was higher with respect to plutonium and cesium. The K_d values were lower for strontium and uranium, which were the active migration species in natural systems. The K_d values for zirconium and niobium were consistent with those for plutonium (actinides). When the material samples with adsorbed radionuclides were brought into contact with the simulated radionuclide-free layer water, not more than 15% of strontium, cesium, and uranium was desorbed, and less than 3% of niobium and zirconium.

The radionuclide distribution velocity at the minimum degree of the clay packing of 1.0 g/cm^3 would be 10^2–10^4 times slower than the velocity of the water distribution. As the degree of the clay packing was increased up to 1.5 g/cm^3, those values were increased by several orders of magnitude. The inspection of the clay packing in the EI-2 PUGR clay barrier showed that the latter value was exceeded. Under the circumstances the radionuclide distribution would occur merely by the diffusion mechanism. As a result, not only radionuclides adsorbed on clays, but the non-adsorbed ones would be also captured in pores of the swelling montmorillonite, when water entered the interlayer space of montmorillonite decreasing the water permeability of the barrier material.

5. Conclusions

Thus, results of the accomplished studies allowed to certify the following:

1) The dynamics of leaching and the absolute leach rates values of ^{93m}Nb, ^{94}Nb, ^{93}Zr, ^{60}Co, ^{137}Cs, ^{90}Sr, and actinide radionuclides identified in the samples of the irradiated technological channels of the E110 zirconium alloy witnessed the high corrosion resistance of the irradiated structural elements and parts of the zirconium alloy. It should be mentioned that the extended radiation and thermal impact on the zirconium alloy during the reactor operation did not result in any significant change in the corrosion stability of the irradiated zirconium alloy.
2) When creating the long-term storage point for the graphite stack and construction steelwork from the PUGR the sorption properties of the barrier material of the mix of clays ensured the localization of radionuclides contained in the parts of the irradiated zirconium alloy. The gross activity of those parts was about 1%, which was less than in the irradiated graphite stack of the reactor.
3) Zirconium radwaste PUGR should be attributed to the special RW category and it is not necessary to retrieve it from the reactor. The burial option is more suitable to solve the zirconium RW PUGR problem.

References

[1] Espartero A.G., Suarez J.A., Rodriguaez G. Radiochemical analysis of ^{93}Zr // *Applied Radiation and Isotopes*. – 2002. – V. 56. – P.41-46.

[2] B.G.Parfenov, V.V.Gerasimov, G.I.Venedictova. Corrosion of Zirconium and its Alloys. – M.: *Atomizdat*, 1967. – 258 p.

[3] B.Lastman, F. Kerze. Metallurgy of Zirconium. Transl. from Eng., under the edition of G.A.Meerson and Yu.V.Gagarinsky. – M.: *Izdatinlit.*, 1959. – 219 p.

[4] Haскer M., Properties of Reactor Materials and Effects of Radiation damage. // *Butterworths*. – 1962. – P.383.

[5] Izmestiev A., Pavliuk A., Kotlyarevsky S. Application of void-free filling technology for additional safety barriers creation during uranium-graphite reactors decommissioning // *Advanced Materials Research*. – 2015. – V.1084. – PP. 613-619.

[6] A.N.Murrey. Sanitary Protection of Water Bodies. M., *Atomizdat*, 1976. – 155 p.

The mechanism for production of beryllium fluoride from the product of ammonium fluoride processing of beryllium-containing raw material

R I Kraydenko, A N Dyachenko, L N Malyutin, I V Petlin

[1]National Research Tomsk Polytechnic University, Institute of Physics and Technology, Department of Rare, Scattered and Radioactive Element Technology, Russian Federation, 634050, Tomsk, 30 Lenin avenue.

E-mail: prof_1990@mail.ru

Abstract. The technique of fluorite-phenacite-bertrandite ores from Russian Ermakovskoe deposit processing by ammonium bifluoride is described. To determine the temperature mode and the thermal dissociation mechanism of ammonium tetrafluoroberyllate (the product of ammonium-fluoride leaching of the ore) the TG/DTA have been carried out. By IR spectroscopy and XRD the semi-products of ammonium tetrafluoroberyllate thermal dissociation have been identified. The hygroscopic low-temperature beryllium fluoride forms higher than 380 °C. The less hydroscopic form of BeF_2 have been produced at 600 °C.

1. Introduction

In nature beryllium is mainly represented by nearly 30 minerals, the most important of them are: beryl, chrysoberyl, behoite, bertrandite, and phenacite [1-3]. In production of metallic beryllium and its compounds on industrial scale only beryl, bertrandite, and phenacite concentrates are used. The largest beryllium deposits with easy-to-break down ores (carbonate-containing bertrandite) are located in the United States: Spor Mountain, Sierra Blanca [4]. Today, beryllium is leached from ores of these deposits only by using an industrial method of hydrometallurgical processing of beryllium raw materials – the sulfuric acid method, without a preliminary preparation of raw materials [5]. There are only three states in the world that carry out hydrometallurgical processing of beryllium-containing raw materials: the United States, Kazakhstan, and China. At the same time, in the territory of the Russian Federation there are unique Orotskoe bertrandite and Ermakovskoe deposits, with the content 0.5-1.0 % of BeO in the ore. The capacity of the Ermakovskoe deposit in terms of commercial beryllium ore amounts to nearly 1400 thousand tons [6]. The Ermakovskoe deposit is represented by fluorite-phenacite-bertrandite ores; a concentrate with the content 24.44 % expressed in terms of Be_2SiO_4 is produced from the ore by flotation concentration (Table 1).

Table 1. Phase composition of the fluorite-phenacite-bertrandite concentrate

Be_2SiO_4	SiO_2	CaF_2	CaO	Fe_2O_3	MgO	Al_2O_3
24.44	13.01	23.65	27.27	1.07	2.51	8.04

Processing of the given raw material using the existing sulfuric-acid scheme is complicated by a number of problems: 1) in order to leach beryllium from the fluorite-phenacite-bertrandite concentrate (FPBC), the concentrate must be subjected to a dual thermal activation (1700 and 950 °C) and fusion with alkali fluxes ($CaCO_3$, Na_2CO_3); 2) a large amount of fluorine, which transitions into a soluble form from fluorite (CaF_2), with further processing of the activated concentrate binds beryllium, which leads to losses of the target component [5]. The described phenomena adversely affect the cost of sulfuric-acid leaching of beryllium.

For efficient and cost-effective processing of FPBC for the purpose of production of metallic beryllium and its compounds the ammonium-fluoride technology has been developed.

2. Technology description

The ammonium-fluoride technology for processing of beryllium-containing raw materials (in this case FPBC) consists in fusion of the concentrate with ammonium bifluoride (NH_4HF_2) at 180-220 °C without a preliminary thermal and chemical treatment of the raw material. Interaction of the concentrate components NH_4NF_2 (except fluorite) proceeds according to chemical equations (1-6) [7,8]:

$$Be_2SiO_4 + 7NH_4HF_2 \rightarrow 2(NH_4)_2BeF_4 + (NH_4)_2SiF_6 + 4H_2O\uparrow + NH_3\uparrow, \quad (1)$$

$$Al_2O_3 + 6NH_4HF_2 \rightarrow 2(NH_4)_3AlF_6 + 3H_2O\uparrow, \quad (2)$$

$$Fe_2O_3 + 5NH_4HF_2 \rightarrow 2(NH_4)_2FeF_5 + 3H_2O\uparrow + NH_3\uparrow, \quad (3)$$

$$CaO + NH_4HF_2 \rightarrow CaF_2 + H_2O\uparrow + NH_3\uparrow, \quad (4)$$

$$MgO + NH_4HF_2 \rightarrow MgF_2 + H_2O\uparrow + NH_3\uparrow, \quad (5)$$

$$SiO_2 + 3NH_4HF_2 \rightarrow (NH_4)_2SiF_6 + 2H_2O\uparrow + NH_3\uparrow. \quad (6)$$

The produced fluorinated sinter is then dissolved in water. Only beryllium and silicon completely go into solution (the degree of leaching is 97 %), iron (~ 55 %) and aluminum (~ 20 %) partially go into solution, calcium and magnesium form insoluble fluorides (the degree of leaching is ~ 0.2 %). The beryllium-containing solution is then purified from impurities using 25 % of ammonia solution. Ammonia hydrolysis of impurities proceeds according to chemical equations (7-9):

$$(NH_4)_2FeF_5 + 3NH_4OH = Fe(OH)_3\downarrow + 5NH_4F, \quad (7)$$

$$(NH_4)_3AlF_6 + 3NH_4OH = Al(OH)_3\downarrow + 6NH_4F, \quad (8)$$

$$(NH_4)_2SiF_6 + 4NH_4OH = Si(OH)_4\downarrow + 6NH_4F. \quad (9)$$

The pH level of precipitation of impurity hydroxides in this case is equal to 8. Such high pH of hydrolysis is related to the presence of an excessive amount of ammonium fluoride (NH_4F) in the production solution. Also, due to the presence of an excess of ammonium fluoride the destruction of the complex ion BeF_4^{2-} practically does not occur, even if raising the pH level to 10. The degree of beryllium hydrolysis at pH 10 does not exceed 5.2 %.

The purified beryllium solution is then evaporated to produce a crystalline ammonium tetrafluoroberyllate ($(NH_4)_2BeF_4$) which is then calcined to produce beryllium fluoride (BeF_2) – the initial compound for magnesiothermal reduction of beryllium [7, 9-11].

3. Experimental part

The aim of this study was to determine the temperature mode and the thermal dissociation mechanism of ammonium tetrafluoroberyllate.

Based on the information from literature sources, the decomposition of $(NH_4)_2BeF_4$ begins at 125 °C, at 230-260 °C ammonium trifluoroberillate (NH_4BeF_3) is formed, at temperatures higher than 270 °C beryllium fluoride is formed [1, 7, 9, 10]. A complete decomposition of ammonium tetrafluoroberyllate at an acceptable rate occurs at 400 °C, but the produced BeF_2 is hygroscopic and is not suitable for production of metallic beryllium, therefore the decomposition is carried out at 900 °C [9, 10]. For accurate identification of the decomposition mechanism of ammonium tetrafluoroberyllate a thermogravimetric and differential-thermal analyses of the mixture BeO-NH_4HF_2 have been carried out (Fig. 1). To carry out the analysis, ammonium bifluoride was taken with a 10 % excess relative to the stoichiometrically required quantity under the reaction (10):

$$BeO + 2NH_4HF_2 = (NH_4)_2BeF_4 + H_2O \quad (10)$$

Experiments were carried out using a combined TGA/DSC/DTA analyzer – derivatograph of the brand SDT Q600 with a software data processing TA Instruments Universal V4.2E. Crucibles: platinum: 40 mcl. Control of the sample atmosphere – argon. The heating rate was 5 °/min.

When mixing pure beryllium oxide and ammonium bifluoride the reaction begins immediately with the release of heat. The moisture allocated prior to loading of the reaction mixture into the analyzer is removed at 53 °C. At 129 °C melting of ammonium bifluoride in the mixture takes place, and the flow of the fluorination process accelerates. The fluorination ends at 160 °C. In the temperature range of 185-255 °C the decomposition of $(NH_4)_2BeF_4$ up to NH_4BeF_3 takes place (Fig.1). The decomposition of NH_4BeF_3 up to beryllium fluoride proceeds in the range of 255-300 °C, while on the TGA curve an inflection at 275 °C is observed. At 549 °C a peak on the DTA curve is observed, which corresponds to melting of the quartz form of beryllium fluoride with subsequent recrystallization and solidification of a new crystalline structure (tridimitic, α-cristobalitic) [11]. Melting of a new form of BeF_2 begins at 811 °C. Since the sample is under argon and there is a constant removal of beryllium fluoride vapors from the surface of the melt, a complete evaporation of the sample takes place when reaching 926 °C.

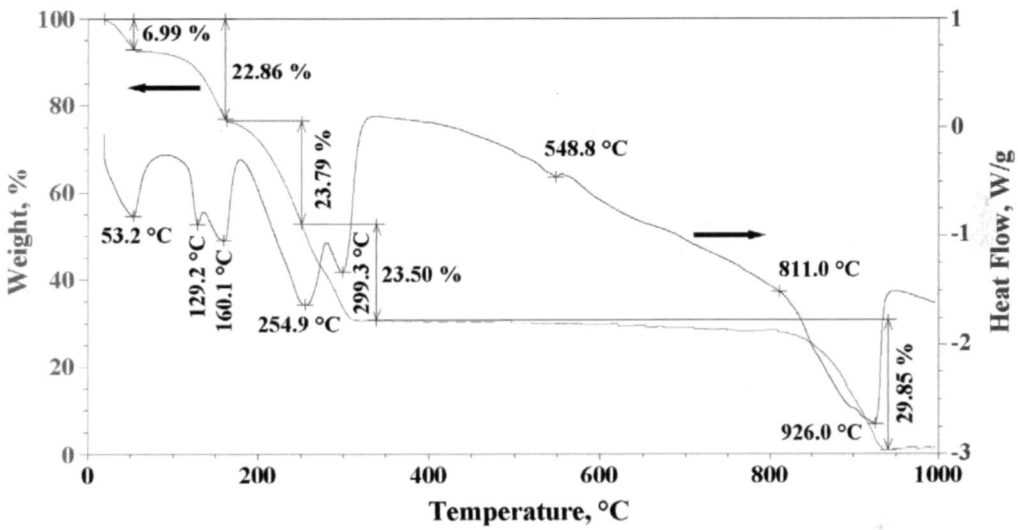

Figure 1. The derivatogram of the system $BeO-NH_4HF_2$

To determine the composition of produced substances during the decomposition, an analysis of products in critical points of the DTA curve has been carried out using methods of infrared (IR) and X-ray spectroscopy (X-ray diffraction analysis – XRD). To prepare the samples 5 g of beryllium oxide and 24 g of NH_4HF_2 were taken. The powders were mixed, abraded, and the mixture was filled into a graphite crucible. The reaction mixture was first heated at 150 °C, then at 250 °C, 280 °C, and 310 °C, in each temperature mode the mixture was incubated for 1 hour. The reaction proceeded under argon.

Analysis of the sample maintained at 150 °C

IR-spectra were recorded on a Fourier spectrometer NICOLET 6700 Thermo Electron Corporation in the wavenumber range of 600 – 4000 cm^{-1}, the acceptable error threshold of the wavenumber scale ± 0.5 cm^{-1}, under nitrogen. For the given range of 4000–600 cm^{-1} the samples were prepared by compression of tablets, the ratio of the investigated material: KBr – 1:300.

In the IR-spectrum of the investigated product in the wavenumber range of 600 – 4000 cm^{-1} there are bands characteristic of the bond N-H (NH_4^+) – 3266, 1427, for the bond of ammonium cation and fluoride-ion – 1101, bands for the bond Be-F are distinct – 793, 552 (Fig. 1). The fluorination product of the mixture $BeO-NH_4HF_2$ at 150 °C consists of a complex fluorine-containing compound of beryllium.

In order to establish a more specific structure of the fluorination product the sample was tested using the method of X-ray spectroscopy on X-ray diffractometer MiniFlex 600. The analysis conditions: X-ray power – 40 kW, 15 mA; detector – SC-70; scanning rate – 2 deg./min; step width – 0.02 deg .; scanning range – from 15.0 to 80.0 degrees.

The fluorination product at 150 °C consists of ammonium tetrafluoroberyllate (Fig. 1).

Analysis of the sample maintained at 250 °C

In the IR-spectrum of the investigated product in the wavenumber range of 600 – 4000 cm^{-1} there are bands characteristic of the bond N-H (NH_4^+) – 3251, 1440, for the bond of ammonium cation and fluoride-ion – 1101, 478, bands for the bond Be-F are distinct – 792, 638, 552, 436 (Fig. 2). The product of thermal dissociation at 250 °C consists of a complex fluorine-containing compound of beryllium.

XRD data indicates that the product of thermal dissociation at 250 °C consists of NH_4BeF_3 and $(NH_4)_2BeF_4$ (Fig. 3). The intensity of trifluoroberyllate bands is much greater than of tetrafluoroberyllate, which indicates an incomplete decomposition of $(NH_4)_2BeF_4$.

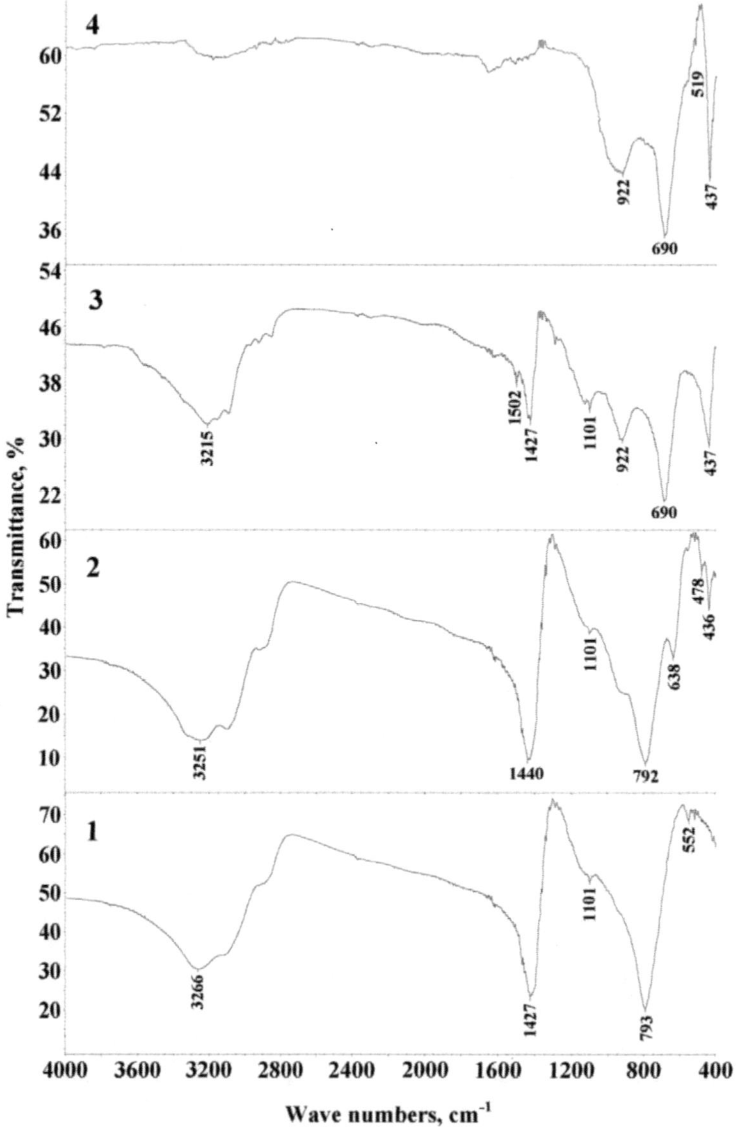

Figure 2. IR-spectra of samples: 1 – maintained at 150 °C; 2 – at 250 °C; 3 – at 280 °C; 4 – at 380 °C.

Analysis of the sample maintained at 280 °C

In the IR-spectrum of the investigated product in the wavenumber range of 600 – 4000 cm^{-1} there are bands characteristic of the bond N-H (NH_4^+) – 3215, 1427, for the bond of ammonium cation and fluoride-ion – 1502, 1101, 478, bands for the bond Be-F are distinct – 922, 690, 437 (Fig. 3). The product of thermal dissociation at 280 °C consists of a complex fluorine-containing compound of beryllium.

XRD data indicates that the product of thermal dissociation at 280 °C (in the region of the TGA curve, where the change in the inclination angle of the line takes place) consists of $NH_4Be_2F_5$ (Fig. 3). The reference on the formation of such a compound during the decomposition of $(NH_4)_2BeF_4$ is present in [11].

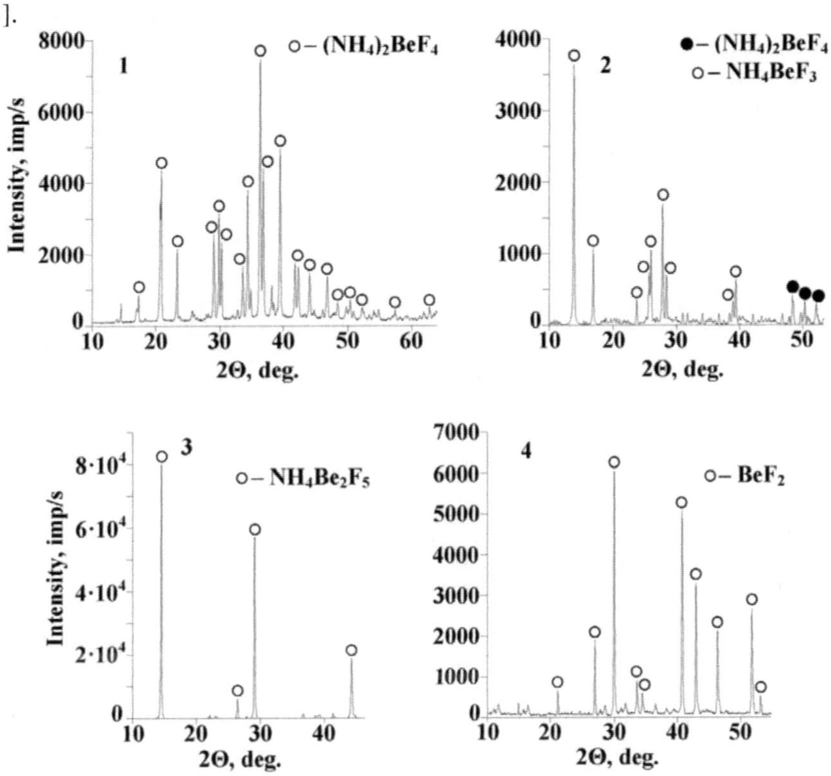

Figure 3. Diffraction patterns of samples: 1 – maintained at 150 °C; 2 – at 250 °C; 3 – at 280 °C; 4 – at 380 °C.

Analysis of the sample maintained at 310 °C

During the analysis of the sample maintained at 310 °C, using methods of IR-spectroscopy and XRD, strips of ammonium cation were found in the sample, indicating an incomplete decomposition of $NH_4Be_2F_5$. Experiments on decomposition of salt in the temperature range of 320-400 °C have been also carried out. Ammonium bands were detected up to 380 °C.

In the IR-spectrum of the given product in the wavenumber range of 600 – 4000 cm^{-1} there are peaks characteristic for the bond Be-F – 922, 690, 519, 437 (Fig. 4). The product of thermal dissociation at 375-380 °C consists of beryllium fluoride.

The produced beryllium fluoride at 380 °C, after exposure to atmospheric environment for 30 minutes, begins to cohere and becomes viscous. This fact indicates high hygroscopicity of the synthesized compound. When BeF_2 was dried in a muffle furnace at 110 °C for 3 hours, a grayish powder was obtained, the IR-spectrum of which does not match with the spectrum of the freshly produced compound. XRD indicated the presence of an oxide component in the dried powder. The produced compound was beryllium oxyfluoride, whose properties are described in [6]. This structure of beryllium fluoride is not suitable for storage and transport, and, consequently, for production of metallic beryllium.

During thermal decomposition of $(NH_4)_2BeF_4$, beginning from 25 °C, a white mass of beryllium fluorocomplex melted at 220 °C and boiled at 310 °C, with allocation of gaseous products. Then the reaction mass solidified forming a transparent amorphous material at 350 °C. With further increase in the temperature up to 380 °C the amorphous mass melted again turning into a thick liquid grayish substance. Upon reaching 500 °C there was an intense boiling of the formed mass with its parallel recrystallization. This phenomenon was accompanied by an intense increase in the volume and splashing of the substance from the reaction zone. Above 560 °C recrystallization of the substance ended, a decrease in the volume of the reaction material took place. The formed mass acquired a crystalline shape and a grayish color at 600 °C.

XRD data indicates that the product of thermal dissociation at 600 °C consists of BeF_2 (Fig. 3). Beryllium fluoride produced at a temperature of ≥ 600 °C, after being exposed to atmospheric environment during one day, does not cohere and does not become viscous. The difference between diffraction patterns of the freshly produced beryllium fluoride and the sample maintained in atmospheric environment has not been revealed.

4 Conclusion

1. As a result of the study, the temperature mode and the thermal dissociation mechanism of ammonium tetrafluoroberyllate up to beryllium fluoride have been established (Table 2).

Table 2. Stages of thermal dissociation $(NH_4)_2BeF_4$

Temperature range, °C	Process
185-255 °C	$(NH_4)_2BeF_4 = NH_4BeF_3 + NH_4F$
255-280 °C	$2NH_4BeF_3 = NH_4Be_2F_5 + NH_4F$
280-380 °C	$NH_4Be_2F_5 = 2BeF_2 + NH_4F$
525-565 °C	Melting and recrystallization BeF_2

2. It is established that beryllium fluoride produced at 380 °C is a hygroscopic substance. Production of beryllium fluoride at a temperature over 600 °C allows forming a less hygroscopic type of substances suitable for storage.

3. The process of thermal dissociation of $(NH_4)_2BeF_4$ is accompanied by liberation of fluorinated gases (NH_4F) and splashing of beryllium-containing compounds. To reduce losses of the target component and toxic contamination of the laboratory it is recommended to use a stage mode of thermal dissociation of $(NH_4)_2BeF_4$.

References

[1] Thorat D.D., Tripathi B.M., Sathiyamoorthy D., *Hydrometallurgy*, **109** (2011), 18-22.

[2] Zaki E.E., Ismail Z.H., Daoud J.A., Aly H.F., *Hydrometallurgy*, **80** (2005), 221-231.

[3] White D.W., Burke J. E. The Metall Beryllium. American Society for Metals, 1955.

[4] Methodological recommendations on application of the Classification of deposits and forecast resources of solid minerals. Beryllium ores. Approved by the decree of the Ministry of Natural Resources of the Russian Federation from 05.06.2007 №. 37-p., Moscow, 2007.

[5] Samoylov V.I, Borsuk A.N. Methods for co-processing of phenacite, bertrandite, and beryl in beryllium hydrometallurgy, Media Alliance, Ust-Kamenogorsk, 2006.

[6] Corporation "Metals of Eastern Siberia". Mining sector. Ermakovskoe deposit. [Electronic resource]. – Access mode: http://mbc-corp.ru/activity/gorsector/ermak/index.wbp. - 28/02/15.

[7] Andreev A.A., Dyachenko A.N., Kraydenko R.I., *Russian Journal of Applied Chemistry*, **81** (2), (2008), 178-182.

[8] Rimkevich V.S., Pushkin A.A., Malovitskii A.A., Dem'yanova Yu.N., Girenko I.V., *Russian Journal of Applied Chemistry*, **82** (1), (2009), 6-11.
[9] Silina G.F., Zarembo Yu.I., Bertina L.E. Beryllium. Chemical technology and metallurgy, Atomizdat, Moscow, 1960.
[10] Chemistry and technology of rare and dispersed elements, part I. Ed. by K.A. Bolshakov, second ed., revised and add., Higher School, Moscow, 1976.
[11] Everest D.A. Chemistry of beryllium. Trans. by I.B. Braverman, Chemistry, Moscow, 1968.

The investigation of hydrogenation influence on structure changes of zirconium with nickel layer

V N Kudiiarov, Yu S Bordulev, R S Laptev, N S Pushilina, E B Kashkarov, M S Syrtanov

Tomsk Polytechnic University, 30 Lenin Avenue, Tomsk, Russian Federation

E-mail: viktor.kudiiarov@gmail.com

Abstract. The results of experimental investigation of hydrogenation influence on structure changes of zirconium alloy (Zr-1%Nb) with thin nickel layer have presented in this work. Nickel layer was formed by magnetron sputter deposition. Hydrogenation was carried out at gas atmosphere at constant temperature. Different hydrogen concentrations were obtained by varying time of hydrogenation. Defect and phase structure was studied by means of X-ray diffraction, glow discharge optical emission spectroscopy, positron lifetime and Doppler broadening spectroscopies. New experimental data about the evolution of the positron annihilation parameters depending on hydrogen concentration in Zr-1Nb alloy with nickel layer was obtained.

1. Introduction

Penetration and accumulation of hydrogen in metal structure has a significant impact on the physical, chemical and mechanical properties of its structure. This problem is particularly topical for zirconium alloys [1-6]. The preparation process of samples with different concentrations of hydrogen and different distribution of hydrogen by volume is an essential part of metal-hydrogen systems experimental studies. It is important to bear in mind that samples are covered with an oxide film, which degrades the process of hydrogen penetration during planning the process of saturation with hydrogen of zirconium alloys.

Increase in the rate of hydrogen sorption is possible by coating the Zr surface with a layer of nickel. Nickel has a high physical and chemical adsorption activity with hydrogen and has a high permeability to hydrogen. Furthermore, the oxide film formed worse on the nickel surface than on the titanium or zirconium surfaces, which promotes to hydrogen absorption. The study of the effect of nickel layer coating on the hydrogen sorption by Zr-1%Nb alloy, and the effect of hydrogenation on the defect structure of zirconium alloy with the nickel layer is carried out in this work.

2. Experimental Procedure

Flat samples of Zr-1%Nb alloy (20×20×0.7 mm) were prepared for this experiment. Samples were mechanically polished and annealed in vacuum (T = 580 °C) for three hours. The deposition of nickel layer on the surface of the samples was carried out by the «Raduga Spectr» technique which was developed in General Physics Department of Tomsk Polytechnic University. Samples were fixed in a vacuum chamber by the samples holder. Ultimate pressure in the chamber was equal to $2.5 \cdot 10^{-3}$ Pa.

Ionic surface cleaning of the samples is of prime importance before the thin protective films sputtering. It is necessary for the removal of surface contamination of various types, as well as oxide layer removal. Ionic surface cleaning method produces virtually atomically clean surface. As a result, the ion cleaning of the surface directly before the coating process improves adhesion of the coating material with the substrate. Surface cleaning was carried out directly in the vacuum chamber before the coating process by argon ion beam, with the parameters shown in Table 1.

Table 1. Parameters of the ionic surface cleaning process

Gas	Voltage, V	Power, W	Current, A	Pressure, Pa	Time, min.
Argon	1800-2000	1000	0.5	0.06	3

The nickel coating was deposited by magnetron sputtering technique with the parameters, presented in Table 2. Sputtering time was varied in order to form nickel layers of different thicknesses. High purity nickel (99,99%) was used as a target for magnetron system.

Table 2. Parameters of nickel magnetron sputtering

Voltage, V	Power, W	Current, A	Pressure, Pa	Time, min.
550	2000	3	0.11	10 ÷ 30

Structural and phase composition was determined by X-ray diffraction using diffractometer Shimadzu XRD 7000 (Japan). Hydrogenation was carried out using special equipment Gas Reaction Controller (USA) [7]. Study of the elemental composition of samples was measured by optical spectroscopy of high-frequency glow discharge performed with Profiler-2 (Horiba, Japan) [8].

The penetration depth of the X-ray beam was varied by changing the angle of incidence while the asymmetric scan by XRD. Decreasing of the angle of incidence was conducted up to complete elimination of the signal from the substrate. Evaluation of the coating thickness was carried out using the geometry of the sliding beam by the relation:

$$h = \frac{\ln(1-R)\sin\alpha \cdot \sin(2\theta - \alpha)}{\mu(\sin\alpha + \sin(2\theta - \alpha))}, \quad (1)$$

where α – angle of incidence, θ – gliding angle (Bragg angle), μ – linear absorption coefficient, R – the proportion of radiation reflected from the layer with the thickness h. When determining the thickness of the nickel coating the value of R has a value of 0.95.

Thicknesses of the nickel layers were additionally determined by analysing the distribution profile of elements on the depth of the samples.

3. Results and Discussion

Figure 1 shows the distribution profiles of elements in samples of zirconium alloy before (a) and after (b) nickel layer deposition. The surface layer of the initial samples (~ 0.3 μm) consists of the oxides and contaminant gases (nitrogen and hydrogen) from the air.

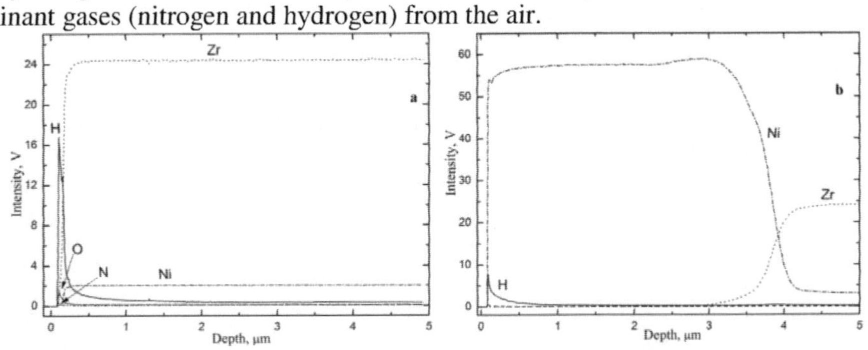

Figure 1. Distribution profiles of the elements in the Zr-1%Nb (a) and Zr-1%Nb with nickel layer (b)

Thus, the initial samples contain hydrogen only in the surface layer. Its concentration is about 0.001 weight %. The nickel layer is formed on the surface of the samples after 30 minutes of magnetron sputtering. The thickness of this layer can be determined from the distribution profile of the elements as a depth at which the intensity of the nickel signal begins rapidly decrease and zirconium signal starts increasing.

Table 3. Results of nickel layer thicknesses determination

Time of magnetron sputtering, min	Thickness (XRD), μm	Thickness (depth profile distribution analysis), μm
10	1.2	1 ± 0.1
20	1.9	1.8 ± 0.1
30	2.6	2.7 ± 0.1

Table 3 shows the results of nickel layer thicknesses determination by different techniques as a function of sputtering time. It has been established that the increase in sputtering time from 10 to 30 minutes increases the thickness of the nickel layer from ~ 1 to 2.7 μm. The results of the thickness determination by X-Ray method correlate with the results obtained by depth profile distribution analysis. Non-destructive XRD analysis allows estimating the nickel layer thickness before hydrogen saturation.

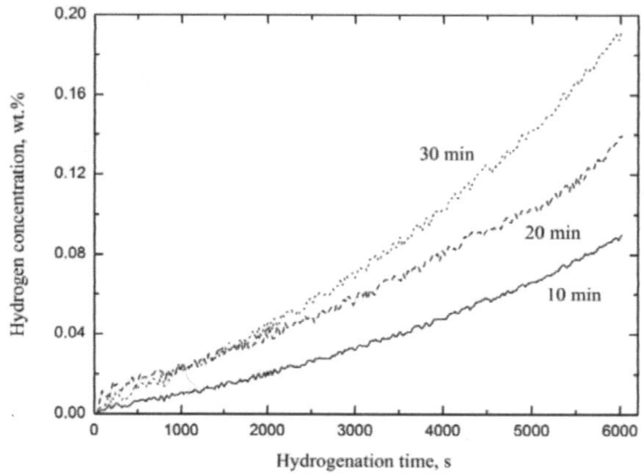

Figure 2. Hydrogen sorption curves by zirconium alloys with nickel layer of different thicknesses

Figure 2 shows that increasing of the nickel layer thickness leads to an increase in the hydrogen sorption rate of zirconium alloy.

Table 4 shows the results of defect structure studies of zirconium alloy with a layer of nickel before and after hydrogenation by positron lifetime and Doppler broadening techniques.

Table 4. Results of the defect structure studies of zirconium alloy with a layer of nickel before and after hydrogenation

Sample	Sputtering Time, min	Hydrogenation Temperature, °C	Hydrogen Concentration, wt.%	τ_1, ps	τ_2, ps	I_2, %	τ_{avg}, ps	Fit
Zr	-	-	-	164±1	-	-	164±1	1.09
Zr + Ni	10	-	-	165±1	-	-	165±1	1.08
Zr + Ni	10	-	-	165±1	-	-	165±1	1.10
Zr + Ni	20	-	-	166±1	-	-	166±1	1.10
Zr + Ni	30	-	-	167±1	-	-	167±1	1.09
Zr + Ni	10	550	2.1	164±1	183±2	53.5±2	176±1	1.12
Zr + Ni	10	450	2.1	164±1	183±2	65.1±2	178±1	1.10
Zr + Ni	10	300	0.15	164±1	183±2	84.5±5	182±1	1.10
Zr + Ni	20	300	0.15	164±1	209±3	10±2	169±1	1.12
Zr + Ni	30	300	0.15	164±1	241±5	7.8±1	170±1	1.07

The positron lifetime in the annealed zirconium alloy is 164 ± 1 ps, which is consistent with a lifetime of positrons zirconium lattice [9], and after coating the lifetime of this component increases slightly. There are no any other components in the positron lifetime spectra. This is apparently due to the use of ^{44}Ti radioisotope as a source of positrons. Since the maximum energy of positrons from this source is 1.47 MeV, information is collected from the volume of material. Using the method proposed in [10], as well as LYS-1 software [11], positron implantation profile in the sample of Zr-1%Nb with a layer of nickel was calculated (Figure 3). The figure shows that more than 95% of the positrons annihilate in the volume of the Zr-1%Nb.

Hydrogenation has a significant impact on the lifetime distribution of positrons in the material. The intensive component (50-85 %) with a lifetime of 183±2 ps is appeared in the samples with the high hydrogen concentration. This positron lifetime corresponds to positron lifetime in a zirconium ZrH_2 [12].

The long-lived component of positron lifetime with a lifetime values of 209±3 ps and 239±5 ps appears in a spectra for the samples with low hydrogen concentrations. The positron lifetime value of 241±5 ps corresponds to the lifetime of positron, trapped in single vacancy [9]. The lifetime of 209±3 ps is significantly higher than the lifetime of positrons, trapped at dislocations (197 ps), but lower than the lifetime in a single vacancy. Apparently, this component corresponds to the positron lifetime in a hydrogen-vacancy defects (mV-nH), formed in the process of hydrogen saturation [6, 7].

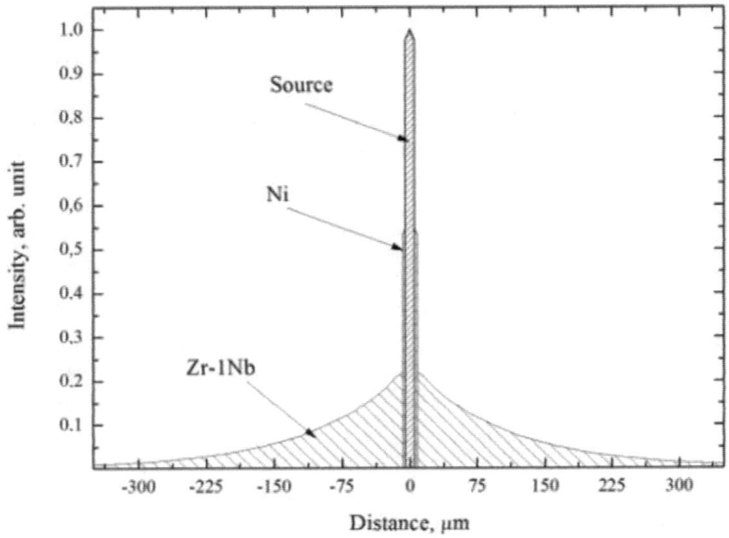

Figure 3. Positron implantation profile in a Zr1%Nb alloy with a layer of Ni for ^{44}Ti positron source

Analysis of Doppler broadening spectra (DBS) confirms these assumptions. The values of the line-shape parameters (S- and W- parameters) of DBS for the samples with nickel layer are virtually identical to the values of these parameters for the initial sample. The relative deviation is less than 1%. Figure 4 shows the dependence of the relative parameters S/S_0 and W/W_0 on the hydrogen content.

The graph shows that at low concentrations of hydrogenation the decrease of the relative parameters S/S_0 and growth of W/W_0 are observed. It also may indicate the formation of hydrogen-vacancy complexes [13-15].

The relative growth of S/S_0 and W/W_0 parameters is observed in the samples saturated up to high concentrations of 1.2 and 2.1 wt.% at the temperatures of 350 °C and 450 °C respectively. It can be possible due to the formation of hydrides, as well as changes in the microstructure, connected with hydride phase formation. In addition to the above, for the samples, that have been saturated up to 2.1 wt.% at 550 °C, S/S_0 parameter increases and W/W_0 parameter is decreases. This is the evidence of the vacancy-type defects formation [16].

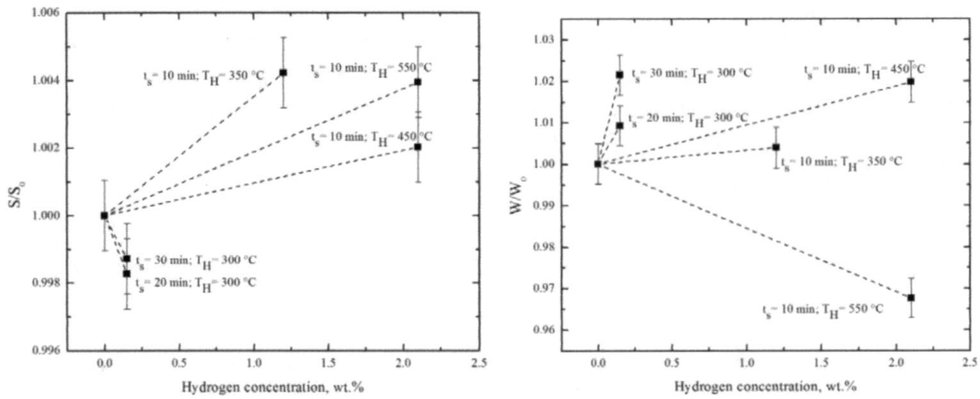

Figure 4. Dependence of S/S_0 parameter (a) and W/W_0 parameter (b) on the hydrogen content. t_s – time of Ni layer sputtering, T_H – temperature of hydrogen saturation

4. Conclusion

Magnetron sputtered nickel layer with different thickness from 1 to 3.4 μm was deposited. It was shown that increasing of the nickel layer thickness leads to an increase in the hydrogen sorption rate of zirconium alloy. The positron lifetime in the annealed zirconium is equal to lifetime of positrons zirconium lattice. Hydrogenation has a significant impact on the lifetime distribution of positrons in the material. The long-lived component of positron lifetime corresponds to the lifetime of positron, trapped in single vacancy appears in a spectra for the samples with low hydrogen concentrations. The intensive component correspond to positron lifetime in a zirconium hydride is appeared in the samples with the high hydrogen concentration.

Acknowledgements

This work was funded within the framework of realization of Strategic Programme on National Research Tomsk Polytechnic University Competitiveness Enhancement in the Group of Top Level World Research and Academic Institutions.

References

[1] Huang J H 1998 *Metallurgical and materials transaction A* **29** 1047.
[2] Kim Y S 2008 *Journal of Alloys and Compounds* **453** 210.
[3] Pushilina N S, Kudiiarov V N, Lider A M, Teresov A D 2015 *Journal of Alloys and Compounds* **645** S476
[4] Ivanova S V 2002 *International Journal of Hydrogen Energy* **27** 819
[5] Pushilina N S, Lider A M, Kudiiarov V N, Chernov I P, Ivanova S V 2015 *Journal of Nuclear Materials* **456** 311
[6] Kashkarov E B, Nikitenkov N N, Tyurin Yu I, Syrtanov M S, Zhang Le 2015 *Iop Conf. Series: Materials Science and Engeneering* **81** 012017
[7] Kudiiarov V N, Gulidova L V, Pushilina N S, Lider A M 2013 *Advanced Materials Research* **740** 690
[8] Lider A M, Pushilina N S, Kudiiarov V N, Kroening M 2013 *Applied Mechanics and Materials* **302** 92
[9] Robles J M C, Ogando E, Plazaola F 2007 *JOP: Cond. Matt.* **19** 176
[10] Dryzek J, Singleton D 2006 *Nucl. Instr. Meth. Phys. Res. B.* **252** 197
[11] http://www.ifj.edu.pl/~mdryzek/page_r22.html
[12] Gainotti A et al. 1968 *Nuovo Cimento B.* **1** 47

[13] Laptev R S, Bordulev Y S, Kudiiarov V N, Lider A M, Garanin G V 2014 *Advanced Materials Research* **880** 134.

[14] Bordulev Y S, Laptev R S, Kudiiarov V N, Lider A M 2014 *Advanced Materials Research* **880** 93

[15] Laptev R, Lider A, Bordulev Y, Kudiiarov V, Garanin G 2014 Hydrogenation-induced microstructure changes in titanium *Journal of Alloys and Compounds* **645** S193

[16] Krause-Rehberg R, Leipner H S 1999 *Positron annihilation in semiconductors: defect studies* (Berlin: Springer) p 324

IOP Conf. Series: Materials Science and Engineering **135** (2016) 012023 doi:10.1088/1757-899X/135/1/012023

Methods of studying the composition of the low-energy ion beams and the surface of deuterated-metal targets

S I Kuznetsov, G N Dudkin, B A Nechaev and I D Bystritsky

Tomsk Polytechnic University, 30 Lenin Ave., Tomsk, 634050, Russia

E-mail: smit@tpu.ru

Abstract. To study the reactions between the light nuclei (dd, pd, d^3He, d^4He) with ultralow collision energies, there is a need to obtain the high-precision experimental results on the purity of the target surface saturated with the hydrogen isotopes (protium, deuterium) and on the number and composition of the accelerated particles falling on the target. To solve this problem, a method has been developed and tested for operational testing the quality of the vacuum system and the cleaning of the metal target surface saturated with deuterium. The paper also presents the measurement results for the true flow of the accelerated ions and neutrals of hydrogen (deuterium), using a multigrid electrostatic energy analyzer. The values of the ion and neutral components of the accelerated particle flow were received for the Hall ion source. The values of the secondary electron emission coefficients were determined for a number of the metal targets (Cu, Ti, Ta, Zr) in the range of the accelerated ion energies of 3–12 keV.

1. Introduction

The modern models of nuclear astrophysics assume that the rates of nuclear reactions between the light nuclei in the stars are increased due to screening the charges of interacting nuclei by negatively charged electrons. Screening the charges at low energies is equivalent to an increase in the kinetic energy of interacting light nuclei by a small quantity U_e that is called the potential of screening [1, 2]. To verify this theoretical assumption, the fusion reactions are studied using the metal targets saturated with deuterium. This is caused by the presence of quasi-free electrons in metals, similar to quasi-free electrons in stellar plasma. To provide the required methodological purity of the experiment for studies of the electron screening effect, the reactions $d(d,p)^3H$ or $d(d,n)^3He$ are usually selected.

The comparison of the results concerning the studies of the mechanisms for proceeding the *dd* -reactions in metals saturated with deuterium [3–12] has shown that some results differ from other results and the results of calculations obtained using the traditional models of nuclear physics [13, 14]. In particular, this statement refers to the results received for the ZrD_2 and TiD_2 targets.

The calculations show that this difference may be connected with the presence of a parasitic adsorbed layer on the surface of the target. Therefore, to analyze the experimental data, there is a need to take into account the ionization losses of the deuteron energy in the adsorbed layer before the introduction of deuterons into the "pure" layer of the deuterated metal.

The layered Auger-spectrometric studies have shown that carbon is the basis of the adsorbed layer deposited in the operating vacuum of $2 \cdot 10^{-7}$ mm Mercury in our experiments. In addition, the surface of the target is significantly contaminated by carbon even in the ultra-high vacuum ($\sim 10^{-9}$ mm Mercury), using the magnetic separation of accelerated ions in the studies of the $d(d,p)^3H$ reactions conducted by other authors [12]. Numerous studies (within the international project ITER) with the

IOP Conf. Series: Materials Science and Engineering **135** (2016) 012023 doi:10.1088/1757-899X/135/1/012023

use of various techniques have shown that the films deposited in tokamaks and various discharges in the presence of hydrogen (including our conditions) are hydrogenated carbon with a weakly ordered structure.

In vacuum, the sources of carbon films are all carbon-containing materials and residual gases contacting with plasma or the beams of accelerated particles. Moreover, the ratio of deuterium and carbon particles in the films may reach a value of D/C = 2 [15, 16]. In the experiments, in real time, the composition and thickness of the films can not be controlled. Consequently, it is necessary to clean the target surface.

This paper presents the results of the study concerning the rate of the material deposition on the surface of the quartz plate (without heating) for different operation conditions of the setup and the experimental results demonstrating the complex effect of the temperature and accelerated deuterium ions on the surface of the target. Quartz of the crystal resonator in the device "Micron – 5" and the titanium film (TiD_2) saturated with deuterium and deposited on the substrate were used as the targets.

In addition, to study the reactions between the light nuclei (dd, pd, d^3He, d^6Li,..) with ultralow collision energies, there is a need to obtain the high-precision experimental results on the number of the accelerated particles falling on the target. To solve this problem, we developed and tested a method for measuring the true flow of the accelerated ions and neutrals of hydrogen (deuterium), using a multigrid electrostatic energy analyzer.

2. Controlling the mass thickness of the adsorbed layer on the target surface

The measurements were carried out in a vacuum chamber of the setup [17] used for the study of nuclear reactions dd, pd, etc. during the interaction of the deuterium (hydrogen) ions accelerated up to the energy of 6 – 12 keV with a metal target preliminary saturated with deuterium.

The plate of the quartz resonator was placed in a vacuum chamber where the target was mounted. The frequency of the quartz resonator was measured consistently for the different operating conditions of the setup with simultaneous controlling the temperature.

The minimum mass contamination of the surface can be recorded at the level of $1.3\cdot10^{-8}$ g/cm^2 under conditions of dynamic equilibrium with the frequency resolution of the recording device 1 Hz. For carbon, about $6.5\cdot10^{14}$ particles/cm^2 (~ 1 monolayer) corresponds to this mass. Such a mass thickness of the adsorbed layer on the target surface is acceptable for the experiments conducted to study the influence of the electron screening in metal (saturated with deuterium) targets on the neutron yield during the d(d,n)^{3}He reaction.

Figure 1 shows a curve that characterizes the rate of the substance desorption from the target surface during the operation of the accelerator. It follows from the figure that the rate of desorption is significantly increased by using the accelerator (*t* = 40 min).

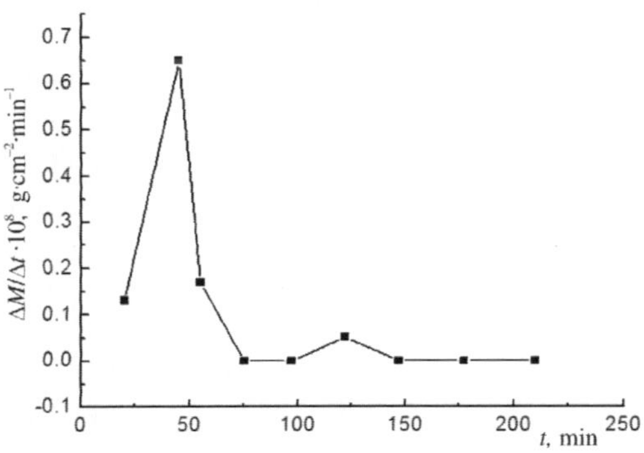

Figure 1. Rate of the substance desorption on the target surface during the experiment.

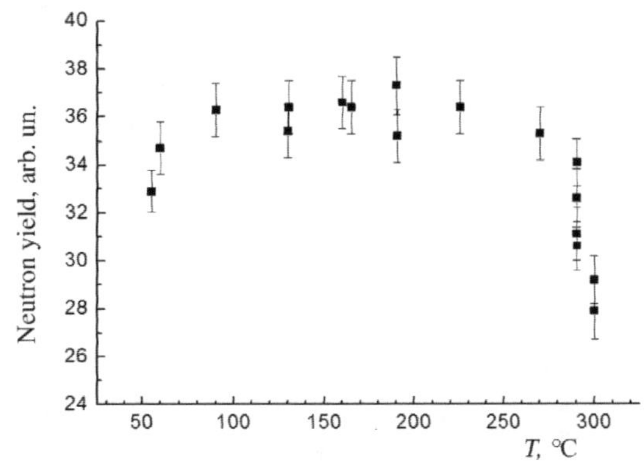

Figure 2. Yield of neutrons versus the target temperature.

IOP Conf. Series: Materials Science and Engineering **135** (2016) 012023 doi:10.1088/1757-899X/135/1/012023

The influence of the temperature factor was checked in the experiments using a titanium target saturated with deuterium (stoichiometry TiD_2). The experiments have shown (figure 2) that the neutron yield is increased during the heating of the target from the room temperature to ≤ 100 °C at the energy of the deuterons falling on the target (12 keV).

The yield of neutrons decreases during the heating of the target above 200 °C due to the decomposition TiD_2 and the subsequent separation of deuterium from the target. This assumption is consistent with the results of the study conducted by other authors and relating to the sublimation of the carbon film and the thermal desorption of deuterium implanted from titanium. [18, 19].

3. Determining the true number of the accelerated particles (ions + neutrals) falling on the metal target

To study nuclear reactions (dd, pd, and others) in the range of low-energy particle collisions, there is a need to measure the number of the accelerated particles falling on the target as accurate as possible. If the metal saturated, for example, with deuterium is used as a target, then such a target can be used as a collector for measuring the integral current of charged particles. At the same time, the current is measured on the entire target area, which is extremely important, in the case if the spatial heterogeneity and dynamic instability of the flow rule out the possibility for local measurement of the small area by using the collector.

However, in this case, to determine the true flux of the accelerated particles, it is necessary to consider the coefficient of the secondary electron emission due to the charged and fast neutral particles which fall on the target and are a substantial part of the integral flow.

In fact, fast neutrals must unavoidably be present in the flow of accelerated particles during the charge-exchange process [20]. Therefore, there is a need in the information on the number of neutrals. The fact of the neutral presence can be recorded during the secondary electron emission when the neutrals interact with the collector material.

The coefficient of secondary electron emission γ is dependent on the collector material, the energy and composition of falling particles, the purity of the collector surface, the temperature, the pressure and composition of residual gas, etc. [21–23]. Thus, it is almost impossible to select the reference values γ for the specific experimental conditions. There is a need to find a method for measuring this coefficient for the specific conditions.

In our experiments we used the capability of the grid electrostatic energy analyzer located behind the target to determine the number of fast neutrals and to control the value γ.

Figure 3 shows the scheme of the measuring unit in the experimental setup that consists of the four-grid analyzer and the metal target with the current recording systems in the digital oscilloscope. The disc target (the area is ≈ 80 cm^2) and the part of the energy analyzer were placed in a copper screen, cooled up to – 170°C.

The cooled screen was used for the local freezing and adsorption of heavy and light hydrocarbons during the evacuation of the operating chamber. These measures were intended to reduce the contamination of the target, since the path of the accelerated deuterium particles in the target materials is < 400 nm at the energy of ~ 12 keV. The residual pressure in the chamber was maintained in the range of $1 \cdot 10^{-5} – 2 \cdot 10^{-6}$ mm Mercury.

The disk target had a hole with the area of ≈1 cm^2, through which a part of the accelerated particle flow fell on the energy analyzer.

The S_2 and S_4 grids of the energy analyzer were grounded. The breaking potential was applied to the grid S1 for cutting off the electrons. The potential cutting off the ions with the energy less U_3 was applied to the grid S_3. The values of the ballast containers C and load resistors R were selected depending on the value of the current recorded.

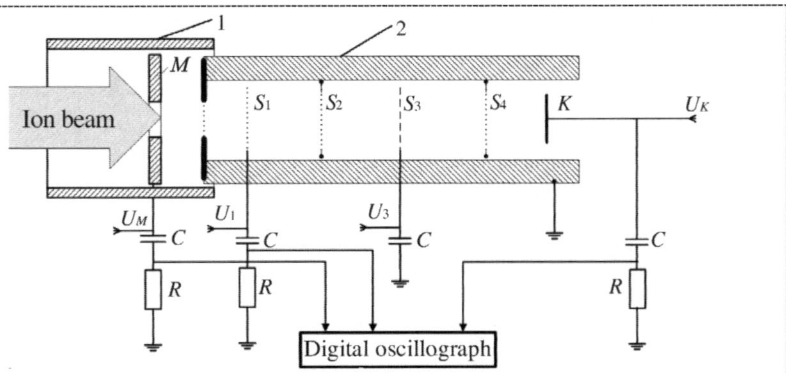

Figure 3. Experimental scheme. 1 is the cooled copper cylinder, 2 is the grid power analyzer, M is the target, K is the collector, S_1–S_4 are the grids.

A series of experiments was conducted to specify the sources (causes) of currents recorded on the target, collector and grids to develop the method for measuring the true flow of the accelerated deuterium particles.

The results of the studies have shown that a positive signal of the collector is determined by the emission of electrons subjected by fast neutrals in the presence of the cutoff voltage on the grid. The signal from the collector is almost completely suppressed after applying the displacement potential > +15 B to the collector (figure 4, oscillogram 4).

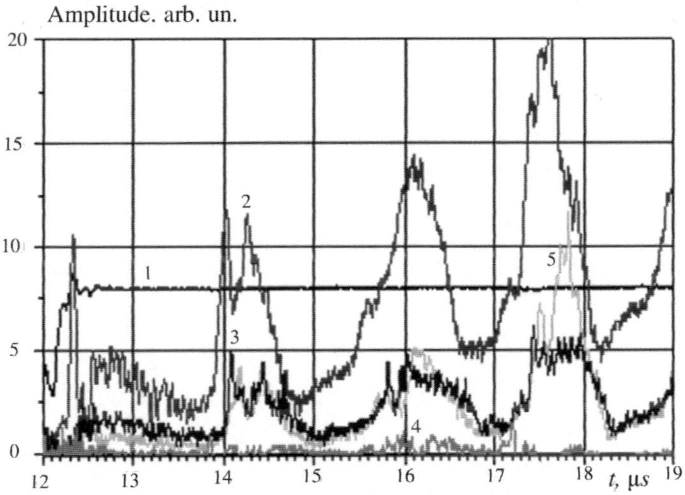

Figure 4. Oscillograms of the current and acceleration voltage. 1 is the acceleration voltage (12 kV), 2 is the target signal, 3 is the signal from the collector (displacement + 9 B), 4 is the signal from the collector (displacement +25 B), 5 is the signal from the first grid of the spectrometer.

Suppression of the positive signal after applying a relatively low positive potential of displacement to the collector of the spectrometer indicates that there are no fast deuterium ions between S_3 and S_4 (grounded).

The secondary electron emission was found not to be essential under the exposure to gamma rays and electrons. In addition, we believe that the coefficient of secondary electron emission is the same for the fast ions and neutrals of atomic deuterium, since the energy of the electrons connected with the neutrals in our range of the acceleration energy is low.

The previous studies did not reveal the accelerated particles of molecular deuterium in a flow, using the time-of-flight technique on the basis of ~ 1 m [24].

Thus, the current I^M on the metal target is caused by the flow of ions and the electrons of secondary emission

$$I^M = I^+ + \gamma \cdot (I^+ + I^0),$$

where γ is the coefficient of electron emission under the exposure to ions and neutrals, I^+ is the flow of ions, I^0 is the flow of neutral particles.

The flow of accelerated particles on the target:

$$I = I^+ + I^0.$$

The values of current integrated in the acceleration range determine the number of the corresponding particles.

The coefficient of the electron emission from the collector depends on the collector material with all other things being equal. Therefore, the collector material should be the same as the target material. Electrical copper, titanium and tantalum saturated with deuterium and deposited on the copper foil ZrD were used as the target materials (and collectors). The copper target and collector were not subjected to any pretreatment. The titanium and tantalum targets and, respectively, the collectors were subjected to evacuation and annealing during the saturation with deuterium. The zirconium target was formed by the magnetron deposition of Zr on a copper substrate in the atmosphere of deuterium.

When voltage is applied to S_3 for cutting off the ions, only neutrals fell on the collector of the energy analyzer. The positive displacement voltage on the collector was selected to suppress the electron emission (under the exposure to neutrals) under these conditions. (The positive potential "returns" the knocked-on electrons on the collector, and the collector current = 0).

Further, in the experiments, the currents were measured under different conditions, using the energy analyzer:

$I_1 = I^+ + \gamma \cdot (I^+ + I^0)$ is the collector current, the sum of currents is determined by the number of charged particles and the number of the electrons knocked-on by ions and neutrals.

$I_2 = \gamma \cdot I^0$ is the collector current determined by the number of the electrons knocked-on by neutrals from the collector.

$I_3 = I^+$ is the collector current caused by the ion current.

The currents I_1, I_2, I_3 were measured by alternating the conditions for the operation of the energy analyzer.

From here it follows (table 1):

$$\gamma = (I_1 - I_2 - I_3)/I_3.$$

$$I^+/(I^+ + I^0) = I_3/(I_3 + I_2/\gamma).$$

$$K = I^M/I = I^+/(I^+ + I^0) + \gamma.$$

Table 1. Results of measurements.

Target (collector)	Cu			Ti + D_2			Ta + D_2		
Energy of deuterium particles (keV)	γ	$I^+/(I^++I^0)$	K	γ	$I^+/(I^++I^0)$	K	γ	$I^+/(I^++I^0)$	K
12				1.1	0.69	1.8			
10	2	0.79	2.78	0.98	0.84	1.8	0.37	0.83	1.2
8	1.9	0.87	2.7						
6	1.92	0.9	2.9				0.45	0.87	1.3

For the target ZrD, the following values were obtained at the energy of ~7 keV: $\gamma \approx 0.17$, $I^+/(I^+ + I^0) \approx 0.64$, $K \approx 0.81$. It should be noted, that the smallest value of the ratio $I^+/(I^+ + I^0)$ was obtained for all targets at a residual chamber pressure of ~ $1 \cdot 10^{-5}$ mm Mercury.

4. Conclusion

Thus, in the experiment there is a possibility for testing the quality of the vacuum system and the different units in the setup to identify and eliminate the factors influencing on the target contamination. The results received for the combined effect on the hydrogenated carbon film on the target surface allow us to develop a procedure for operating with the titanium target saturated with deuterium.

If the target is produced from the material identical to the collector material of the energy analyzer, then the correction coefficients *K* obtained by the energy analyzer can be used for determining the true flow of the accelerated particles falling on the target.

This work was supported by the RFBR grant No. 14–22–03039.

References

[1] Rolfs C, Rodney W S Cauldrons in the Cosmo University of Chicago Press Chicago 1988
[2] Assenbaum H J, Langanke K and Rolfs C 1987 *Z. Phys.* A **327** 461
[3] Greife U, Gorris F, Junker M et al. 1995 *Z. Phys.* A **351** 3107
[4] Aliotta M., et al. 2001 *Nucl. Phys.* A **690** 790
[5] Kasagi J, Yuki H, Baba T et al. 2002 *J. Phys. Soc. Jpn.* **71** 288.
[6] Czerski K et al. 2002 *Nucl. Instr. Meth. in Physics Research* B **193** 183
[7] Raiola F. et al. . 2004 *Eur. Phys. J.* A **19** 283
[8] Raiola F et. al. 2006 *Eur. Phys. J.* A **27** 79
[9] Huke A, Czerski K and Heide P 2007 *Nucl. Instr. Meth.* B **256** 599
[10] Czerski K et al. 2004 *Europhys. Lett.* **68** 363
[11] Bonomo C. et al. 2003 *Nuclear Physics* A **719** 37
[12] Czerski K et al. 2008 *J. Phys.* G: *Nucl. Part. Phys.* **35** 014012
[13] Ichimaru S 1993 *Rev. Mod. Phys.* **65** 252
[14] Czerski K et al. 2006 *Eur. Phys. J.* A **27** 83
[15] Kolbasov B, Guseva M I, Gureev V M et al. 2006 *VANT* Ser. *Termoyaderniy sintes* **2** 50
[16] Svechnikov N Yu, Stankevich V G, Suhanov L P et al. 2009 *Poverhnost* **6** 5 (In Russian)
[17] Bystritsky V M, Gerasimov V V, Krylov A R, Parzhitskii S S, Anan'in P S, Dudkin G N et al. 2008 *Eur. Phys. J.* A **36** 151
[18] Neklyudov I M, Morosov A N, Voevodin V N, Kulich V G 2004 *Int. Scient. J. for Alternative Energy and Ecology* **5** 13
[19] Suvorov A L, Guseva M I, Korshunov S N, SaluzhniyA G, Lasarev N E 2000 *VANT* **4** 6. (In Russian)
[20] Darkalov E E, Veselovsorov A N, Pogorelov A A 2008 *ПТЭ* **2** 113 (In Russian)
[21] Abroyan I A, Eremeev M A, Petrov N I 1967 *UFN* **92** 1 105 (In Russian)
[22] Brusilovskyi B A, Kineticheskaya ionno-elektronnaya emissiya M. Energoatomisdat 1990 (In Russian)
[23] Pleshivzev N V, Bajin A I Fizika vozdeyistviya ionnih puchkov na materiali M. Vuzovskaya kniga 1998 (In Russian)
[24] Veretelnik V I, Dudkin G N, Nechaev B A et al. 2008 *15th Int. Symp. on High Current Electronics*: *Proc.* (*Tomsk, Russia, 21–26 September 2008*): Publishing House of the IAO SB RAS 2008 pp 151–153

Comparative analysis of using natural and radiogenic lead as heat-transfer agent in fast reactors

R A Laas, R V Gizbrekht, P A Komarov and V N Nesterov
National Research Tomsk Polytechnic University, Tomsk, Russia

E-mail: nesterov@tpu.ru

Abstract. Fast reactors with lead coolant have several advantages over analogues. Performance can be further improved by replacement of natural composition lead with radiogenic one. Thus, two main issues need to be addressed: induced radioactivity in coolant and efficient neutron multiplication factor in the core will be changed and need to be estimated. To address these issues analysis of the scheme of the nuclear transformations in the lead heat-transfer agent in the process of radiation was carried out. Induced radioactivity of radiogenic and natural lead has been studied. It is shown that replacement of lead affects multiplication factor in a certain way. Application of radiogenic lead can significantly affect reactor operation.

1. Introduction

Concept of the BREST reactor assumes replacement of sodium with lead – a chemically passive and poorly activated element with high boiling temperature (1740 °C). In addition, combination of the lead heat-transfer agent and dense nitride fuel creates possibility for the following main positive changes for ensuring safe operation of fast reactors [1 – 4]:

- at breeding ratio (BR) of the core above unity reactivity margin at fuel burn-out may be reduced down to the level not exceeding ratio of the delayed neutrons. As a result, power excursion on prompt neutrons is not possible at any emergencies;
- big emissions of radioactive substances into environment are not possible even if the reactor vessel and NPP containment are damaged;
- lead has lower moderation ability than sodium; that's why emergency removal of lead from the core will affect neutron spectrum less severely and, as a result, change of the reactivity will be lower than in case of the sodium moderator.

Isotopic composition of the lead used in the BREST reactors are to be estimated. Two opposite concepts exist which assume increase the ration of:

- light isotopes, which should reduce induced radioactivity in the heat-transfer agent;
- heavy isotopes, which will allow reducing absorption of neurons in the heat-transfer agent thus improving breeding properties of the core and increase value of the nuclear fuel reproduction factor.

Besides natural lead with sufficiently high content of light isotopes, there is also radiogenic lead, which is formed in uranium and thorium ores. Isotopes Pb^{206}, Pb^{207}, Pb^{208} are final products of the radioactive decay chains of U^{238}, U^{235} and Th^{232}, respectively. So, isotopic composition of the radiogenic lead depends significantly on elemental composition of the ore and age of the deposit [5-7]. For example, radiogenic lead from thorium or uranium-thorium ores with low uranium fraction contains mainly Pb^{208}, which is interesting because of the following reasons:

- isotope Pb^{208} has unique neutron-physical properties. Nucleus of Pb^{208} is double magic with closed neutron and proton shells. Levels of excitation of nuclei Pb^{208} are in the area of higher energies than levels of excitation of other natural isotopes of lead. That is why Pb^{208} has lower moderation power.
- isotope Pb^{208} has extremely low section of radioactive capture of neutrons over the whole range of energies.

This means that use of the radiogenic lead with high content of Pb^{208} as moderator in fast reactors may improve their neutron-physical characteristics. Poor absorption of neutrons will allow using uranium-plutonium fuel with smaller plutonium fraction. Poor moderation of neutrons will cause reduction of spectral component of the void reactivity coefficient (VRC). [8]

2. Change of the lead-based moderator nuclide composition

Lead moderator will be irradiated with neutrons, which will lead to production of radionuclides and thus increased activity. [9].

Figure 1 represents scheme of main nuclear reactions in the lead moderator, which change nuclide composition and accumulate radioactivity.

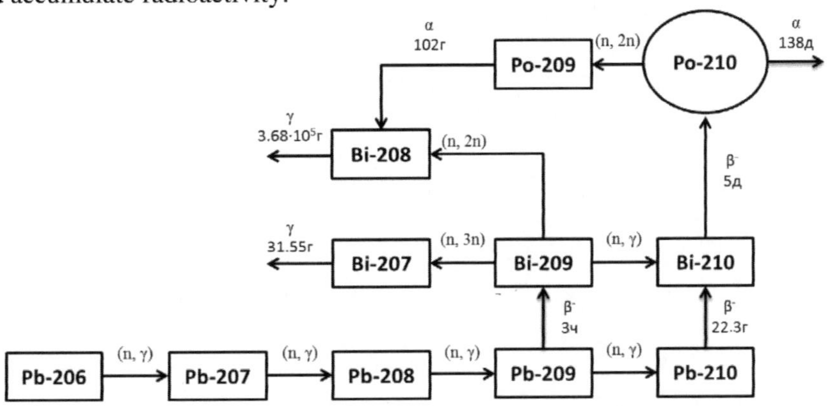

Figure 1. Process of lead moderator transmutation

For determining dependence of the concentration values of different elements upon time of the moderator operation, it was necessary to determine values of microscopic cross sections of interaction of the neutrons of each group with the moderator nuclides. The TENDL-2012 database was used for the calculation.

The values of microscopic sections from the TENDL-2012 database were approximated by different functions, obtaining dependences of the microscopic cross sections on energy of neutrons for 26 groups. Then mean values of cross section in each group were determined taking into account spectrum of the neutron flow in a group using the ratio:

$$\sigma = \frac{\int_{E_1}^{E_2} \sigma_{\text{app}}(E) \cdot \Phi(E) \cdot dE}{\int_{E_1}^{E_2} \Phi(E) \cdot dE}, \tag{1}$$

where E_1 and E_2 are the boundary values of energies for each group of neutrons; $\sigma_{\text{app}}(E)$ are approximated dependences of the microscopic sections on energy of neutrons; $\Phi(E)$ is dependence of the neutron flow density on energy of the neutrons inside the considered group. These dependences were determined using the ratios:

a) Watt spectrum for neutrons from 1st to 11th group inclusively:

$$\Phi(E) = e^{-E} \cdot sh\left(\sqrt{2E}\right) \cdot \sqrt{E} = \begin{cases} (0.913 + 0.75 \cdot E) \cdot \sqrt{E} \\ \left(0.944 + 0.0644 \cdot E^{-0.608}\right) \cdot \sqrt{E} \end{cases}, \quad (2)$$

for $0.01 \leq E < 0{,}25$ MeV,

b) Fermi spectrum for neutrons from 12th to 23d group inclusively:

$$\Phi(E) \sim \frac{1}{E}, \quad (3)$$

for $0.25 \leq E < 10$ MeV;

As the BREST reactor is fast neutron reactor, we may not consider low energy groups. The calculation is limited by first 17 groups inclusively, because fraction of neutrons in subsequent groups has negligibly low value.

Cross sections for different nuclides of moderator are shown given Table 1.

Table 1. Approximated values of microscopic cross sections from the TENDL-2012 database of micro-constants

No of group	Pb^{204}	Pb^{206}	Pb^{207}	Pb^{208}	Pb^{209}	Bi^{209}		Po^{210}
	σ_c, b	σ_c, b	σ_c, b	σ_c, b	σ_c, b	σ_c, b	$\sigma_{(n,2n)}$, b	$\sigma_{(n,2n)}$, b
1	0.00201	0.00089	0.00131	0.00080	0.00107	0.00100	0.41887	0.50067
2	0.00256	0.00045	0.00060	0.00072	0.00063	0.00100	0	0
3	0.00495	0.00056	0.00025	0.00188	0.00110	0.00150	0	0
4	0.00981	0.00078	0.00017	0.00125	0.00068	0.00200	0	0
5	0.01092	0.00344	0.00014	0.00086	0.00054	0.00250	0	0
6	0.00989	0.00604	0.03195	0.00384	0.00049	0.00250	0	0
7	0.02481	0.00632	0.03425	0.00275	0.00051	0.00250	0	0
8	0.10906	0.00612	0.02150	0.00560	0.00067	0.00250	0	0
9	0.67265	0.00810	0.00412	0.00117	0.00019	0.00200	0	0
10	0.06275	0.00801	0.03089	0.00001	0.00131	0.00200	0	0
11	0.08517	0.01679	0.01541	0	0.00219	0.00200	0	0
12	0.23902	0.00004	0.00094	0	0.00028	0.00500	0	0
13	0.25621	0.02949	0.13245	0	0.00042	0.02300	0	0
14	1.24659	0.00010	0.00252	0	0.00122	0.00150	0	0
15	0.04955	0.00015	0.00380	0	0.03667	0.19000	0	0
16	0.00611	0.00023	0.00573	0	0.52077	0.00070	0	0
17	0.00875	0.00035	0.00837	0	0.02953	0.00060	0	0
$\overline{\sigma}$	**0.13726**	**0.00659**	**0.02175**	**0.00237**	**0.00212**	**0.00378**	**0.00019**	**0.00022**

3. Results and discussion

Isotopic composition of the heat-transfer agent influences absorption of neutrons in it. Radiation capture of neutrons goes in direct proportion with fraction of light isotopes, since they have greater capture cross sections. Relative increase of k_{eff} at transition from natural to radiogenic lead will be about 2% ($k_{\text{eff}}^{\text{rad}} / k_{\text{eff}}^{\text{nat}} = 1.021$), thus significantly increasing reactivity margin in the reactor only due to use of the radiogenic lead.

Formation of radionuclides in the moderator causes increase of general radioactivity in it. In Figures 2–4 dependences of specific radioactivity in natural and radiogenic lead upon time of its operation in nuclear reaction and after the reactor shutdown are shown.

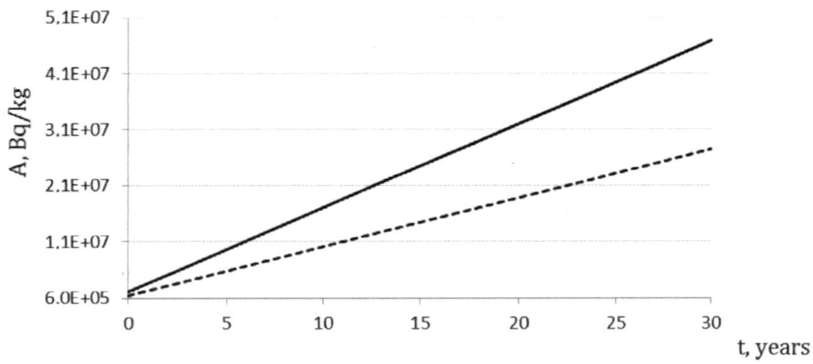

Figure 2. Change of specific alfa-activity of heat-transfer agent at 30-year operation of reactor: —— – radiogenic; - - - - - – natural

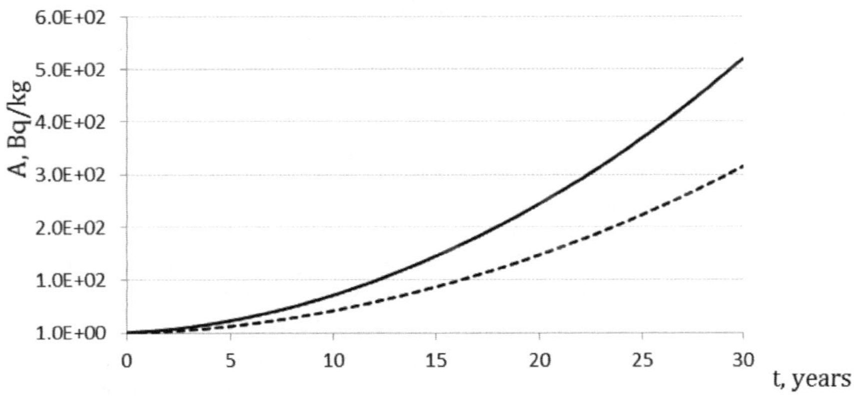

Figure 3. Change of specific gamma-activity of heat-transfer agent at 30-year operation of reactor: —— – radiogenic; - - - - - – natural

Specific beta-activity of heat-transfer agent sharply increases during first hours of the rector operation and preserves its value at constant level: $7.6 \cdot 10^{12}$ Bq/kg and $1.3 \cdot 10^{13}$ Bq/kg for natural and radiogenic lead, respectively.

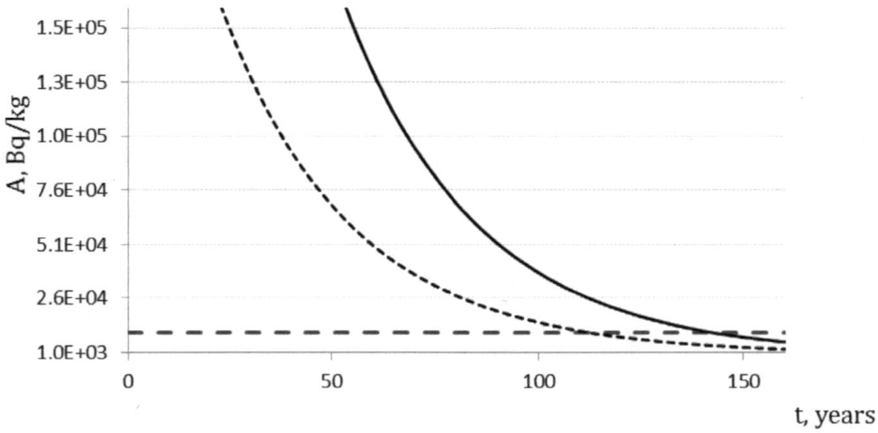

Figure 4. Change of specific beta-activity caused by Pb^{209}, Pb^{210}, Bi^{210} after shutdown of reactor: ——— – radiogenic; - - - - - – natural

Specific gamma-activity of the heat-transfer agent will be equal to 316 and 521 Bq/kg for natural and radiogenic lead, respectively. It will remain constant because of long half-life of Bi208, which makes main contribution into gamma-activity - about $3 \cdot 10^5$ years.

Conclusion

Radiogenic lead has greater induced activity compared to natural lead. This leads to increase of holding time. The results show that beta-activity is of primal importance. Due to accumulated beta-activity holding times are about 100 and 150 years for natural and radiogenic lead respectively.

However, the authors consider that increase of the reactivity margin by about 0.02 is more significant advantage of radiogenic lead in comparison with its disadvantage – the induced radioactivity. Since design value of reactivity margin in fast neutron reactors with lead heat-transfer agent does not exceed efficient share of the delayed neutrons (about 0.0063 for U^{235}), radiogenic lead may significantly affect implementation of the nuclear fuel cycle and nuclear safety in the course of the reactor plant operation.

References

[1] Apse V A, Apsje V A, Shmelev A N and Sirotkin A M 2010 *Izvestija VUZov Serija: Jadernaja jenergetika.* **4** 15
[2] Orlov V V 2000 Trudy Mezhdunarodnogo seminara Bystryj reaktor i toplivnyj cikl estestvennoj bezopasnosti dlja krupnomasshtabnoj jenergetiki. Toplivnyj balans, jekonomika, bezopasnost', othody, nerasprostranenie [In Russian]
[3] Bedenko S V and Shamanin I V 2011 Osnovy uchjota i kontrolja deljashhihsja materialov v proizvodstve: Uchebnoe posobie 91 [In Russian]
[4] Bedenko S V, Knyshev V V, Jakovlev M V and Plevaka M N 2015 *Izvestija Tomskogo politehnicheskogo universiteta. Inzhiniring georesursov* **4** 99
[5] Godoy J M, Godoy M L and Aronne C C 2007 Journal of Brazilian Chemical Society **18** 12.
[6] Nier A O, Tompson R W and Murphey B F 1941 *Physical Review* **60** 112.
[7] Scherbak N P 1978 Katalog izotopnyh dat porod ukrainskogo shhita
[8] Adamov E O and Orlov V V 1999 Tjazhelye zhidkometallicheskie teplonositeli v jadernyh tehnologijah **1** 25
[9] Shmelev A N, Kulikov G G and Glebov V B 1992 Atomnaja jenergija **73** 450

Control of changes in the defect structure of titanium saturated with hydrogen

V V Larionov, A M Lider and R S Laptev

National Research Tomsk Polytechnic University, Tomsk, Russia

E-mail: lvv@tpu.ru

Abstract. The hydrogenated samples of technical titanium were investigated using the EPA method and the measurements of the thermal electromotive force for these samples saturated with a different amount of hydrogen. The structure of the hydrogenated samples was studied by the X-ray diffraction method. The results have shown that the hydrogenated titanium structure starts changing at the same time, depending on the amount of added hydrogen. The intensity of the annihilation process increases with the increase in the hydrogen concentration in α-titanium up to the values of 4% wt and does not change up to the values of 5% wt (α + δ) - titanium. At the same time, the value of the thermal electromotive force decreases in this range of values. The annihilation intensity is stabilized for the values of 5% wt, and the value of the thermal electromotive force is increased. The inflection point for the thermal electromotive force versus the hydrogen concentration corresponds to the formation of δ - hydrides. The increase in the positron lifetime starts in the concentration range of 6-8% and moves to the stable level up to the concentrations of 21-22%. In this range, there is a transition from the (α + β) to the (α + δ) phase. The lifetime of positrons and the number of defects are increased, the value of the thermal electromotive force is reduced (up to the concentration of 24%), then there is a stabilization mode for all these parameters up to the values 32% wt.

1. Introduction

The development and improvement of the materials used in the nuclear industry or under extreme conditions should be controlled to meet such conditions. As a rule, these materials absorb hydrogen and become the hydride-forming ones [1]. In this case, hydrogen plays both a negative and positive role [2]. To control hydrogen, there is a need to use the methods providing the minimum radiation, the minimum toxic substances, and a wide temperature range. The development of control methods is closely connected with the study of physical processes accompanying the hydrogenation of metals [3]. Many of them, namely, the nuclear magnetic resonance, quasi-elastic neutron scattering, the Gorsky and Mössbauer effects require the complex equipment, and the received results are essentially dependent on the concentration gradient of added hydrogen, the hydrogenation method, and the form of the samples under study. Hydrogen is known to have a high diffusion mobility in metals and alloys and can form complex systems which include the vacancy-type defects, impurity atoms, dislocations, self interstitial atoms, as well as the grain boundaries [4]. Hydrogen interacts actively not only with the existing structural defects [1], but also causes the formation of a large number of new defects [5-8]. To create operational production control methods [1], there is a need to study the degassing processes [9], the nature and formation of hydrogen defects.

In the work we used the classical control methods for the measurement of eddy currents, the thermal electromotive force and electrical resistance [10-12], as well as the modern method of electron-positron annihilation (EPA) [13]. The EPA method is quite effective for studying the dynamics of appearance, transformation and disappearance of defects in metals, and for determining the concentrations and dimentions of defects during the hydrogen saturation [5, 14].

The simultaneous use of different experimental techniques allows us to obtain the comprehensive quantitative and qualitative data on the structure and modification of the material and determine the defects in a wide temperature range for any hydrogenation method. This combination is useful not only for hydrogenation, but for degassing of hydrogenated samples.

2. Experiment

Preparation of samples. The commercially pure titanium samples with the linear dimensions of 30×30×1 mm were mechanically polished and annealed in vacuum at the temperature of 650 °C for 1 hour. The samples were saturated with hydrogen from the gaseous phase at a high pressure and temperature according to the Sieverts' method [12], using the automated complex "Gas Reaction Controller" [13, 14] (Advanced Materials Corporation). The samples were placed in a chamber that was evacuated to a pressure of 10^{-6} atm. Then the samples were heated in vacuum to a temperature of 600 °C at the heating rate of 4 °C/s. The slow rate of heating removes completely the desorbed gases during heating. Then the chamber was quickly filled with hydrogen to a pressure of 0.66 atm. The retention time of the samples in the atmosphere of hydrogen was varied to obtain the different hydrogen concentrations. The hydrogen concentrations were determined by the volume (during hydrogenation) and gravimetric method. When the H-concentration in the sample reaches a certain value, the system automatically removes hydrogen from the samples, and then slowly (1.5 °C/min) decreases the temperature of the sample up to the room temperature in vacuum. The hydrogen concentration values obtained in this experiment were in the range of 0.9-25.1 atm %. This means that, in accordance with the Ti-H diagram, the new phases can be formed during hydrogenation.

3. Results and discussion

Methods. Spectra of PL in the samples were collected using the spectrometer described in details in [6]. The time resolution of this spectrometer is 240 ps. Three spectra of $5\cdot10^6$ events each were collected for every sample. They were mathematically fitted by the sum of 1 gaussian and several exponential functions according to the two-state trapping model by the LT10 software [18]. After the background and the source contribution subtracting, decomposition of the spectra was carried out by two exponential components. Each component is characterized by the positron lifetime (τ) and intensity (I) values [19-21]. Figures 1-2 show the results of measurements for the intensity I and the positron lifetime (τ) versus the hydrogen concentration in titanium.

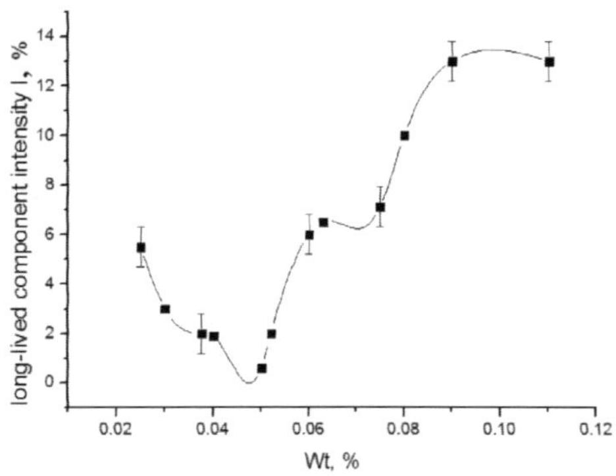

Figure 1. Intensity of EPA versus the concentration of hydrogen in titanium.

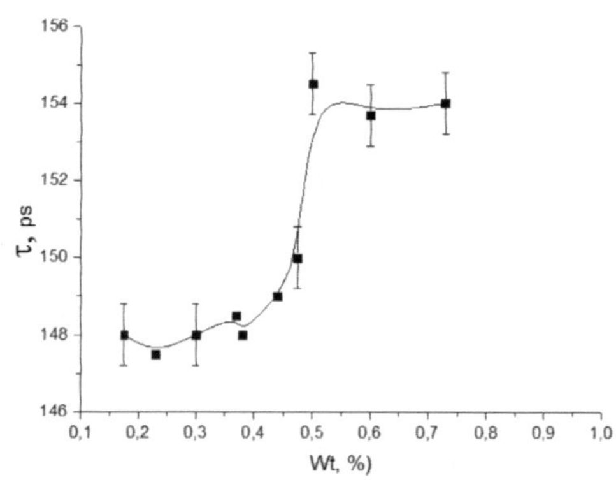

Figure 2. Positron lifetime in titanium versus the concentration of hydrogen in the BT1-0 titanium samples.

After the formation of defects, the curve *I* (wt) reaches the plateau region when the concentrations of hydrogen in titanium are in the range of 0-5 wt%. The increase in the concentration is also characterized by the increase in the lifetime of positrons (Figure 2). It should be noted that there is the inflection point on the curve of the thermal electromotive force for the same concentration (Figure 3).

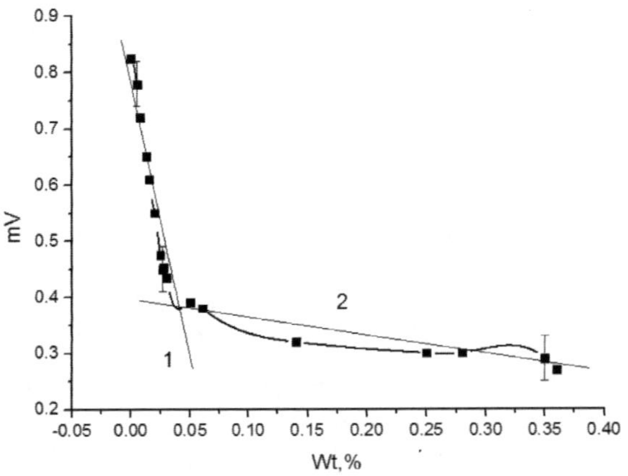

Figure 3. Thermal electromotive force versus the concentration of hydrogen in titanium BT1-0.

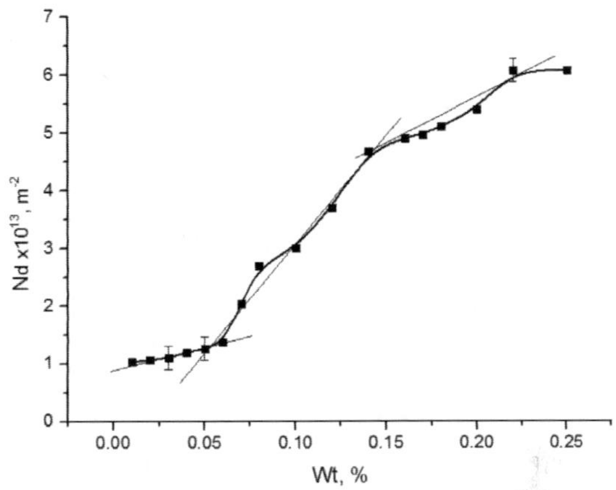

Figure 4. Number of defects versus the weight content of hydrogen (w_t) in the titanium alloy.

The relative change in the absolute coefficient of the thermal electromotive force S is determined by the formula [23] (1, 2):

$$\frac{\Delta S}{S_0} = \frac{3\pi^2\hbar Q}{e^2 k_F^2 \rho}\left(\frac{\Delta x}{x} - 1\right) \quad (1)$$

$$x = -\frac{k_F}{2}\left(\frac{d\ln}{dk}\rho_0\right)_{k=k_F}$$

$$\Delta x = -\frac{k_F}{2Q}\left(\frac{dQ}{dk}\right)_{k=k_F} - 1. \quad (2)$$

In the formulas (1-2), *Q* is the transport cross-section of electron scattering in a defect, *Kp* is the wave vector on the Fermi surface, *p* is the specific resistance of the metal, *e* is the electron charge, *h* is the Planck's constant. This value depends on the value of the transport cross-section of electron scattering *Q* in the defect, the specific resistance of hydrogenated titanium, i.e., on the number of defects and the shape of the cross-section.

It follows from the formula (1) that the value *ΔS/S* is determined by the value *Δx/x*, the cross section of scattering *Q*, and a possible change in the sign of the derivative *dQ/dk* caused by the presence of defects. Figure 3 shows the inflection point for the thermal electromotive force versus the concentration of hydrogen in titanium, which is caused by the transition of titanium from one structural state to another one.

A further increase in the hydrogen concentration does not lead to the changes in the thermal electromotive force and is completely dependent on the transport cross-section of electron scattering in the defect, i.e., on the number of defects and the shape of the cross-section and is indirectly confirmed by the measurements shown in Figure 4. This fact is typical when there are the changes in the structure of the titanium alloy. The further increase in the hydrogen concentration does not lead to the changes in the thermal electromotive force. The jump in the change of the thermal electromotive force can be

IOP Conf. Series: Materials Science and Engineering **135** (2016) 012025 doi:10.1088/1757-899X/135/1/012025

explained by the formation of a chemical compound (hydrogen with titanium), the presence of the phase transition $\alpha \rightarrow \gamma$ and grains, as well as the different material in intergranular zones. If to draw a tangent to the curves and drop a perpendicular to the axis of concentrations, then the obtained value of the hydrogen concentration in the metal corresponds to the concentration of titanium during the transition in the different structural state.

The number of defects in titanium N_d versus the hydrogen concentration is shown in Figure 4. In this work, the dislocation density of N_d is determined by formula (3) using the method described in [22]:

$$N_d = \pi\beta^2 ctg^2\Theta / 16b^2, \quad (3)$$

where β is the broadening of X-ray lines caused by microdeformation of the lattice, Θ is the angle corresponding to the peak of the X-ray line, b is the Burgers vector.

There are three regions (0.01-0.05) wt%, (0.5-0.14) wt% and 0.14- 0.24 wt%. The results for these regions are similar to the results received by the EPA method (Figure 1 and 2). If to draw a tangent to each region and drop a perpendicular from the point of intersection, then we obtain the concentration of hydrogen in titanium, the value of which corresponds to the change in the hydrogenation of the sample. A comparison of Figures 1-4 shows an identical character of the dependences with a characteristic inflection point on the curve of the hydrogen concentration that is equal to 0.05 wt% for the thermal electromotive force and a good agreement between the EPA (Figure 2) and N_d (Figure 4) values. This effect can be used to estimate the transition of the titanium alloy from one state to another one in order to control the material and allows this technique to be introduced for the laboratory and industrial control.

4. Conclusion

This method proposed for controlling the changes in the structure of titanium saturated with hydrogen is based on the simultaneous use of electron-positron annihilation and the measurement of the thermal electromotive force and electrical resistance. The results have shown that there are two characteristic regions in which hydrogen in titanium has an effect on the value of the thermal electromotive force. The characteristic graphic inflection found for the thermal electromotive force versus the hydrogen concentration allows us to determine the concentration of hydrogen in titanium, the value of which corresponds to the structure rearrangement of the titanium alloy. The transition from one structure to another one is shown to be observed for the value of the hydrogen concentrations in titanium equal to about 0.5 wt%. A comparison of the structural-phase state for the titanium alloy such as TiH_2: 4.04 wt% TiH: 2.02 wt% $TiH_{0.5}$: 1.01 wt% allows us to control the compounds of titanium with hydrogen using the measurements of the thermal electromotive force obtained for the thermal electromotive force versus the concentration of hydrogen in titanium.

This work was supported in the framework of the state task Science No. 1524, project 0.1325.2014.

References

[1] Geld P.V., Ryabov R.A., Kodes E.S. Hydrogen and Imperfections of the Metal Structure, Metallurgia, Moscow, 1979.
[2] Eliezer D., Eliaz N., Senkov O. N., Froes F.H. Positive effects of hydrogen in metals, 2000 Materials Science and Engineering A280 220.
[3] Kolachev B.A., Talalaev D.V. Toutoch Titanium, third ed., Aviation Publishing House, Moscow, 1993.
[4] Wert C.A. Hydrogen in metals. Topics in applied physics. 29 (1978), pp. 305-330.
[5] Cizek J., Prochazka I., Becvar F., Kuzel R., Cieslar M., Brauer G., Anwand W., Kirchheim R., Pundt A. Phys. Rev. B 69 (2004) 224106.
[6] Socher S., Lavrov E.V., Weber J. Phys. Rev. B 86 (2012) 125205.

[7] Sakaki K., Yamada T., Mizuno M., Araki H., Shirai Y. Mater. Trans. 43 (11) (2002) pp. 2652-2655.

[8] Cizek J., Prochazka I., Kuzel R., Becvar F., Cieslar M., Brauer G., Anwand W., Kirchheim R., Pundt A., Mater. Sci. Forum 60 (2004) pp. 145-146.

[9] Evard E.A., Gabis I.E., Voyt A.P. Study of the kinetics of hydrogen sorption and desorption from titanim, J. Alloys. 404-406 (2005) pp. 335-338.

[10] Larionov V.V., Lider A. M., Garanin G. V. Eddy Current Analysis for Nuclear Power Materials Adv. Mater. Res. 1085 (2015) pp. 335-339.

[11] Karolik A S, Luhvich A A 1994 J. Phys: Condens mater 6 873.

[12] Karolik A S 1988 Fizika Metal. I Metallovedenie65 463.

[13] Y. Shirai, H. Araki, T. Mori, W. Nakamura, K. Sakaki: J. Alloys Compd. 330 (2002), pp. 125-131.

[14] J. Cizek, I. Prochazka, S. Danis, M. Cieslar, G. Brauer, W. Anwand, R. Kirchheim, A. Pundt: Journal of Alloys and Compounds. 446-447 (2007), pp. 479-483.

[15] Y.S. Bordulev, R.S. Laptev, V.N. Kudiyarov, A.M. Lider: Advanced Materials Research. Vol. 880, 93 (2014).

[16] V. N. Kudiiarov, A. M. Lider, S.Y. Harchenko, Hydrogen Accumulation in Technically Pure Titanium Alloy at Saturation from Gas Atmosphere, Adv. Mater. Res. 880 (2014) 68-74.

[17] R.S. Laptev, Y.S. Bordulev, V.N. Kudiiarov, A.M. Lider, G.V. Garanin, Investigation of Defects Accumulation in the Process of Hydrogen Sorption and Desorption, 880 (2014) Adv. Mater. Res. 134-138.

[18] Laptev R. S. , Lider A. M. , Bordulev Y. S. , Kudiyarov V. N. , Garanin G. V. , Wang W., Kuznetsov P. V. Investigation of Defects in Hydrogen-Saturated Titanium by Means of Positron Annihilation Techniques // Defect and Diffusion Forum. - 2015 - Vol. 365. - p. 232-236.

[19] Y. Chen, X. Wan, F. Li et al.: Materials Science and Engineering: A. 466 (2007), pp. 156-159. [20] V.N. Kudiiarov, L.V. Gulidova, N.S. Pushilina, A.M. Lider: Advanced Materials Research.740 (2013), pp. 690-693.

[21] Kuznetsov P. V., Mironov Y. P. , Tolmachev A. I., Bordulev Y. S. , Laptev R. S., Lider A. M., Korznikov A. V. Positron Spectroscopy of Defects in Submicrocrystalline Nickel after LowTemperature Annealing // Physics of the Solid State. - 2015 - Vol. 57 - . 2. - p. 219-228 [22] D. [22] G. Malychin, V.V. Korneeva, The calculation of the X-ray diffraction of the field distortion of edge dislocations, Bulletin of the Kharkov University. Series Physical. 887 (2010) 115-119.

[23] A.S. Karolik, Electrical resistivity and thermopower in compact submicrostructure. Copper.Mater.Sci. (Materialovedenie) (in Russian).4(2011)15-21.

The study of titanium oxynitride coatings solubility deposited by reactive magnetron sputtering

L A Leonova, E L Boytsova, A A Pustovalova
Tomsk Polytechnic University, Institute of Physics and Technology,
Lenin Avenue 30, Tomsk, 634050, Russia

E-mail: leonovala@tpu.ru

Abstract. To improve hemocompatibility of cardiovascular stents the coatings based on titanium oxides and oxynitrides were used. In the present work the morphology, surface properties (wettability and surface energy), and in vitro solubility of the ternary system Ti-N-O coating were investigated. Experimentally, low dissolution rate of the coating in saline NaCl (0,9%) was confirmed. Instrumental methods of quantitative analysis (XRF, AES) revealed that the Ti-N-O coating is chemical-resistant and does not change the qualitative and quantitative composition of body fluids.

1. Introduction

Surface biocompatibility is a significant problem of medical implants used in cardiovascular surgery. About 10% of patients suffer the re-narrowing of blood vessel at the stent site. Hemocompatible coatings are formed on the implant surfaces to prevent the aversive response of human organism [1]. Titanium oxides and oxynitride coatings have attracted interest due to their excellent biocompatibility, high chemical stability, corrosion resistance, and nontoxicity [2]. Complex (Ti-N-O) films are promising coatings for coronary and vascular stents because of ability to prevent thrombus formation and fibrinogen adhesion to the implant surface [3].

It is well known fact [4] that this coating type protects the metal base of the implant from the corrosive effects of body fluids (blood, lymph, interstitial fluid).

The aim of this work was to investigate the solubility of titanium oxynitride coatings in NaCl saline (0.9%).

2. Experimental part

Titanium oxynitride coatings were deposited by reactive magnetron sputtering using midrange magnetron sputtering installation "UVN-200MI" (TPU, Tomsk). As the substrate for one-side coatings the NaCl crystal plate with dimensions of 10×10 mm and a thickness of 1 mm was used. Sputtering parameters [1]: the cathode material is Ti, the working chamber pressure is 0.1 Pa, power is 1 kW, current is 3 A, working gas leakage rate is 5 ml/min, distance between the substrate and the magnetron is 100 mm. The ratio of the partial gases pressure O_2 and N_2 in the magnetron chamber was equal unit. The substrate temperature during deposition was 120° C, the deposition time was one hour.

Coating morphology was obtained by scanning electron microscope (SEM) Quanta 400 FEG, FEI.

The experimental samples were dissolved in NaCl saline (0.9%, pH = 7) during 30 days. Condition of NaCl solution volume depends on the area of the specimen and were determined according to ISO

10993-12 recommendation [5]. The solvent acidity changed to pH=6 with the maximum exposure time.

Further, the samples were removed, and the solution was filtered through a "blue tape" filter and analyzed to detect elements of the coatings. The sample with the same coating not immersed in NaCl solution was served for the control. For these purposes, we used the X-ray fluorescence (XRF) analysis based on the X-rays interaction with the examined substance and the atomic-emission spectroscopy (AES). The AES is based on determining the elemental composition of substance using optical line spectra of atoms and ions irradiation of the sample excited by the light sources. The spectra obtained with XRF spectrometer Thermo Electron QUANT'X (EDXRF, USA) are shown on figures 1, 2.

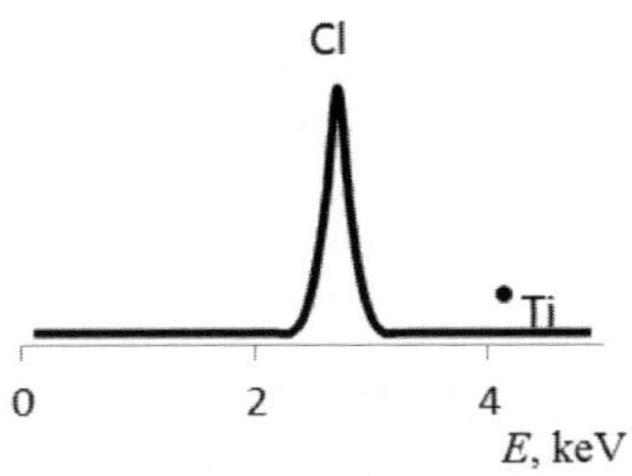

Figure 1. XRF spectrum of the solvent

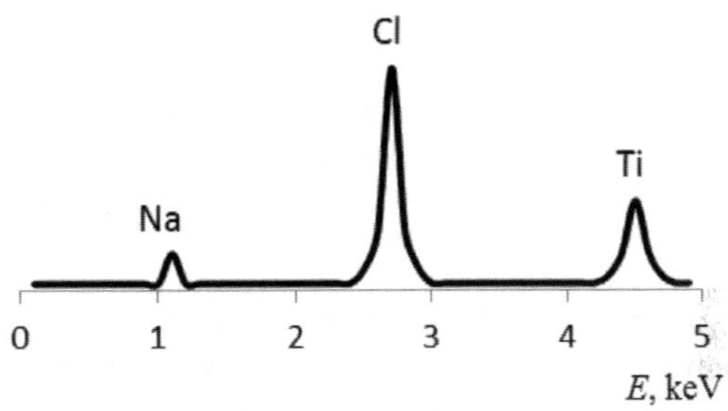

Figure 2. XRF spectrum of the Ti-N-O coating

AES analysis was carried out using an atomic emission spectrometer ICAP 6300 Duo (USA). As the standard reference solutions the state standard MES-1 and MES-2 samples were used. Obtained results are shown in table 1.

Table 1. The results of AES analysis of the samples after 30 days

No of measurement	The elements concentration in the test solution, mg/L	
	Na	**Ti**
1	119600	0.0005
2	120500	0.0006
average value	120000	0.0005

Surface properties such as wettabillity and surface free energy were studied for determination of Ti-N-O coating solubility.

The contact angles of the coating were measured by the OCA 15 (Plus DataPhysics Instruments GmbH) with sessile drop method, where liquid drop is set on a surface and the contact angle is measured from the drop shape [6]. The wetting angle from the interfacial surface tensions was calculated by Young's equation (figure 3). It describes the interactions between the cohesion and adhesion forces referred as surface energy [7].

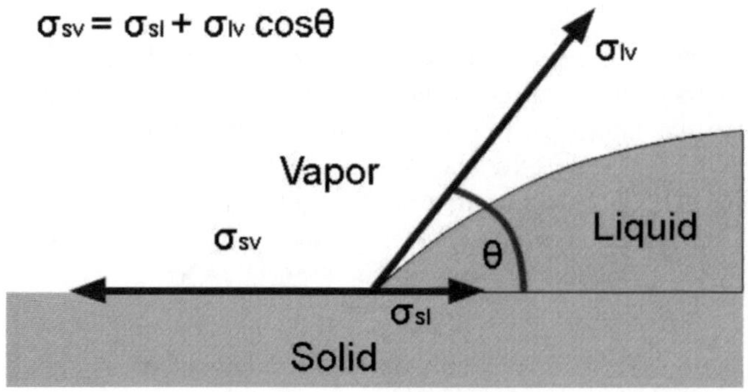

Figure 3. Schematic representation of the Young's equation (θ is the contact angle, σ_{lv} is the surface tension of the liquid, σ_{sl} is the solid/liquid interfacial tension, and σ_{sv} is the surface free energy of the solid, [mN/m = mJ/m^2])

The Owens, Wendt, Rabel and Kaelble method (the OWRK) was used for calculating the surface free energy using several liquids such as deionized water (θ_w), glycerol (θ_g), and diiodomethane (θ_d). The OWRK method calculates the surface free energy and its polar (σ^P) and dispersion (σ^D) parts using the equations 1-3 [8]:

$$\sigma_{sv} = \sigma_{sv}^D + \sigma_{sv}^P \quad (1)$$

$$\sigma_{lv} = \sigma_{lv}^D + \sigma_{lv}^P \quad (2)$$

$$\sigma_{sl} = \sigma_{sv} + \sigma_{lv} - 2\left(\sqrt{\sigma_{sv}^D \sigma_{lv}^D} + \sqrt{\sigma_{sv}^P \sigma_{lv}^P}\right) \quad (3)$$

The linear regression method is used to obtain the surface energy of solid, which is described by equation 4:

$$\frac{1 + cos\theta}{2} \frac{\sigma_{lv}}{\sigma_{lv}^D} = \sqrt{\sigma_{sv}^P} \sqrt{\frac{\sigma_{lv}^P}{\sigma_{lv}^D}} + \sqrt{\sigma_{sv}^D} \quad (4)$$

These interactions along the boundary of both phases are important for solving wetting and adhesion.

3. Results and discussion

3.1. Surface morphology of Ti-N-O coating

SEM images show that the coating has fine-grained surface morphology (figure 4*a*). Figure 2*b* demonstrates the homogeneous compact coating structure consisting of spherical elements (grains). Grains size is in the range of (6–75) nm [9].

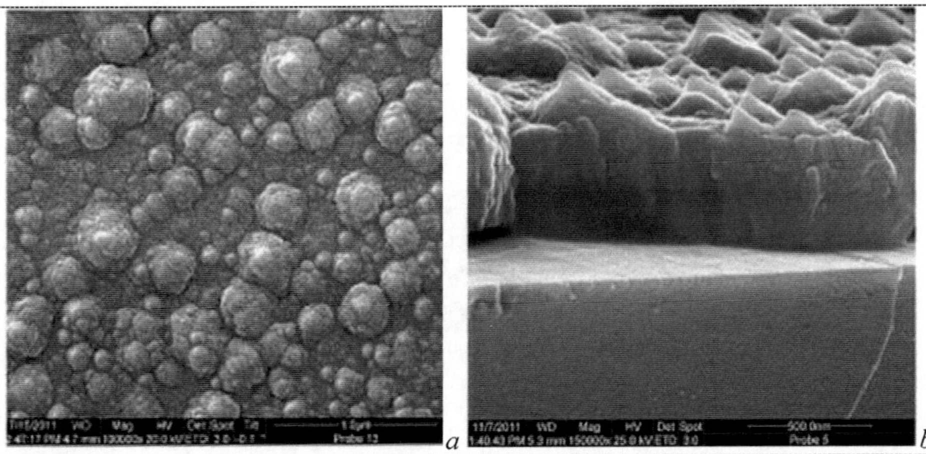

Figure 4. SEM images of Ti-N-O coating: (*a*) surface morphology and (*b*) cross-section [9]

3.2. Interpretation of the XRF spectra

Identification of elements from XRF spectra is based on the conformity of the spectral lines energy of element characteristic radiation and their relative intensities with tabulated values. The energy maximum of the signal was determined. The position the maximum is 2.8 keV that corresponds to Cl element: from NaCl solution (see in figure 1) and the crystal NaCl (see in figure 2). The maximum corresponding to Ti element with energy of 4.5 keV was not detected on the spectrogram (figure 1). While, on the XRF spectrum of the Ti-N-O coating the characteristic radiation titanium line is sufficiently intense (see on figure 2). The N and O elements cannot be detected using XRF spectrometer.

3.3. Interpretation of the AES spectrum

AES analysis allows determining not only qualitatively but also quantitatively elemental composition of the sample solution. It is obvious that with decreasing of an element concentration in a series of samples with the same composition, the spectral lines intensity in the corresponding spectra will decrease provided invariance excitation parameters and recording of spectra [10]. Since the spectral lines corresponding to an element have different intensities, the weaker lines will disappear in the spectrum faster than more intensive lines. As a result by the absence in the spectra of the certain lines and presence more intense of the element line we may estimate its concentration in the sample.

Using AES experimental data we identified the ions presence of Na and Ti elements in solution. AES results as well as XRF indicate expected high concentration of Na in the solution, and trace amounts of Ti confirm good chemical resistance of titanium oxynitride coating.

3.4. Measurements of wettability and surface energy

The wetting angle measurements (table 2) show that Ti-N-O coating is hydrophobic because the water contact angle is over 90 degree (θ_w), and the surface free energy is low [7]. Currently, most implant surfaces used in clinical application are hydrophobic, according to studies [6]. Hydrophobic surfaces are prone to entrap of air bubbles due to hydrocarbon contamination that can prevent the protein adsorption and cell receptor adhesion [6].

Table 2. Data of contact angle and surface energy measurements of Ti-N-O coating

Coating	θ_w (deg.)	θ_g (deg.)	θ_d (deg.)	σ (mN/m)
Ti-O-N	116±0.2	108±0.4	70±0.2	20±0.8

Conclusions

A low degree of dissolution of the titanium oxynitride coating in saline NaCl (0,9%) was confirmed experimentally. By means of modern spectroscopic methods of analysis (XRF, AES) it is proved corrosion-resistant properties of the Ti-N-O coating and does not change the qualitative and quantitative composition of the saline. The results obtained using two methods of analysis are in good agreement. Low solubility and hydrophobicity of Ti-N-O coating deposited on the implant surface can prevent the metallic surface corrosion and also stop the foreign ions yield from steel stents surface.

Acknowledgement

The authors wish to thank Prof. V. Pichugin, and their colleague M. Konishchev for the help in preparing the samples. SEM analysis and contact angle measurements were performed in framework of DAAD program at University of Duisburg-Essen, Germany. Authors express their gratitude to Prof. M. Epple and Prof. M. Ulbricht.

References

[1] Pichugin V F, Pustovalova A A, Konishchev M E, Khlusov I A, Ivanova N M, Sun Zh, Gutor S S 2016 *J. Surf. Investig. X-ray, Synchrotron Neutron Tech.* **10** 282–91

[2] Yin Z F, Wu L, Yang H G and Su Y H 2013 *Phys. Chem. Chem. Phys.* **15** 4844–58

[3] Rizzo A, Signore M A, Tapfer L, Piscopiello E, Cappello A, Bemporad E and Sebastiani M 2009 *J. Phys. D. Appl. Phys.* **42** 115406

[4] Huang N, Leng Y X, Yang P, Chen J Y, Sun H, Wang J, Wan G J, Zhao A S and Ding P D 2006 *Nuclear Instruments and Methods in Physics Research B* **242** 18

[5] ISO 10993-12:2012(en) 2012 *Biological evaluation of medical devices* 20

[6] Gittens R A, Scheideler L, Rupp F, Hyzy S L, Geis-Gerstorfer J, Schwartz Z and Boyan B D 2014 *Acta Biomater.* **10** 2907

[7] Guy O J and Walker K-A D 2016 *Silicon Carbide Biotechnology* 85

[8] Owens D K and Wendt R C 1969 *J. Appl. Polym. Sci.* **13** 1741–7

[9] Konischev M E, Kuzmin O S, Pustovalova A A, Morozova N S, Evdokimov K E 2014 *Russ. Phys. J.* **56** 1144–9

[10] Bekker Yu. Spectroscopy 2009 *M.: Technosphera* 528

IOP Conf. Series: Materials Science and Engineering **135** (2016) 012027 doi:10.1088/1757-899X/135/1/012027

The Study of ^{99m}Tc-Aluminum Oxide Using for Sentinel Lymph Nodes Detection in Experiment

A.A. Medvedeva [1, 2], I.G. Sinilkin [1, 2], R.V. Zelchan [1, 2], V.I. Chernov [1, 2], A.Yu. Lyapunov[1], O.D. Bragina[2], N.V. Varlamova[2], V.S. Skuridin[2], A.P. Dergilev[3]

[1] Tomsk Cancer Research Institute, Kooperativny Street 5, Tomsk, Russia
[2] Tomsk Polytechnic University, Lenin Avenue 30, Tomsk, Russia
[3] Novosibirsk State Medical University, Krasny Prospect 52, Novosibirsk, Russia

E-mail: Chernov@oncology.tomsk.ru

Abstarct: The purpose of the study was a comparative research of the possibility of using the radiopharmaceuticals ^{99m}Tc-Al_2O_3 and ^{99m}Tc-Nanocis for visualizing sentinel lymph nodes in the experiment. Measurement of the sizes of ^{99m}Tc-Al_2O_3 and ^{99m}Tc-Nanocis colloidal particles was performed in seven series of radiopharmaceuticals. Pharmacokinetics of ^{99m}Tc-Al_2O_3 and ^{99m}Tc-Nanocis was researched on 50 white male rats. The possibility of the use of ^{99m}Tc-Al_2O_3 and ^{99m}Tc-Nanocis for lymphoscintigraphy was studied in the experiments on 12 white male rats. Average dynamic diameter of the sol particle was 52-77 nm for ^{99m}Tc-Al_2O_3 and 16.7-24.5 nm for ^{99m}Tc-Nanocis. Radiopharmaceuticals accumulated in the inguinal lymph node – at 1 hour after administration the average uptake of ^{99m}Tc-Al_2O_3 was 8.6% in it and the accumulation of ^{99m}Tc-Nanocis was significantly lower - 1,8% ($p < 0,05$). In all study point average uptake of ^{99m}Tc-Al_2O_3 in the lymph node was significantly higher ^{99m}Tc-Nanocis accumulation. The results dynamic scintigraphic studies in rats showed that ^{99m}Tc-Al_2O_3 and ^{99m}Tc-Nanocis actively accumulated into the lymphatic system. By using ^{99m}Tc-Al_2O_3 inguinal lymph node was determined in 5 minutes after injection and clearly visualize in all the animals in the 15th minute, when the accumulation became more than 1% of the administered dose. Further observation indicated that the ^{99m}Tc-Al_2O_3 accumulation reaches a plateau in a lymph node (average 10.5%) for 2 hour study and then its accumulation remains practically at the same level, slightly increasing to 12% at 24 hours. In the case of ^{99m}Tc-Nanocis inguinal lymph node visualized in all animals for 15 minutes when it was accumulated on the average 1.03% of the administered dose. Plateau of ^{99m}Tc-Nanocis accumulation in the lymph node (average 2.05%) occurred at 2 hour of the study and remained almost on the same level (in average 2.3%) to 24 hours. Thus, the experimental study of a new domestic radiopharmaceutical showed that the ^{99m}Tc-Al_2O_3 several times actively accumulates in the lymph nodes as compared to imported analogue and its practical application will facilitate intraoperative identification of sentinel lymph nodes.

1. Introduction

In recent years radionuclide techniques have proven themselves to identify the sentinel lymph nodes (SLN) - first lymph nodes on the path of lymph drainage from a malignant tumor. These nodes, filtering afferent lymph, become a "trap" for cancer cells, and that is why SLN biopsy with followed by histological examination is an objective diagnostic criterion for the lymphatic metastasis detection. It is believed that if the SLN are not affected by metastatic disease, all other regional lymph nodes remain intact [1-3, 5-7].

The world practice has considerable experience in relation to radionuclide visualization of SLN at melanoma and breast cancer. At tumors in other locations (lungs, head, neck, gastrointestinal tract, uterus) the effectiveness of this method is studied in scientific research [1-3, 6].

Optimal radiopharmaceuticals to identify SLN is colloid labeled with technetium-99m [1,6]. In Russian Federation, to identify the SLN has been applied colloidal rhenium sulphide (^{99m}Tc-Nanocis, CIS bio International). The main disadvantage of this radiopharmaceutical is the low level of accumulation in the SLN (1.5-2% of the administered dose) and redistribution to lymph nodes 2 and 3 orders, which reduces the specificity of SLN visualization [1, 3, 6, 7].

Currently, in the Russian Federation there are no registered radiopharmaceuticals for SLN imaging. In this regard, the Tomsk Cancer Research Institute and Tomsk Polytechnic University developed by the original radiopharmaceutical based on labeled with technetium-99m aluminum oxide (^{99m}Tc-Al_2O_3) (the project № 16.N08.12.1011 «Preclinical studies of new lymphotropic radiopharmaceutical labeled by technetium-99m aluminum oxide" Federal Program "Development of the Russian Federation, the pharmaceutical and medical industry period up to 2020 and beyond ").

The aim of present investigation was a comparative study of the possibility of using the radiopharmaceuticals ^{99m}Tc-Al_2O_3 and ^{99m}Tc-Nanocis for visualizing sentinel lymph nodes in the experiment.

2. Material and Methods

Measurement of the sizes of ^{99m}Tc-Al_2O_3 and ^{99m}Tc-Nanocis colloidal particles was performed in seven series of radiopharmaceuticals on apparatus NICOMP 380 ZLS (company PSS NICOMP, USA).

Pharmacokinetics of ^{99m}Tc-Al_2O_3 and ^{99m}Tc-Nanocis was researched on white male rats "Wistar" weighing 200-250 g; radiopharmaceuticals was injected between the first and second fingers of the hind paw subcutaneously in a dose of 30 MBq. Before the injection of radiotracer animals were anesthetized with ethyl ether. The volume of the injected dose was 0.1 ml (volume activity is 300 MBq/ml). The animals were decapitated in groups of 10 individuals (5 rats for each radiopharmaceutical) in 1, 2, 3, 5 and 24 h after injection of radiopharmaceuticals.

The recovered organs and tissues were packed into vials for weighing and direct radioactivity measurement. Standard radiopharmaceutical and bioassays have the same volume (after weighing vials filled with distilled water up to the same level) and the geometric form. Radioassay of blood and internal organs (the contents of the radiopharmaceutical as a percentage of the entered radioactivity per 1 ml of blood or 1 g of tissue or organ) performed on the radiometer RIS A1 (Russian Federation). For radiometric studies differential discriminator was set up on the photon peak of 140 keV, with a window width of 20%. According to the results of radioassay also determined the level of the radiopharmaceutical level in inguinal lymph node and injection site.

The possibility of the use of ^{99m}Tc-Al_2O_3 and ^{99m}Tc-Nanocis for lymphoscintigraphy was studied in the experiments on 12 white male rats "Wistar" (6 rats for each radiopharmaceutical) weighing 300-350 g. Before a single subcutaneous injection of radiopharmaceuticals between the first and second fingers of the hind paw at a dose of 18-20 MBq and during scintigraphic studies animals were anesthetized using ethyl ether. All scintigraphic studies were performed on the gamma-camera E-CAM-180 (Siemens, USA) with a setting up of a differential discriminator on a photon peak of 140 keV, with a window width of 20%, using a parallel low-energy high-resolution collimator. During the study animals were put with their ventral surface to the detector of the gamma camera so that the whole body of the animal appeared in picture. Since the introduction, kinetics of the distribution of radiopharmaceuticals through the organs and tissues was fixed by step-frame recording within 15 min (1 frame per minute), in the matrix of 64 * 64 pixels. Static scintigraphy was performed in 1, 2, 3 and 24 hours in the front and rear projections in the matrix of 256 x 256 with a set of 500 pulses per position. According to the results of scintigraphic studies the percent of radiopharmaceuticals accumulation in the inguinal lymph node and the injection site from administered dose were determined.

Keeping and participation of the animals in the experiment were performed in accordance with the rules adopted by the "European Convention for the Protection of Vertebrate Animals used for Experimental and other Scientific Purposes" (Strasbourg, 1986).

The data received were processed by the method of variation statistics using "STATGRAF" numeric package. Between-group comparisons were done by using the Wilcoxon rank-sum test.

3. Results and Discussion

Measurement of the sizes of radioactive particles of $^{99m}Tc\text{-}Al_2O_3$ and ^{99m}Tc-Nanocis in 7 batches showed that the average dynamic diameter of the sol particle was 52-77 nm for $^{99m}Tc\text{-}Al_2O_3$ and 16.7-24.5 nm for ^{99m}Tc-Nanocis, which, according to the literature, is quite acceptable for scintigraphic visualization of lymph nodes [1, 3, 6].

Radiometry of rats organs showed that $^{99m}Tc\text{-}Al_2O_3$ and ^{99m}Tc-Nanocis actively withdrew subcutaneous injection spot - after 1 hour in the spot was about 2/3 of the administered dose (Table 1). After 24 hours at the injection site remained about half of the dose. Leaving the injection site, the radiopharmaceuticals accumulated in the inguinal lymph node – in 1 hour after administration the average accumulation of $^{99m}Tc\text{-}Al_2O_3$ was 8.6% in it and the accumulation of ^{99m}Tc-Nanocis was significantly lower - 1,8% ($p < 0{,}05$). In the second hour of study average accumulation of $^{99m}Tc\text{-}Al_2O_3$ in inguinal lymph node has exceeded 10% and in a day gradually increased to 12.8%. Accumulation of ^{99m}Tc-Nanocis after 2 hours reached 3.6% and slightly fluctuated at this level up to 24 hours of observation (Table 1). In all study point average accumulation of $^{99m}Tc\text{-}Al_2O_3$ in the lymph node was significantly higher ^{99m}Tc-Nanocis accumulation.

Received through the thoracic duct into the blood radiopharmaceuticals, very actively uptake in liver and spleen. The hepatic level of accumulation of $^{99m}Tc\text{-}Al_2O_3$ rise from 3.8% in 1 h after injection to 21.1% in 24 hours of the study, for ^{99m}Tc-Nanocis these values were 6.6% and 18.3%, respectively. In the spleen, the magnitude accumulation of $^{99m}Tc\text{-}Al_2O_3$ ranges from 1.5% (in 1 hour) to 4.5% (in 24 hours), the level of accumulation of ^{99m}Tc-Nanocis grew from 1.3% (in 1 hour) to 3.6% (in 24 hours). It was noted a slight content of radiopharmaceuticals in the heart, lungs and blood, which recorded less than 1% of the injected dose.

The results dynamic scintigraphic studies in rats showed that $^{99m}Tc\text{-}Al_2O_3$ and ^{99m}Tc-Nanocis actively accumulated into the lymphatic system from the injection site (Table. 2). After 2 hours at the injection site was 57% of the administered dose of $^{99m}Tc\text{-}Al_2O_3$ and 50.5% of ^{99m}Tc-Nanocis. After 24 hours the injection spot was kept less than half the dose of the radiopharmaceuticals.

Table 1. The content of ^{99m}Tc in the organs and tissues of rats at different times after subcutaneous injection of radiotracer $^{99m}Tc\text{-}Al_2O_3$ and ^{99m}Tc-Nanocis according to direct radiometry.

Time		**1h**	**2h**	**3h**	**5h**	**24h**
				$^{99m}Tc\text{-}Al_2O_3$		
Injection site	(%)	68.3±5.3	57.0±8.7	53.6±5.0	51.1±5.2	46.7±6.7
Lymph node	%	8.6± 1.1*	10.5± 1.6*	11.2± 1.6*	11.8± 1.9*	12.8± 1.4*
Blood	%/ml	0.91±0.01	0.26± 0.08	0.28± 0.08	0.30± 0.07	0.31±0.08+
Liver	%	0.33± 0.15	0.38± 0.11	0.45± 0.14	0.43± 0.12	0.35± 0.11
	%/g	0.13± 0.06	0.15± 0.05	0.18± 0.06	0.18± 0.05	0.14± 0.07
Stomach	%	3.8±1.2	9.6±1.1	12.1±2.4	16.8±1.8	21.1±1.6
	%/g	0.32±0.1	0.8±0.09	1.0±0.14	1.4±0.15	1.8±0.14
Spleen	%	1.5± 0.4	2.6± 0.6	4.1± 0.9	4.3± 0.8	4.5± 0.9

	%/g	0.74±0.2	1.4±0.29	2.1±0.43	2.2±0.4	2.2±0.42
Heart	%	0.41±0.11	0.29±0.12	0.67±1.08	0.6±0.07	0.9±0.1
	%/g	0.34±0.1	0.24±0.09	0.56±0.11	0.61±0.06	0.76±0.07
Lung	%	0.34±0.12	0.26±0.09	0.15±0.06	0.13±0.02	0.1±0.08
	%/g	0.12±0.05	0.09±0.04	0.06± 0.03	0.05± 0.04	0.04±0.3
99mTc-Nanocis						
Injection site	(%)	66.3±3.15	59.0±4.1	55.5±5.7	53.2±4.9	51.2±4.3
Lymph node	%	1.8± 1.2	3.6± 2.4	3.6± 1.8	3.6± 1.7	3.2± 1.0
Blood	%/ml	0.2±0.1	0.24± 0.04	0.27± 0.03	0.27± 0.07	0.3±0.1
Liver	%	1.2± 0.2	1.5± 0.2	1.5± 0.4	1.5± 0.7	2.2± 0.4
	%/g	0.48± 0.1	0.60± 0.15	0.62± 0.22	0.63± 0.31	0.88± 0.38
Stomach	%	6.6±0.3	10.5±0.9	13.9±2.07	16.3±1.9	18.3±1.4
	%/g	0.55±0.2	0.87±0.21	1.15±0.24	1.35±0.31	1.52±0.29
Spleen	%	1.3± 0.2	2.1± 0.3	2.8± 0.5	3.3± 0.7	3.6± 0.6
	%/g	0.65±0.1	1.1±0.15	1.4±0.14	1.65±0.2	1.80±0.2
Heart	%	0.3±0.1	0.3±0.06	0.31±0.09	0.3±0.11	0.4±0.15
	%/g	0.25±0.09	0.24±0.08	0.25±0.08	0.24±0.06	0.23±0.07
Lung	%	0.6±0.1	0.6±0.12	0.61±0.15	0.59±0.14	0.54±0.2
	%/g	0.22±0.09	0.22±0.1	0.23± 0.08	0.21± 0.06	0.20±0.1

* - p <0.05% compared to ^{99m}Tc-Nanocis

Table 2. The content of ^{99m}Tc at the injection site and inguinal lymph node after subcutaneous injection of radiotracer $^{99m}Tc\text{-}Al_2O_3$ and ^{99m}Tc-Nanocis according to dynamic scintigraphy.

Localization	Time after the injection	Average accumulation of $^{99m}Tc\text{-}Al_2O_3$ (%)	Average accumulation of ^{99m}Tc-Nanocis (%)
Injection site	5 minutes	76.8±3.7	82.7±5.0
	15 minutes	71.3±4.2	67.7±3.0
	1 h	68.2±5.3	55. 7±4.8
	2 h	57.0±8.6	50.5±5.1
	3 h	53.5±4.9	47.7±4.6
	24 h	46.7±6.8	35.33±3.3
Inguinal lymph node	5 minutes	0.9±0.2	0.98±0.22
	15 minutes	1.19±0.13	1.03±0.22
	1 h	8.6±1.1	1.4±0.86
	2 h	10.5±1.6	2.05±1.67
	3 h	11.1±1.2	2.18±1.57
	24 h	12.0±1.4	2.31±1.45

* - p <0.05% compared to ^{99m}Tc-Nanocis

By using $^{99m}Tc\text{-}Al_2O_3$ inguinal lymph node was determined in 5 minutes after injection and clearly visualize in all the animals in the 15th minute, when the accumulation became more than 1% of the administered dose. Further observation indicated that the $^{99m}Tc\text{-}Al_2O_3$ accumulation reaches a plateau in a lymph node (average 10.5%) for 2 hour study (Fig. 1) and then its accumulation remains practically at the same level, slightly increasing to 12% in 24 hours. In the case of ^{99m}Tc-Nanocis inguinal lymph node visualized in all animals for 15 minutes when it was accumulated on the average 1.03% of the administered dose.

Plateau of ^{99m}Tc-Nanocis accumulation in the lymph node (average 2.05%) occurred in 2 hour of the study (Fig. 1) and remained almost on the same level (in average 2.3%) to 24 hours.

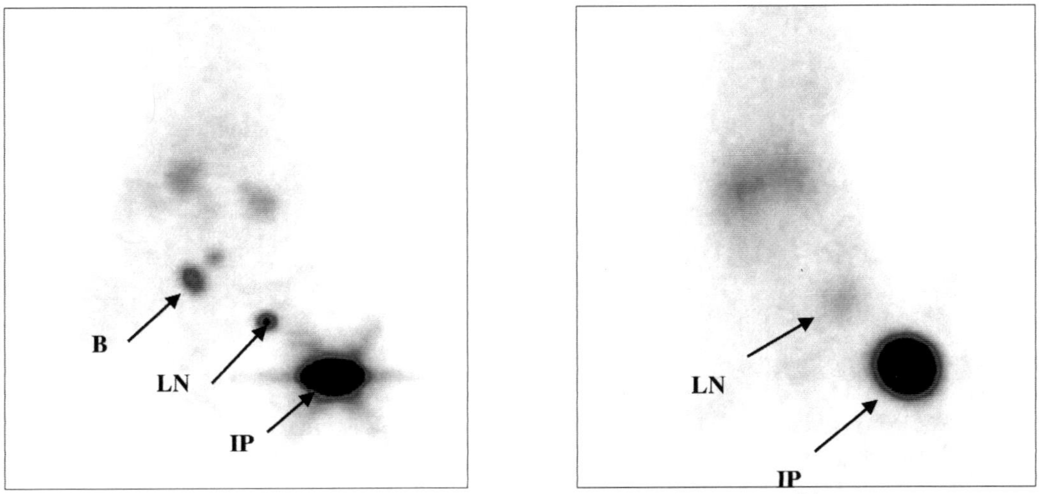

(a) (b)

Figure 1. Scintigrams rats performed through 2 hours after injection of $^{99m}Tc\text{-}Al_2O_3$ (a) and ^{99m}Tc-Nanocis (b). A - The level of radioactivity in the injection point (IP) – 56.3% of the injected dose, in the inguinal lymph node (LN) – 11.2%. B - bladder. B - The level of radioactivity in the injection point (IP) – 61.3% of the injected dose, in the inguinal lymph node (LN) – 3.5%.

In oncology practice spot of radiopharmaceutical injections are often located in close proximity to the sentinel lymph node, making it difficult to visualize [1, 6].

We have established a multiple higher accumulation of $^{99m}Tc\text{-}Al_2O_3$ in SLM in comparison with import analogue. Considering this fact the clinical use of new domestic radiopharmaceutical will facilitate intraoperative identification of such nodes.

4. Conclusions

Thus, the experimental study of a new domestic radiopharmaceutical showed that the $^{99m}Tc\text{-}Al_2O_3$ several times actively accumulates in the lymph nodes as compared to imported analogue and its practical application will facilitate intraoperative identification of SLN.

References

[1] Chernov VI, Sinilkin IG, Choynzonov EC, Chijevskaya SY, Titskaya AA, Zelchan RV, Bragina OD, Lyapunov AY, Skuridin VS 2015 *Eur J Nucl Med Mol Imaging* **42** (Suppl 1) 704

[2] Chernov VI, Sinilkin IG, Shiryaev SV. 2010 *National leadership on nuclear medicine* Ed. Lishmanov YB, Chernov VI Tomsk STT B **2** 336

[3] Afanasyev SG, Avgustinovich AV, Chernov VI, Sinilkin IG 2009 Siberian Journal of Oncology 34 27

[4] Jimenez IR, Roca M, Vega E, García ML, Benitez A, Bajén M et al. 2008 *Nucl Med Commun* **29** 166

[5] Paredes P, Vidal-Sicart S, Zanón G et al. 2005 *E.J.Nucl. Med.* **32** 1283

[6] Schauer AJ et al. 2005 The Sentinel Lymph Node Concept Springer Berlin Heidelberg New York P 565

[7] Sinilkin I, Chernov V, Zelchan R, Titskaya A, Skuridin V 2014 *Eur J Nucl Med Mol Imaging* **41** (Suppl 2) 518

IOP Conf. Series: Materials Science and Engineering **135** (2016) 012028 doi:10.1088/1757-899X/135/1/012028

Influence of temperature gradient on diffracted X-ray spectrum in quartz crystal

A R Mkrtchyan[1,2], A P Potylitsyn[2], A V Vukolov[2], A I Novokshonov[2], A S Gogolev[2], R V Amiragyan[1], A E Movsisyan[1]

[1]Institute of Applied Problems of Physics, Yerevan, Armenia

[2]Tomsk Polytechnic University, Tomsk, Russia

E-mail: novokshonov@tpu.ru

Abstract. In this work characteristics of hard X-ray (with energy higher than 30 keV) were investigated. In the experiment we measured spectra of X-ray reflected by a quartz mono-crystal in Laue geometry under influence of the temperature gradient. The measurements were made by the spectrometer BDER-KI-11K with 300 eV resolution on the 17.74 keV spectral line of Am241 and the spectrometer XR-100CR with 270 eV resolution on the same spectral line. An existence of temperature gradient leads to increasing of the diffracted beam intensity. The intensity was measured dependently on the temperature of one of the edge of the crystal.

1. Introduction

Designing of new methods of monochromatic hard X-ray beams producing with tuned parameters such as angular distribution and monochromaticity is stimulated by requirements of new, more sensitive and universal methods for providing of investigations in different fields of science and engineering. It is well known that an existence of acoustic field or temperature gradient in crystal monochromators allows to control parameters of diffracted X-ray in space and time [1, 2]. In the works [3, 4] the first time it was observed a phenomenon of full X-ray transfer from the direction of propagation to the reflection direction in mono-crystal of quartz under the influence of temperature gradient or ultrasonic waves. In the works [5] a sharp decreasing of X-ray absorbing coefficient under the temperature gradient perpendicular to atomic reflective planes was observed in Laue geometry. In the work [6] a high-resolution diffractometer based on the proposed techniques was made. In the work [7] it was experimentally shown that under the influence of high-temperature gradient an angular width of full X-ray transfer increases proportionally to a crystal width. For designing of X-ray optic elements with controlled parameters which are based on proposed techniques it is very important to know what deformation exactly is in the crystal influenced by temperature gradient or acoustic waves (see, for instance, works [8, 9]). The results of investigation of reflected X-ray intensity depending on the temperature gradient value are presented in our work.

2. Experimental setup

As investigated samples there were used rectangular plates of quartz mono-crystal with 6, 8, 9 mm thickness. One of the crystal edges was heated up. This edge of the plates was parallel to atomic reflective planes of the crystal $(10\bar{1}1)$, i.e a temperature gradient was applied

IOP Conf. Series: Materials Science and Engineering **135** (2016) 012028 doi:10.1088/1757-899X/135/1/012028

perpendicularly to the reflective planes. The temperature gradient vector and the diffraction vector were anti-parallel to each other. In the experiment a continuous X-ray spectrum was used. This spectrum was generated by the X-ray tube Mo BSV-29 with 45 kV voltage and 10 mA current.

A dependence of reflected X-ray intensity on a temperature gradient was experimentally investigated. The dependence was investigated with 25, 30, 35 keV energies of the X-ray. The schematic of the experiment is shown in figure 1. In the figure $L1 = L2 = 25$ cm, distance between the last slit and the crystal is 8 cm, angles of observation are $\theta = 6° \div 8°$. Angular divergence of the beam was less than 2 mrad.

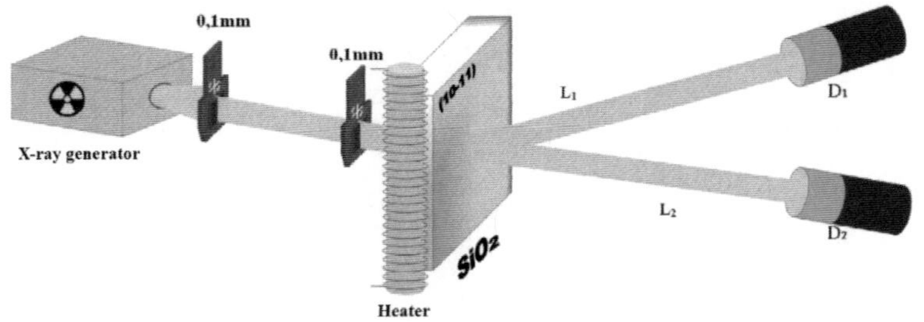

Figure 1. The schematic of the experiment.

The temperature of the crystal was measured in four points, these points are shown in figure 2 ($t1 = 0$ mm, $t2 = 4$ mm, $t3 = 15$ mm, $t4 = 30$ mm). The crystal was heated up from the side of $t1$ point. The beam was passing through the crystal between the points $t2$ and $t3$. One of the temperature distributions is shown in figure 3.

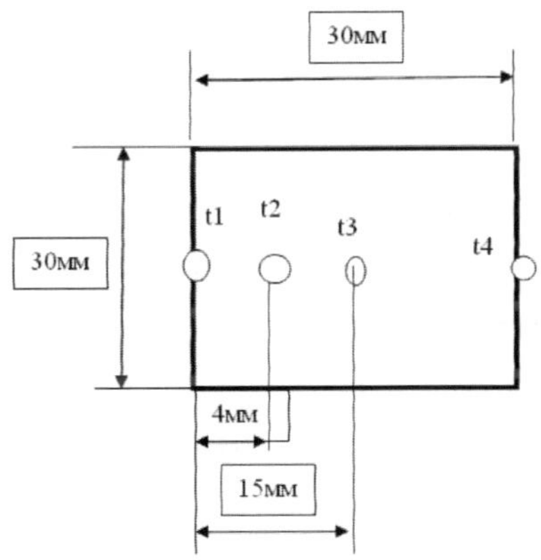

Figure 2. The schematic of the temperature measuring points location.

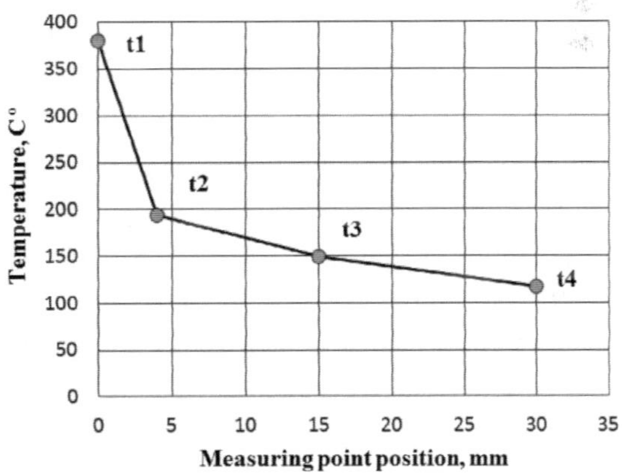

Figure 3. The temperature distribution.

In figure 4 a dependence of temperature in the points on the heater current for 6 mm thickness quartz crystal is shown. Maximal temperature in the point $t1$ was 380°, further heating leads to destroying of the crystal.

IOP Conf. Series: Materials Science and Engineering **135** (2016) 012028 doi:10.1088/1757-899X/135/1/012028

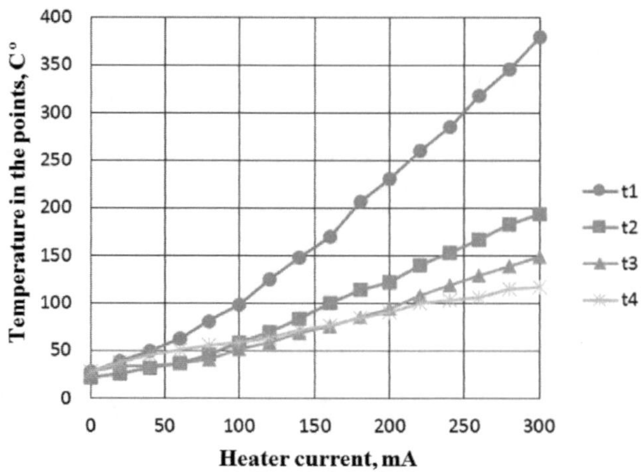

Figure 4. Dependence of temperature in the points on the heater current.

3. Results

In figure 5 there are shown spectra of diffracted in 9 mm thickness crystal X-ray for different temperature gradients. As one can see from the figure that there take place a multiple increasing of the diffracted X-ray intensity, a widening of the spectrum and a misplacing of maximum in low energy region within the increasing of temperature gradient. The widening of spectrum is caused by a bending of the crystal atomic reflective planes and the misplacing of maximum is a result of increasing of distances between the reflective planes.

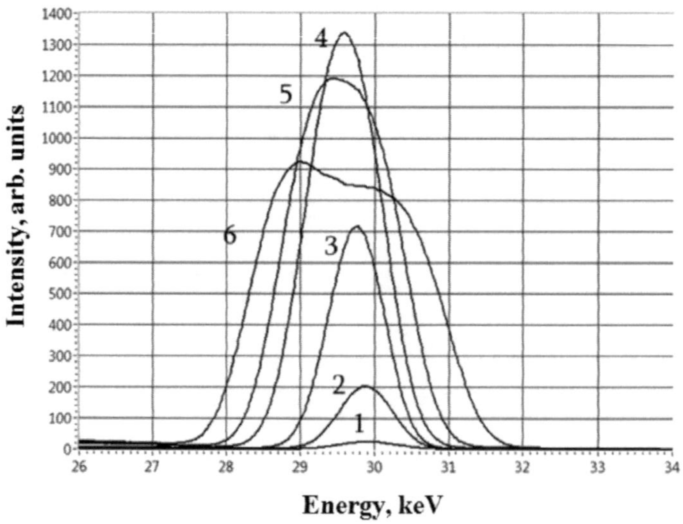

Figure 5. The spectra of diffracted in 9 mm thickness crystal X-ray for different temperature gradient:
1) $\Delta T/\Delta x = 0\ C^{\circ}/cm$; 2) $\Delta T/\Delta x = 10\ C^{\circ}/cm$;
3) $\Delta T/\Delta x = 100\ C^{\circ}/cm$; 4) $\Delta T/\Delta x = 150\ C^{\circ}/cm$;
5) $\Delta T/\Delta x = 200\ C^{\circ}/cm$; 6) $\Delta T/\Delta x = 250\ C^{\circ}/cm$.

In figures 6, 7 the same as in figure 5 spectra are shown but for the 8 mm and 6 mm thickness crystals respectively. As one can see from figures 5 - 7 the beam spectral line is widening within increasing of the crystal thickness.

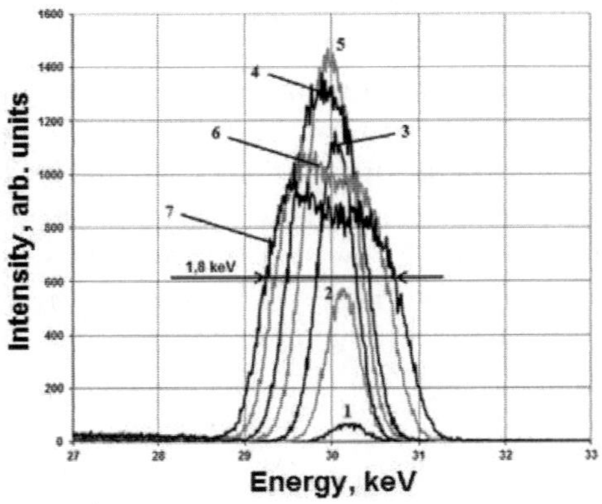

Figure 6. 8 mm crystal:
1) $\Delta T/\Delta x = 0\ C^{\circ}/cm$;
2) $\Delta T/\Delta x = 10\ C^{\circ}/cm$;
3) $\Delta T/\Delta x = 100\ C^{\circ}/cm$;
4) $\Delta T/\Delta x = 150\ C^{\circ}/cm$;
5) $\Delta T/\Delta x = 200\ C^{\circ}/cm$;
6) $\Delta T/\Delta x = 250\ C^{\circ}/cm$;
7) $\Delta T/\Delta x = 300\ C^{\circ}/cm$.

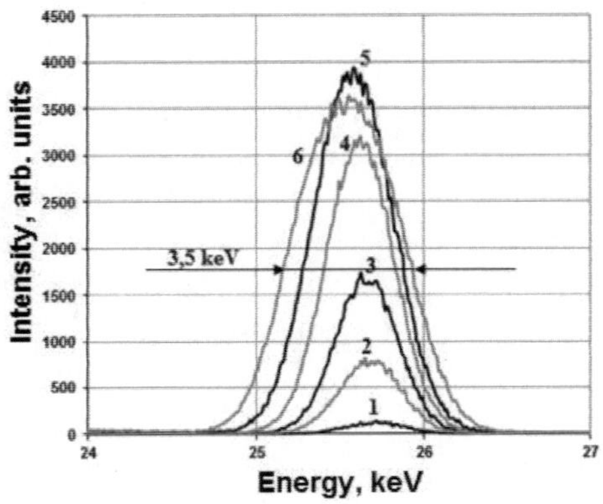

Figure 7. 6 mm crystal:
1) $\Delta T/\Delta x = 0\ C^{\circ}/cm$;
2) $\Delta T/\Delta x = 10\ C^{\circ}/cm$;
3) $\Delta T/\Delta x = 100\ C^{\circ}/cm$;
4) $\Delta T/\Delta x = 150\ C^{\circ}/cm$;
5) $\Delta T/\Delta x = 200\ C^{\circ}/cm$;
6) $\Delta T/\Delta x = 250\ C^{\circ}/cm$.

There are shown the dependencies of diffracted X-ray intensity on the value of temperature gradient for energies 30 keV and 40 keV in figure 8. The Left figure is for 6 mm thickness crystal and the right one is for 9 mm thickness crystal.

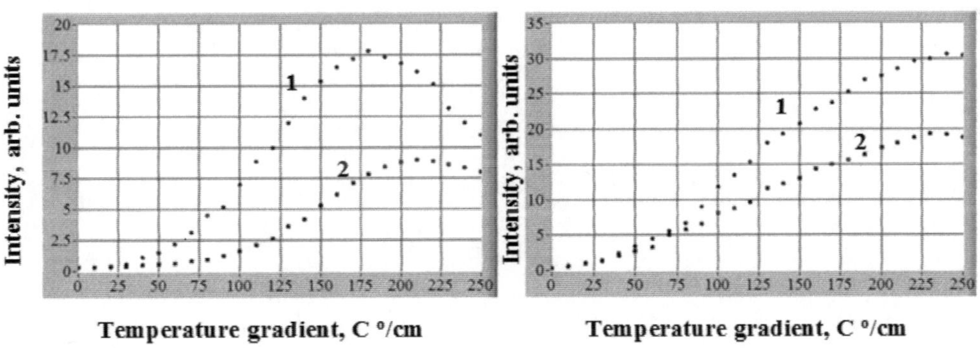

Figure 8. The dependencies of diffracted X-ray radiation intensity on the value of temperature gradient. 1 - the dependence for 30 keV energy; 2 - the dependence for 40 keV energy. Left figure - 6 mm thickness crystal; Right figure - 9 mm thickness figure.

One can see from the left figure that it takes place increasing of intensity more than in 60 times for 30 keV energy and more than in 45 times for 40 keV energy. For 9 mm thickness of the crystal (right figure) the intensity increases more than in 100 times for 30 keV energy and about 80 times for 40 keV energy.

The multiple increasing of X-ray radiation intensity is caused by the phenomenon of full X-ray radiation transfer from the propagation direction to the reflection direction.

The saturation and the decreasing of reflected X-ray intensity within the increasing of temperature gradient is a result of that within large deformations of the crystal a length of extinction becomes many times larger than effective region of diffraction for every radiation mode which takes part in the diffraction. It means that since the length of extinction for 40 keV energy photons is larger than for 30 keV photons so in the last case the saturation takes part within smaller temperature gradient.

Thus, it is experimentally demonstrated that the intensity of diffracted X-ray can increase on the order of 2 when there is the temperature gradient in the crystal. Also it is shown that within the temperature gradient increasing the intensity of reflected beam increases and the spectral line width within the maximal intensity are determined by the crystal thickness. Further increasing of temperature gradient leads to increasing of the spectral line width and decreasing of the intensity.

References

[1] A R Mkrtchyan, M A Navasardyan, R G Gabrielyan, L A Kocharyan, R N Kuzmin, *Solid State Communication*, **59**, 147-149, 1986.

[2] Kocharyan V, Mkrtchyan A, Gogolev A, Khlopuzyan S, Grigorian P, *Advanced Material Research*, **1084**, 107-110, 2015.

[3] Mkrtchyan A R, Navaserdyan M A, Mirzoyan V R, *Letters to JTPh*, **8**, 677, 1982.

[4] Mkrtchyan A R, Navasardyan M A, Gabrielyan R G, *Letters to JTPh*, **11**, 1181, 1983.

[5] Kocharyan V R, Aleksanyan R Sh, Truni K G, *Journal of Contemporary Physics*, **4**, 190-194, 2010.

[6] Mkrtchyan A R, Mkrtchyan A H, Kocharyan V R, Movsisyan A E, Dabagov S B, Potylitsyn A P, *Journal of Contemporary Physics*, **3**, 141-143, 2013.

[7] Noreyan S N, Mirzoyan V K, Kocharyan V R, *News of NAN Armenia. Physics*, 124-130, 2004.

[8] Noreyan S N, Mirzoyan V K, Kocharyan V R, *Surface. X-ray, Synchrotron and Neutron Investigations*, **1**, 18-21, 2004.

[9] Kocharyan V R, Gogolev A S, Movsisyan A E, Beybutyan A H, Khlopuzyan S G, Aloyan L R, *Journal of Applied Crystallography*, **48**, 853-856, 2015.

IOP Conf. Series: Materials Science and Engineering **135** (2016) 012029 doi:10.1088/1757-899X/135/1/012029

LAWS OF THE OXIDATION OF CARBON ISOTOPES IN PLASMA PROCESSES UNDER MAGNETIC FIELD

V F Myshkin[1], E V Bespala[1], V A Khan[1,2], S V Makarevich[1]
[1]Tomsk polytechnic university, Tomsk, Russia
[2]V.E. Zuev Institute of Atmospheric Optics Siberian Branch of the Russian Academy of Science, Tomsk, Russia

E-mail: bespala_evgeny@mail.ru

Abstract. From law of quantum mechanics it follows that spin precession phase of unpaired electron in external magnetic field cannot be determined. It uncertainty necessary take into account in different physical and chemical processes. The expression of the rate constant of a chemical reaction based on the number of discrete spin states was obtained. The equations of chemical kinetics of plasma oxidation of carbon isotopes in the magnetic field were given.

1. Introduction

Most of the chemical element represented by two or more isotopes with different nuclear spin. Known magnetic isotope effect (MIE) - separation of isotopes with spin and spinless nuclei in chemical reactions between radicals in the liquid occurring in an external magnetic field [1]. Studies concerning the effect of the spin nuclei isotope ions изотопов ^{25}Mg, ^{57}Fe, ^{67}Zn on the rate of biochemical processes, are also developing. For example, by reacting Mg+ ion and ATP molecules may induce radical pairs and changes the rate of chemical reactions involving isotopes with paramagnetic nuclei [2].

Without a magnetic field, the spatial orientation of the spin is not determined and all the spin orientations are equally probable. Valence electron spin precession relative to the direction of the magnetic field is observed in the magnetic field. This reduces the number of possible combinations of mutual spatial orientation of the colliding radicals. Also, the rate of change of physical and chemical processes is observed [3]. On the conditions of formation of a chemical bond between the two radicals influenced by two factors: the energy and spin [1]. The influence of magnetic field on the gas-phase chemical reactions received little attention.

The purpose of the study is the analysis of paramagnetic effects in low-temperature plasma, placed in an external magnetic field.

2. Paramagnetic phenomena in the gas system

When the diffusion of radicals into the liquid in a magnetic field and it is possible to capture radicals in "cage." At the same time there are 4 states: one singlet and three triplet. Spin pair multiplet can be configured in 3 ways: oppositely directed, co-current up or down. Possible combinations of spins are shown in Figure 1. Due to the cage effect in the liquid main factor controlling the reactivity of radicals in the external magnetic field is a dynamic spin pairs.

Approaches used for liquid-phase reactions, but for gas-phase reactions are not suitable. Currently, the ability to influence paramagnetic phenomena on the gas-phase chemical reactions, because of the much greater value of the conversion time of spin pairs compared to the time of initial contact radicals are not considered.

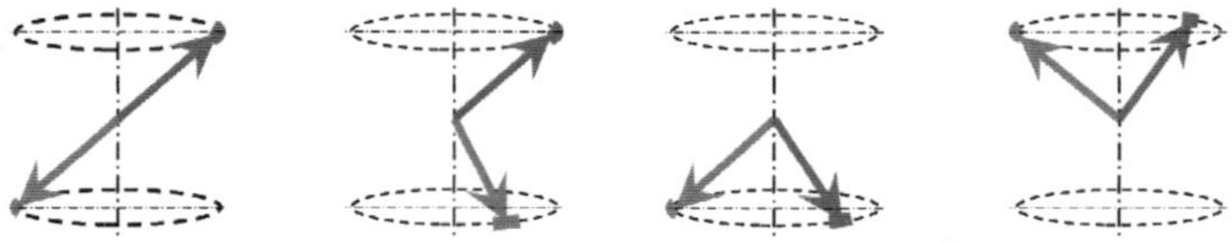

Figure 1. Scheme of forming different states of multiplet

It is considered that due to the statistical independence of the spatial orientation of the spins of the colliding radicals diffusion couples (uncorrelated) the probability of different states of a spin multiplet is identical and equal to 25% [4]. This is due to the indistinguishability of various phase states of the precession of the spin with the direction of +1/2 or -1/2 with respect to the external magnetic field, such as in the experiments of Stern and Gerlach. The interaction of the back of the multiplet considered in parallel planes or in one plane.

In quantum mechanics, the electron state, for example, spin "up" +1/2 denoted as state vector $|+1/2\rangle$, which involves degeneration of the condition. The states of the multiplet can be formed spins of electrons in different planes (having different phases of the valence electron spin precession) (see Figure 1). At the same time singlet state can only occur in the case of coincidence of the two planes of the spins and the triplet state - for any combination of these planes spins. It is obvious that in gas phase the probability of formation of various states of the multiplet depends on difference realizations of angle dispositions with spin precession of forming spin pairs ($|+1/2\rangle$, $|+1/2\rangle$).

The uncertainty relation for S_x and S_y projection of the spin operator can be written in the form of expression $\Delta S_x\, \Delta S_y \geq \hbar/2\, |\langle S_z\rangle|$. Under the condition $\langle S_z\rangle = \hbar/2$ with dispersion $\Delta S_z = 0$ we obtain $\Delta S_x\, \Delta S_y \geq (\hbar/2)^2$. Under the same conditions, the dispersion of the spin operator of projection on the axis OX and OY ($\Delta S_y \approx \Delta S_x$) we get $\Delta S_x \geq \hbar/2$.

Estimation algorithm $\Delta\varphi$ values of trigonometric ratios demonstrated by Figure 2.

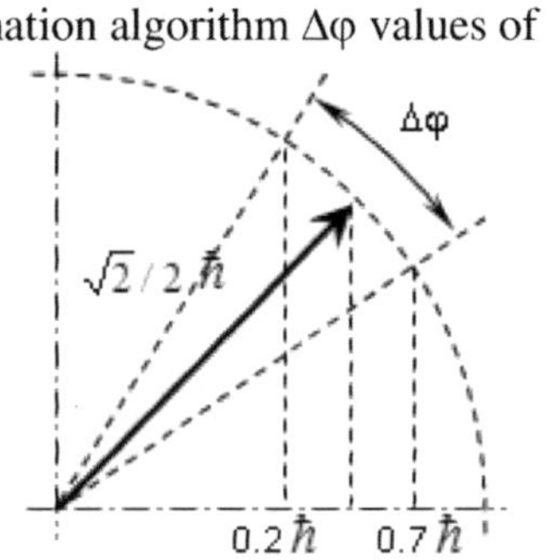

Figure 2. Scheme of estimation uncertainty phase of the spin precession.

Estimation algorithm $\Delta\varphi$ values of trigonometric ratios demonstrated by Figure 2. The value $\Delta S_x \geq \hbar/2$ at different limit values can be implemented as: $\Delta S_x = (0.7\text{-}0.2)\hbar$ and $\Delta S_x = [0.25\text{-}(-0.25)]\hbar$. Accordingly, the uncertainty (discrete) $\Delta\varphi$ the angle of spin precession in the range 73.4°-41.4°. Can be distinguished $n = 2\pi/\Delta\varphi$ ranges phase at the valence electron spin precession radical realized in various physical and chemical processes as "discrete" states. In the range $\Delta\varphi \approx 41.4°\div73.4°$ mean value is 57°; and the mean value of n is $n \approx 6.3$.

Consideration uncertainty of phases of spin precession, for one of the radicals can be detected n singlet states S at its collision with the second radical with axisymmetric spin direction. At the same time state S can be considered as a special case of the state of T_0. Triplet states can be implemented as follows: $T_0 - n(n\text{-}1)$; but T_{-1} and T_{+1} states as combinations of n «up» with n states «down» – n^2 spin position of colliding radicals. The probability of formation of the singlet state at the next collision of all possible sets of interacting spins is:

$$\delta S = \frac{[S]}{[S + T_0 + T_{-1} + T_1]} = \frac{n}{n + n(n-1) + 2n^2} = \frac{1}{3n}. \quad (1)$$

Therefore, the proportion of singlet states of all possible set of interactions of the spins of radicals is 1/3*n* or 5.3%, but not 25%.

3. The gas-phase chemical reaction between the radicals in the magnetic field

In the gas-phase processes each *3n* radicals collision with a 100% probability may be terminated form reaction products. The frequency of the formation of the singlet state in a collision radicals in the external permanent magnetic field, taking into account the frequency of collisions of particles per unit volume of gas can be written as (2):

$$\nu_S = \frac{1}{3n}\frac{\sqrt{2}}{2}\bar{\upsilon}\,\sigma\,[R1][R2] \quad (2)$$

where $R1$, $R2$ – concentrations of the two types of radicals, υ – the mean relative velocity of thermal motion, σ – effective cross section of collision of two types of radicals.

However, equation (2) does not consider the possibility of collapse resulting in an excited state of a new molecule. When taking into account short lifetime of the radiative relaxation of excited states, the equation (2) can be considered valid.

The equation of the second order chemical reaction is proportional to the product of the concentrations of the reactants involved:

$$\frac{d[R1]}{dt} = \nu_s = k\,[R1][R2]. \quad (3)$$

From a comparison of the expressions (2) and (3) shows that the rate constant of a chemical reaction can be written as expressions:

$$k = \frac{1}{3n}\frac{\sqrt{2}}{2}\bar{\upsilon}\,\sigma\ . \quad (4)$$

To estimate the rate constant of the oxidation of carbon in the low-temperature plasma will take the relative velocity at 1500°C $\bar{\upsilon} \approx 3\ 10^5$ cm/s and the effective collision cross section is $\sigma \approx 5\ 10^{-15}$ cm^2. As a result, we obtain the estimate $k \approx 16.8\ 10^{-11}$ cm^6/s. This value is close to the experimental value (5.22 10^{-11} cm^6/s) for the reaction of carbon oxidation at 1500 K at pressures of 0.5-1.2 atm [5]. The difference in the reaction constant may be related to the oxidation reaction of carbon by several ways, and we estimate inflated sectional collision.

In general, the isotope effect of radical processes in the gas phase is determined by the kinetic energy and the distribution of the excited levels for reactions with different thermal effect. The collision frequency determined by the arithmetic average rate determined from the Maxwell velocity distribution of particles of the thermal motion:

$$\Delta N = N\frac{4}{\sqrt{\pi}}\frac{\upsilon^2}{\upsilon_{prob}^2}\frac{\Delta\upsilon}{\upsilon_{prob}}\exp\left[-\frac{\upsilon^2}{\upsilon_{prob}^2}\right], \quad (5)$$

where υ – particle velocity, k – Boltzmann constant, T – temperature.

If you exceed the reaction threshold of faster particles collide more often. Therefore, the formation of chemical bonds have a higher probability. Without a magnetic field, both isotopes have the same and constant in time the probability of the formation of singlet pairs. The molecules formed from the fast particles have a greater excess of energy, and consequently have a higher probability of spontaneous collapse.

Expression (2) can be written for each pair of components of the Maxwell velocity distribution.

For the analysis of rates of chemical reactions with the participation of different isotopes, having different velocity distribution of thermal motion can be use the expression for the formation frequency of singlet states, compiled for a discrete analogue of the Maxwell distribution of velocities:

$$\nu_s = \frac{\sqrt{2}}{6n}\sum_{i=1}^{n}\left(\sum_{j=1}^{m}\sigma_{12}^{j}\upsilon_{12}^{j}\,N_1^{ij}\,N_2^{ij}\right), \tag{6}$$

where m – the number of discretization of the Maxwell velocity distribution, n – the number of discretization phase of the precession of the spins of valence electrons is the same for both radicals, υ_{12} – the relative velocity of the colliding radicals, N_1^{ij}, N_2^{ij} – radical concentration with the given values phase precession of the electron spin and the velocity of the radical, σ_{12}^{j} – cross section of collisions of particles with a relative velocity υ_{12}^{j}.

4. The kinetics of plasma oxidation of carbon isotopes under the magnetic field

Under the conditions of the plasma for describing of chemical kinetics used Pauli equation that takes into account the distribution of the atoms of the reactants electronically excited states. [6]. Given the high content of atoms in excited states is necessary to take into account the contribution of the chemical reaction of the excited states, differing in the value of the g-factor. In a magnetic field, the spin precession frequency, except the g-factor, also determines the spin of the nucleus.

In [7-9] it showed enrichment ^{13}CO in products of incomplete oxidation of carbon plasma in a magnetic field. Oxidation ^{12}C and ^{13}C in a low-temperature plasma, using equation (6) can be described by the following equations.

1. The oxidation of the carbon isotope ^{12}C (^{13}C) by molecular (or atomic) oxygen:

$$\frac{d[^{12}CO]}{dt} = \frac{\sqrt{2}}{6n}\sum_{i=1}^{n}\left(\sum_{j=1}^{m}\left(\sigma_{12C-O2}^{j}\upsilon_{12C-O2}^{j}\,N_{O2}^{ij}(t)\,N_{12C}^{ij}(t)\right)\right), \tag{7}$$

2. Equation for computation of decline rate of molecular (atomic) oxygen in the oxidation of carbon isotopes ^{12}C, ^{13}C:

$$\frac{d[O_2]}{dt} = -\frac{\sqrt{2}}{6n}\sum_{i=1}^{n}\left(\sum_{j=1}^{m}\left(\sigma_{13C-O2}^{j}\upsilon_{13-O2}^{j}\,N_{O2}^{ij}(t)\,N_{13C}^{ij}(t)\right)\right) - \frac{\sqrt{2}}{6n}\sum_{i=1}^{n}\left(\sum_{j=1}^{m}\left(\sigma_{12C-O2}^{j}\upsilon_{12C-O2}^{j}\,N_{O2}^{ij}(t)\,N_{12C}^{ij}(t)\right)\right), \tag{8}$$

where N_{O2}, N_{13C}, N_{12C} – the concentration obtained by discretization of the Maxwell distribution of m velocities of and n discrete states precession phases for O_2, ^{13}C, ^{12}C.

3. On isotopic selectivity of CO oxidation by molecular (atomic) oxygen, the magnetic field has no effect (CO molecule has no unpaired electrons):

$$\frac{d[CO_2]}{dt} = \frac{\sqrt{2}}{6n}\sum_{i=1}^{n}\left(\sum_{j=1}^{m}\left(\sigma_{CO-O2}^{j}\upsilon_{CO-O2}^{j}\,N_{O2}^{ij}(t)\,N_{CO}^{ij}(t)\right)\right). \tag{9}$$

Uniform phases of spin precession state is maintained collisions which do not result in the formation of chemical bonds. The spin-spin relaxation occurs when collisions with atoms of plasma-forming gas.

Isotopes of same elements differ from g and a. Therefore for each isotopes it is typical the time of occurrence of valence electron of radical in one of range of uncertainty $\Delta\varphi$ in the angles of the precession:

$$\Delta t = \frac{h}{n\,g\mu_B(H+a)}. \tag{10}$$

Depending of collision frequency, which depend on pressure and temperature, radicals of two reactants collide step-by-step in different precession phases of their valency electron spin. Besides probability of formation of singlet stage as a result of each collision of two radicals change according to harmonic law [10]. Oscillation frequency of chemical activity is difference between collision frequency of radicals and frequency of paramagnetic resonance of two radicals. Two reactants undergo a chemical reactiony every *3n* collision with probability 1.0. Under certain conditions for two isotopes possible isotopes effects of carbon oxidation in external field. For example, equality collision frequency of one isotopes of carbon with oxygen with difference frequency of paramagnetic resonance of it isotope and oxygen.

5. Conclusion

Spin precession in a magnetic field should be considered for discrete angular coordinate during consideration different physical and chemical processes. Therefore constant of chemical reaction between radicals in external permanent magnetic field can calculate from collision frequency, which depend on gas kinetic parameters of radicals, temperature and pressure of gas. Besides theoretical assessment of constant of oxidation rate of each carbon isotopes and isotope effect of chemical reaction in external permanent magnetic field.

The reported study was funded by RFBR according to the research project No 16-08-00246.

6. Reference

[1] Sagdeev R Z, Salikhov K M, Molin Yu N 1977 *Russian Chemical* **46/4** pp 569–601
[2] Buchachenko A L 2014 *Russian Chemical* **83/1** pp 1–12
[3] Hans E., Lundager M 2007 *Journal of Crystal Growth* **305** pp 271–277
[4] Wallikewitz B H, Kabra D, Gelinas S and Friend R H 2012 Ph*ysical Review B* **85** pp 1–10
[5] Dean A J; Davidson D F; Hanson R K 1991 *J. Phys. Chem.* **95** pp 183–191
[6] Polak L S, Ovsyannikov A A, Slovetskiy D I, Vurzel F B 1975 *Moscow: Science* p 304
[7] Myshkin V F, Khan V A, Plekhanov V G, Izhoykin D A, Bespala E V 2015 *Russian Physics Journal* **57/10** pp 1442-1448
[8] Myshkin V F, Izhoykin D A, Ushakov I A, Shversov 2014 *Advanced Materials Research* **880** pp 128–133
[9] Myshkin V F, Plekhanov V G, Izhoykin D A, Khan V A 2014 *Advanced Materials Research* **1040** pp 360–366
[10] Ivankov Yu V, Ivanova O A, Ivankova E Yu, Levin M N Vestnik VGU **1** pp 35–38.

IOP Conf. Series: Materials Science and Engineering **135** (2016) 012030 doi:10.1088/1757-899X/135/1/012030

Description and development of the means of a model experiment for load balancing in distributed computing systems

A E Nagiyev[1], A I Sherstnyova[1], I A Botygin[1], N Y Galanova[2]

[1] Tomsk Polytechnic University, Tomsk, Russia
[2] Tomsk State University, Tomsk, Russia

E-mail: andrew_nagiev09@mail.ru, sherstneva@tpu.ru ,bia@tpu.ru

Abstract. The results of the statistical model experiments research of various load balancing algorithms in distributed computing systems are presented. Software tools were developed. These tools, which allow to create a virtual infrastructure of distributed computing system in accordance with the intended objective of the research focused on multi-agent and multi-threaded data processing were developed. A diagram of the control processing of requests from the terminal devices, providing an effective dynamic horizontal scaling of computing power at peak loads, is proposed.

1. Introduction

Modern trends of technology tendencies in the IT industry (the concept of cloud computing, big data concept, the concept of Internet-things, the concept of fog computing, the concept of mobilization of calculations, etc.) not only exponentially increase the volume of transmitted data, but also uniquely determine the need to use in computing systems distributed multiprocessor and multicomputer systems (MCS). As it is noted by many experts, if the physical infrastructure of MCS complexity is reduced (an era of universal MCS architectures begins), the logical complexity of such software systems only increases [1-3].

Note, that even the necessity of increasing of efficient use of existing computing power and communication resources leads to system solutions for load balancing, optimization of network traffic. This requires the development of special software, capable quickly and flexibly to optimize the management of the information network and computing infrastructure. Also it is necessary to take into account that theoretically possible execution parallelism is not always obvious for many tasks. Because of this imbalance (communication imbalance, platform imbalance, etc.) of load computing nodes of distributed MCS is inevitable.

Thus, one approach to ensure the required efficiency of functioning of distributed IDC is clearly a load balancing components [4-6]. That load balancing (adjustment) optimizes the calculations in a distributed architecture, information and computing resources of complex system and increases their failover. The problem of distribution perform of dynamic requests (load balancing problem) between the components of MCS, which dynamically change topology, relates to the field of basic research in the field of computer technology and belongs to the section of Artificial Intelligence. Therefore, universal solution for load balancing problem does not exist. It should be taken into account in each case of development: character of input data, intensity of their revenue streams, characteristics of

IOP Conf. Series: Materials Science and Engineering **135** (2016) 012030 doi:10.1088/1757-899X/135/1/012030

processing algorithms, and much more but from the field of human intellectual activity. It can not only eliminate the load imbalance, but also can formulate new goals and objectives of really intelligent distributed MCS.

2. Program experiments

In the present work program experiments have been carried out to test and refine the technological scheme of such a method of load balancing, as a method of dynamic connection of resources (DCR).

During the program experiment first assessment of effectiveness of the classic server solution was carried out, when load is determined randomly by formed applications Poisson flow (flow homogeneous and independent from each other's events). Processing applications obeyed to an exponential law (the absolute continuity of the distribution). After that the effectiveness of the distributed computing system (DCS) was evaluated. The interaction of the components of DCS was performed by the method of the DCR. The obtained results were compared with the results of the classic server solutions.

During the experiment classic software server solution consisting of the management server and dynamically connecting to it terminals of requests was created.

Software experiment showed the ineffectiveness of the server solution. Increasing of the number of requests, which received from the generator, leads to the failure of the entire system. In particular, the simulation server by the law of uniform distribution with the interval 10 ms failure occurred at 850 processing requests. Based on this fact, conclusion can be made, that an increase in the frequency of generating query system will run much slower, because the server needs to receive the request, process and send it back to the terminal. An increasing of the number of received requests in different streams did not increased the server data processing speed. Flows began to compete with each other that led to failure of the system.

According to the authors, a method of dynamic connection of resources was used to improve system performance, which leads to acceleration of processing requests. In the proposed functional structure multicomputer complex (figure 1) the main changes have affected the processing server. Its structure has been improved and represents the main control center (MCC). Also special load controllers have been added, with its help there was a dynamic connection of additional control centers (ACC) and buffer storage of incoming requests. Generator of requests is unchanged.

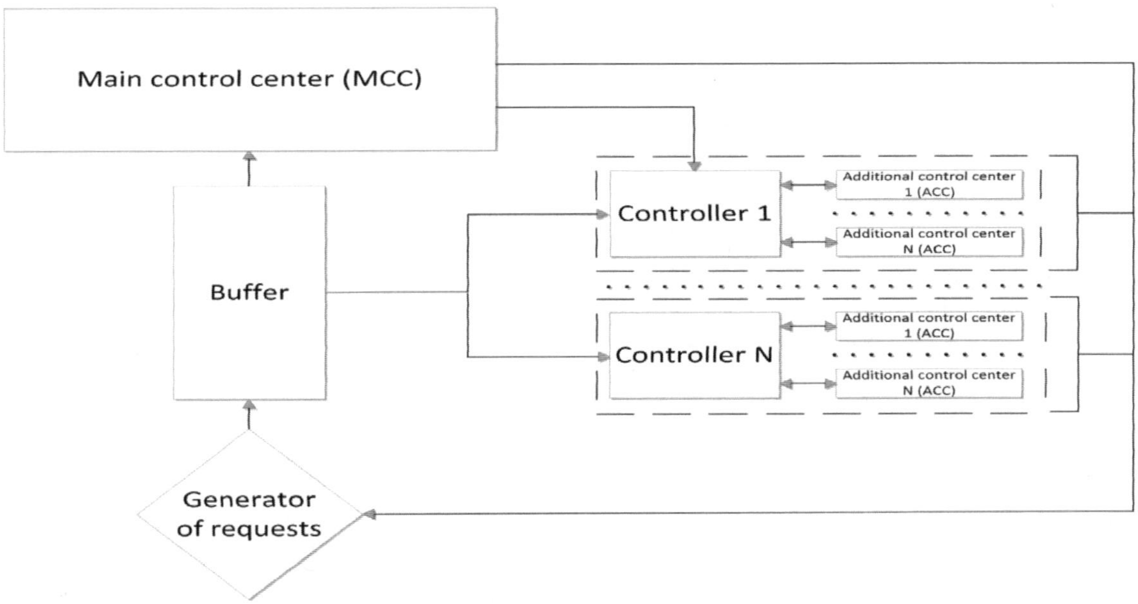

Figure 1. The functional structure of the simulated MCS

Buffer of requests of MCC intends for storage of incoming requests to the processing of the generator of requests (figure 2). After processing the request for calculation, removal from the buffer occurred.

With the increasing the amount of requests incoming from the generator, the main control center, like a server in a classic server solution, does not have time to process all the requests. However, unlike the classical server solution where there was failure of the system, in MCS using DCR, incoming requests accumulated in the buffer. Load controllers were connected when a certain number of requests for processing in a buffer was reached by the software (figure 3). Controllers in separate streams started to run the ACC. The number of connected ACC is limited only by speed of queries and receipt of the physical features of the system (presence of additional computing power).

Each controller starts running with the buffer in a separate thread. Also, each additional control center runs in a separate thread. ACC closed after processing all the requests submitted via the load controller

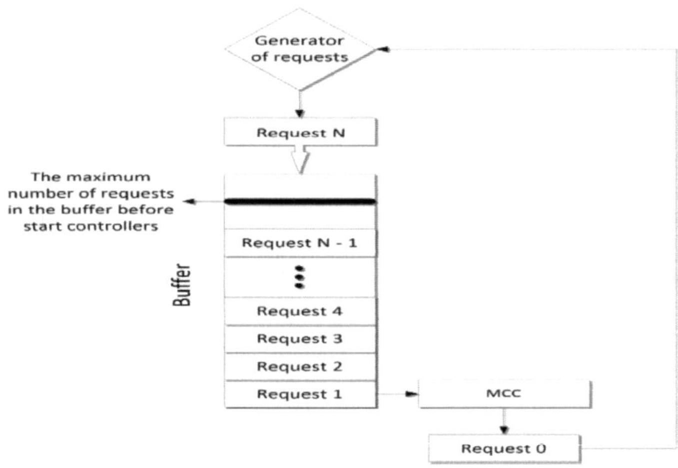

Figure 2. The scheme of processing of incoming requests in MCC

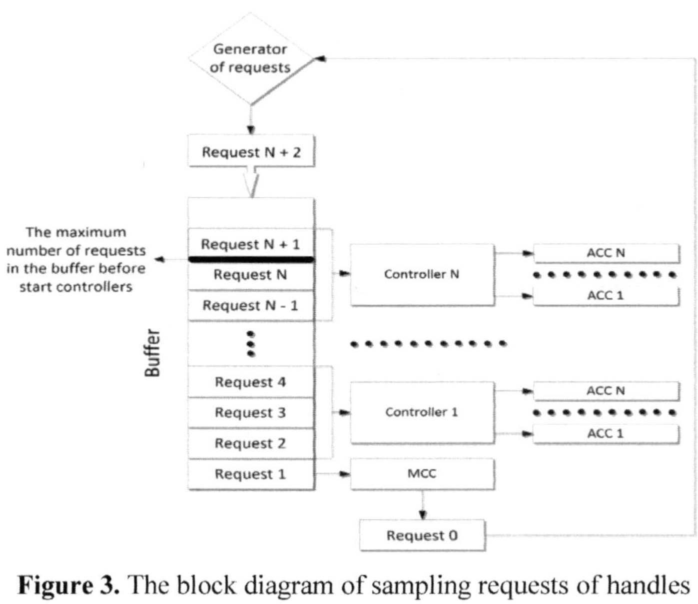

Figure 3. The block diagram of sampling requests of handles controllers

IOP Conf. Series: Materials Science and Engineering **135** (2016) 012030 doi:10.1088/1757-899X/135/1/012030

3. Software modelling

Software modeling of the experiment illustrated in figures 4, 5. In figure 4 there is a curve showing the number of threads running currently in the generator of requests. The main control center loads central processing unit (CPU) (figure 5). As you can see, ACC has time to receive, to process and to send out different threads back into the generator of requests at a certain low frequency of incoming requests which are set by software. However, the number of processing threads is increased with speed of incoming requests. This is due to the fact that the MCC had not got a time to process all the incoming requests, and they accumulated in the buffer of requests. At that moment the classic server solution refused because of lack of buffer, so threads began to compete for server resources. In the proposed structure of MCC load controller connected when a certain number of threads specified by the software. This controller connected the ACC and it increased the computing power system. This led to the fact that the increase of the number of requests in the buffer was stopped. There was a slow decrease of the number of requests in view of their distribution between the additional control centers.

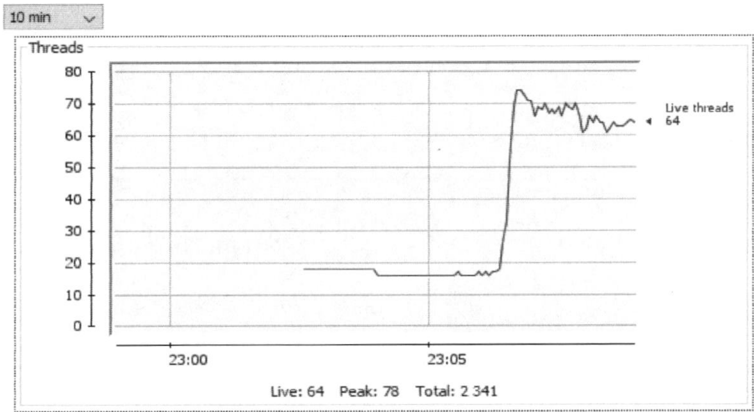

Figure 4. The graph of number of active flows in a distributed computer system.

It can be seen in figure 5 the difference between the CPU load before sending the requests from the generator and during processing of requests. The increase of frequency of receiving requests increases CPU usage by the main control center, which should be faster to handle incoming requests. Then, load growth is ensured by connecting the ACC. This was accompanied by a uniform distribution computing between the CPU cores based on pseudorandom numbers.

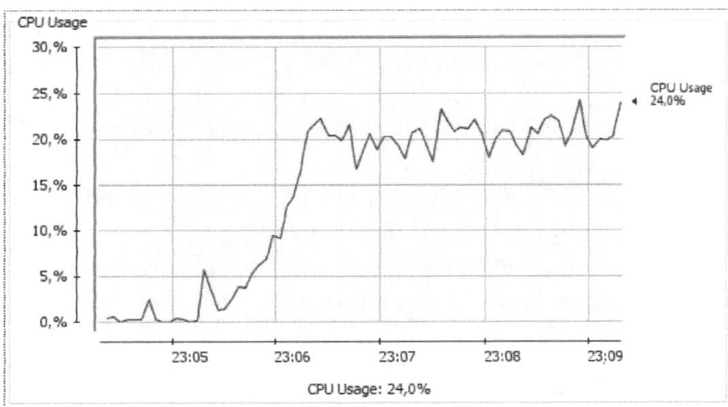

Figure 5. The graph of increasing the CPU load.

There is a processing of the remaining requests in a buffer and switching off the dispatchers and ACC after sending requests from the generator. At the same time a decrease in CPU usage to its original condition, as shown in figure 6, 7.

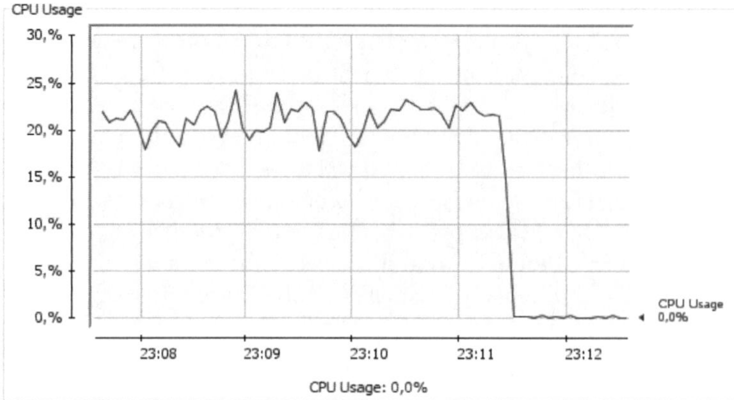

Figure 6. The graph of decreasing the CPU load.

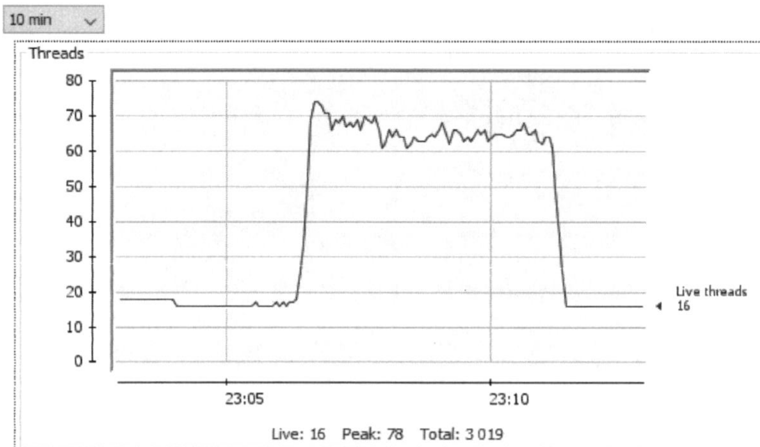

Figure 7. The graph of decreasing threads of the CPU.

4. Conclusion

As a result of experiments, it can be said that the time of connection of terminals and processing requests to MCC was reduced by 57%, compared to the classical version. And in the same time to the main control center is connected and handled on 58.5% more terminals than at the MCC without connecting of ACC. When the second ACC is connected, processing speed increased by 13% compared to the system with a single ACC. When the third ACC is connected, processing speed is also increased by 24% compared to the system with two ACC.

Experiments were carried out by connection of different numbers of requests terminals (from 10 to 10 000) with uniform distribution in the interval from 0 to 100 arbitrary time units.

Despite the fact that the experiment took place only on a single computer with a quad-core processor, a simple generator of requests with a uniform distribution was enough to experimentally show the effectiveness, flexibility and speed of the proposed method dynamic connection of resources. This provides horizontal scalability of processing power, limited only by the physical ability to connect additional nodes.

On the basis of a model experiment can be said that the considered method of dynamically connect resources will be also effective in a significant increase in the number of connected nodes and zoom in architecture.

References

[1] I. A. Botygin, V. N. Popov, V. A. Tartakovsky, V. S. Sherstnev Architecture of scalability file system for meteorological observation data storing // Proc. of SPIE, 21st International Symposium Atmospheric and Ocean Optics: Atmospheric Physics. – 2015. – vol. 9680. – pp. 96800J-1– 96800J-4. – doi: 10.1117/12.2205749.

[2] Sherstnyov V. S., Sherstnyova A. I., Botygin I. A., Kustov D. A. Distributed Information System for Processing and Storage of Meteorological Data // High Technology: Research and Applications 2015. Key Engineering Materials. – Trans Tech Publications, Switzerland, 2016. – Vol. 685. – pp. 867-871.

[3] Popov V.N., Botygin I.A., Koshelev N.V. Uniform Model of Representation Heterogeneous Data Hydrometeorological Observations // High Technology: Research and Applications 2015. Key Engineering Materials. – Trans Tech Publications, Switzerland, 2016. – Vol. 685. – pp. 925-929.

[4] A.V. Boukhanovsky, S.V. Ivanov, Features of optimization of distribution computational load in tasks of parallel processing of information and simulation modelling, IMMOD-2003, Saint-Petersburg., 2003, pp. 69-75.

[5] Intuit. Threads and multithreading. URL: http://www.intuit.ru/studies/courses/641/497/lecture/11284?page=2 (the date of circulation 06. 10. 2015).

[6] Load balansing: basic algorithms and methods // Habrahabr. URL: http://habrahabr.ru/company/selectel/blog/250201/ (дата обращения 10. 11. 2015)

Radioactive γ/β tracer to explore dangerous technogenic phenomena

P M Nagorsky[1], V S Yakovleva[2], E O Makarov[3], P P Firstov[3], A G Kondratyeva[2], A A Stepanenko[2]

[1]Institute of Monitoring of Climatic and Ecological Systems Siberian Branch of the Russian Academy of Sciences, 10/3 Academichesky avenue, Tomsk, 634055, Russia
[2]National Research Tomsk Polytechnic University, 30 Lenin Ave., Tomsk, 634050, Russia
[3]KamchatkaBranch of Geophysical Survey, Russian Academy of Sciences, bul. Piipa 9, Petropavlovsk-Kamchatskiy, 683006, Russia

E-mail:npm_sta@mail.ru,vsyakovleva@tpu.ru, firstov@emsd.ru

Abstract.A radioactive γ/β tracer to explore dangerous technogenic phenomena has been proposed: the ratio of the measured flux density of β- and γ-radiations in the surface layer of the atmosphere. The time dependence analysis of the ratio of β- and γ-pulse count rate has been carried out. A significant increase of the γ/β ratio was recorded under the cyclone passing through Japan (Fukushima) to Kamchatka. The proposed γ/β tracer can be a very sensitive indicator of non-stationary processes related to hazardous natural and technogenic phenomena.

1. Introduction

The study of artificial radioactivity caused by nuclear weapons testing and technological processes and accidents [1–3] is still dominant in atmospheric radioactivity studies. The Automated Radiation Monitoring System of Russian is aimed to solve these problems. Its main activity is γ-background monitoring. The γ-radiation is chosen due to its high penetrating ability. Monitoring of the atmospheric β-radiation is mainly replaced by monitoring of β-activity of individual radionuclides. Monitoring of the atmospheric α-radiation is carried out sporadically for scientific purposes.

The level of γ-background is known to be significantly higher in precipitation [2–6]. According to [3–6] rain and snow play a major role in the atmospheric γ-background variations and cause a short-term abrupt increase (spikes) in the recorded characteristics of γ-radiation field of 125% [6] and even up to 7 times [5]. To describe this phenomenon, a special term "radon washout" was introduced abroad to describe the washout of γ-emitting decay products of radon and thoron onto the Earth's surface by precipitation. However, the expected significant correspondence between the intensity of precipitation and the level of γ-radiation dose has not yet been found [5, 6].

2. Statement of the problem

The methods to analyze the back trajectories of air masses, transport rate of air masses and information about the emanation of radon from the lithosphere are used to estimate the level of ionizing radiation [2, 5, 7]. It is difficult to predict γ-radiation increase reliably due to precipitation, since the concentration of radon decay products in precipitation depends on the concentration of radon in the atmosphere, the formation and transport of clouds and the type of precipitation [2, 5]. The data on γ-radiation is only used in papers devoted to the assessment of ionizing radiation level. It is known that the decay processes of radon and its decay products form not only γ-background, but also α- and β-background. However, this factor is not considered to analyze the level and variations of the background radiation components.

One more important fact should be emphasized: the direct impact of ionizing radiation on the environment, biological objects and human beings is determined by the type of radiation. The fluxes of short-range α-particles in the air can only affect the skin and lungs. The fluxes of β-particles are capable of passing through the air from tens of centimeters to several meters. In biological tissues the path of these particles lies within the range from parts to units of centimeters (affects the skin and subcutaneous tissue). The γ-radiation flux has a high penetrating power and can pass more than 1 km in the air (affects the entire body).

Our previous studies [2, 8] proved that variations of β- and γ-background are not completely consistent within the intra-day and inter-day intervals and within the longer intervals as well. The determined difference in the dynamics of γ- and β-background is the basis for the current research, which includes the investigation of the dynamics of γ/β tracer, which is the ratio of the flux density of γ- and β-radiation measured in the surface layer of the atmosphere.

3. The results of monitoring and analysis

Monitoring of the analyzed variables was performed by the method described in [8] in Tomsk city in Russia (aseismic region with a continental climate) and in Paratunka, Kamchatka, Russia (seismically active region with maritime climate). The monitoring data results were supplemented with current data on space weather (solar radio wave, Wolf number, indices of geomagnetic activity, cosmic rays intensity, etc.).

As indicated previously [2, 4, 5, 7], synoptic processes play a significant role in the dynamics and level of atmospheric emissions of β- and γ-radiation and these processes can lead to the increase of background level by several times. This can be clearly seen in Figure 1, which shows the data obtained in seismic regions (Paratunka, Kamchatka). In Figure 1: on the left – variations of meteorological and radiation quantities for the period 2 September 2011 – 31 December 2011; on the right –the enlarged fragment (11 November 2011 – 01 December 2011). The data on γ- and β-flux densities variations and their ratio (right column) were smoothed by a bandpass filter with the band periods 4 hours – 10 days and then normalized. Waveletspectrograms were also constructed for obtained data. As can be seen in Figure 1, air pressure fall caused by the passage of cyclone is followed by rain (or snow) and corresponding spikes in β- and γ-background. It can be seen that the spikes in the levels of ionizing radiation component background (β and γ) caused by the next cyclone do not result in a significant change in their ratio. The exceptions are variations of the atmospheric β- and γ-radiation in the second half of November 2011 during the passage of cyclone through the area of Fukushima (Japan) to Kamchatka. The scheme of the cyclone route is shown in Figure 2.

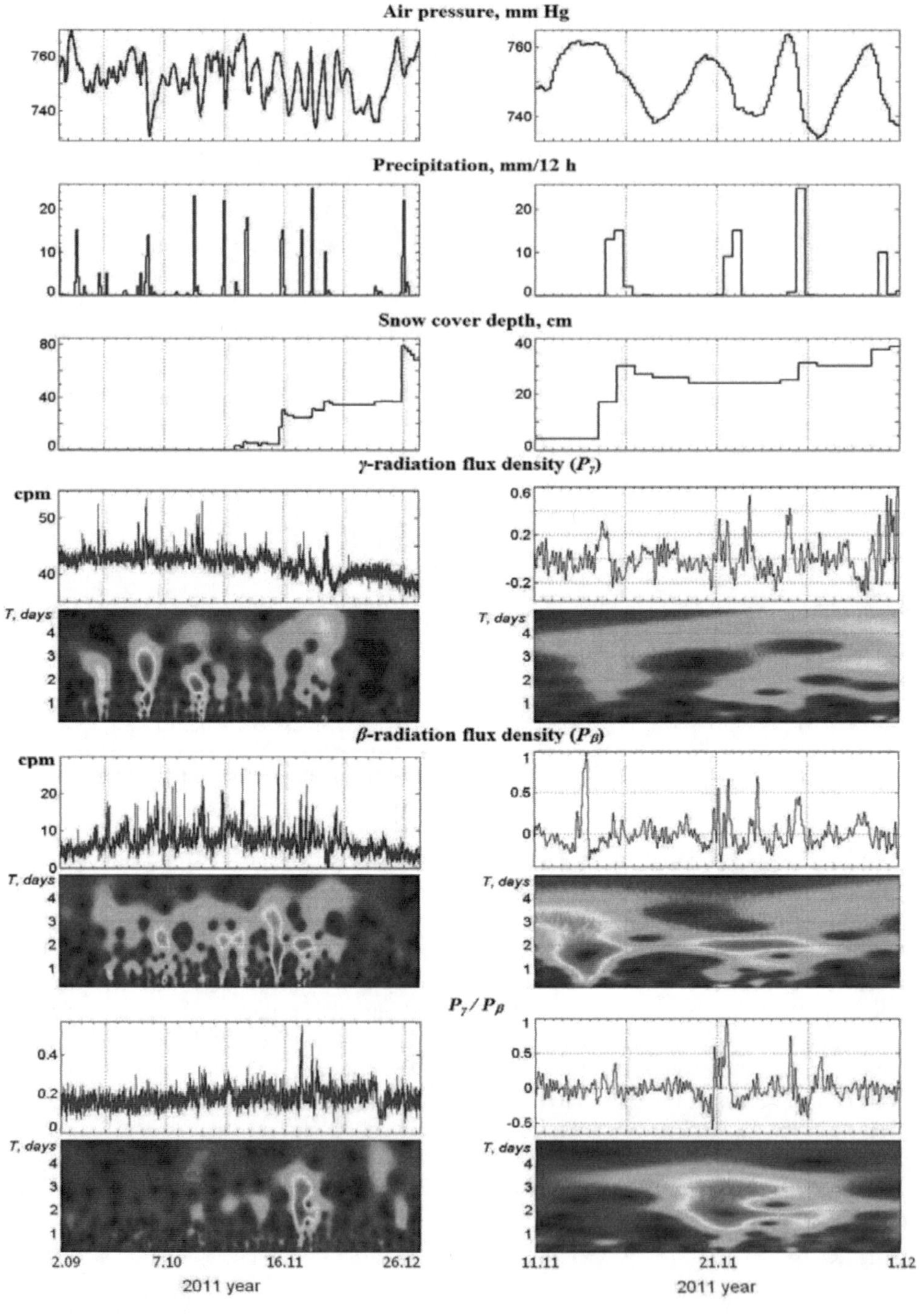

Figure. 1. Variations ofatmospheric pressure, precipitation, snow cover depth, number of pulsesγ-andβ-radiation flux densities measured at 5m heightand their normalizedratio.

Figure. 2. The data of IR channel from NOAA weather satellite about the cyclone at the territory of Kamchatka and Japan.

The capture of radioactive aerosols of technogenic origin by the cyclone in the region of Fukushima nuclear power plant is a possible cause of spikes in the time dependence ratio of β- and γ-count rates. The presence of radioactive aerosols of technogenic origin in the atmosphere must be followed by the formation of new radioactive elements or corresponding changes in the variations of atmospheric β- and γ-radiation flux densities and their ratio.

4. Conclusion

The proposed γ/β tracer can be a very sensitive indicator of non-stationary processes related to hazardous natural and technogenic phenomena. This owes to the fact that technogenic, lithosphere and atmospheric processes result in qualitative changes of the radionuclide composition in the atmosphere and as a consequence in the change of the γ/β ratio. The variations of the γ/β tracer cause changes in the mechanisms of ionizing radiation impact on the environment, which is an important factor to assess the environment.

Acknowledgments

This work is funded by Russian Federal Target Program № 14.575.21.0105 (RFMEFI57514X0105) of Ministry of Education and Science of Russian Federation.

References

[1] Utkin VI and Yurkov AK 1997*Volcanology and Seismology***4** 82–94
[2] Firstov PP, Yakovleva VS, Shirokov VA et al. 2007 *Annals of Geophysics***50** 547–556
[3] EURADOS report 1999 *Radiation protection***106** URL: http://ec.europa.eu/energy/nuclear/radiation_protection/doc/publication/rp106.pdf
[4] Cortes G, Sempau J and Ortega X 2001 *Nukleonika***46** 161–164
[5] Mercier J-F, Tracy B L, d'Amours R, Chagnon F, Hoffman I, Korpach E P, Johnson S and Ungar R K 2009 *Journal of Environmental Radioactivity***100**527–533
[6] Burnett J L, Croudace I W and Warwick P E2010 *Journal of Radiological Protection***30**525–532
[7] Greenfield M B, Ito N., Kubo K et al. 2008 *J. of App. Phys.***104** 074912
[8] Yakovleva V S, Karataev V D, Vukolov A V, Ippolitov I I, Kabanov M V, Nagorsky P M, Smirnov S V, Firstov P P and Parovik R I 2009 *ANRI***4** 55–60

IOP Conf. Series: Materials Science and Engineering **135** (2016) 012032 doi:10.1088/1757-899X/135/1/012032

Modeling of operating history of the research nuclear reactor

A Naymushin[1], Yu Chertkov[1], M Shchurovskaya[2], M Anikin[1], I Lebedev[1]
[1]National Research Tomsk Polytechnic University, TPU, 634050 Russia, Tomsk, Lenin avenue 30.
[2]National Research Nuclear University MEPhI (Moscow Engineering Physics Institute), NRNU MEPhI, 115409, Russia Moscow, Kashirskoe shosse 31.

E-mail: agn@tpu.ru

Abstract. The results of simulation of the IRT-T reactor operation history from 2012 to 2014 are presented. Calculations are performed using continuous energy Monte Carlo code MCU-PTR. Comparison is made between calculation and experimental data for the critical reactor.

1. Introduction

The major research tool for the processes taking place in the reactor is experiment. However, modern state of computational technology and calculation methods allow completion and partial replacement of experimental methods with simulation of processes using computer codes. Nuclear technologies is a field of increased responsibility, therefore, compliance rationale of calculation data with experimental data is needed so that computational software could be used for the simulation of real prototypes.

MCU-PTR code was chosen for the purpose of IRT-T reactor simulation [1, 2]. This code uses Monte-Carlo method and is oriented on the calculations for research reactors. Created model of the IRT-T reactor core and the reflector simulates material composition and reactor geometry in details and allows neutronics calculations for the different states and operating regimes of the reactor.

2. Simulation of burnup process

The operating regimes of the reactor recorded in the operating logs were used as input data for the computation of the fuel burnup process. Weekly cycles were taken into consideration, as well as control rods movement and reloading of fuel assemblies between the reactor operating cycles. Typical weekly operation of the reactor consists of a 4-day operation at the nominal power level and a 3-day shutdown. Fuel reloading takes place when the excess reactivity is running low.

Initial information about the reactor burnup was obtained from the results of the simulation of the reactor operation history from 1984 to 2012 using diffusion code TIGRIS [3]. The simulation of the simplified operation history of the reactor using MCU-PTR code was used to estimate the poisoning of the beryllium reflector blocks with products of high absorption cross-sections as a result of the interaction with neutrons for the past years.

The detail simulation of the reactor weekly operation with actual control rod positions for the past seven cycles from 2012 to 2014 was carried out using MCU-PTR code. It was assumed that in the horizontal plane, all fuel tubes of one fuel assembly have the same burnup (i.e., they burn as one material). The burnup for six axial layers for each fuel assembly was calculated.

Neutronics characteristics (criticality, reactivity worth of control rods, excess reactivity, shutdown margin) of the investigated operating cycles were determined and analyzed. Thermohydraulic parameters of the reactor cores were also analyzed [3].

3. Computation of critical states

The reactor critical states corresponding to the beginning and the end of the operating cycle are the most important. Computation of the Xe-free critical states was carried out for the beginning of operating cycle (at IRT-T reactor start-up) and for the end of operating cycle before fuel reloading (Table 1). The value shown in the last column of the Table 1 is the calculated reactivity for the measured critical states (the measured reactivity equals zero for the critical state).

Table 1. Calculated reactivity for the measured critical states at the beginning and at the end of operating cycles (standard deviation 0.00015)

Date	Energy generation, MW·h	Position of control rods, cm CR-1	CR-2	AR	k_{eff}	ρ,%Δk/k
01.10.2012	0	42	42	34	1,0118	1,17
12.11.2012	2496	33	33	32	1,0284	2,10
12.11.2012	0	54	53	33	1,0142	1,30
20.02.2013	5390	34	34	33	1,0201	1,97
20.02.2013	0	54	54	31	1,0086	0,85
22.04.2013	4328	36	36	23	1,0096	0,95
24.04.2013	0	50	50	41	1,0058	0,57
17.06.2013	3372	36	36	22	1,0133	1,31
17.06.2013	0	60	57	32	1,0084	0,83
26.09.2013	1588	10	12	60	1,0177	1,74
26.09.2013	0	32	31	60	1,0092	0,92
13.01.2014	7558	38	38	24	1,0182	1,78
14.01.2014	0	55	56	31	1,0091	0,90
24.03.2014	3722	32	33	25	1,0212	2,08

Calculated reactivity for the critical states ranges from 0.6 to 1.3 %Δk/k and from 0.95 to 2.1 %Δk/k for the beginning and for the end of the cycles, respectively. Comparing critical states for the beginning and the end of operating cycles, it can be seen (Figure 1) that at the beginning of operating cycles the calculated reactivity is always less. For example, the calculated reactivity during the cycle #99 increased from ~1% Δk/k to ~2% Δk/k.

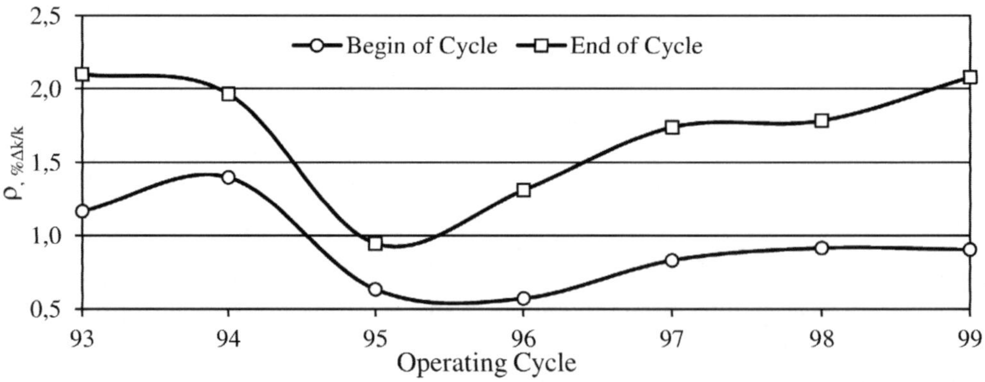

Figure 1. Calculated reactivity for the measured critical states at the beginning and at the end of operating cycle (line connecting calculated points serves as an eye-guide only

IOP Conf. Series: Materials Science and Engineering **135** (2016) 012032 doi:10.1088/1757-899X/135/1/012032

The average calculated reactivity for the measured critical states is 0.93 %Δk/k and 1.7 %Δk/k for the beginning and for the end of the cycles respectively. After the replacement of the spent fuel assemblies with the fresh ones, the calculated reactivity decreases. It can be explained by the fact that the actual energy generation is larger than the energy generation used in the computation.

4. Control rods reactivity worth

Automatic regulator rod (AR) is calibrated using the asymptotic period method. According to this procedure, the reactor is brought to critical state at power level of 5 kW with specified position of AR control rod (in the beginning of calibration AR is fully inserted). After that, AR is partially withdrawn, leading to input of positive reactivity. Asymptotic period can be defined by measuring time needed for doubling the power. Inserted reactivity is calculated using in-hour equation. After measuring the reactivity worth of inserted part of AR rod, the reactor is brought to critical state by means of shim rods, then, the next part of AR rod is calibrated. Shim rods (CR-1 CR-2 and CR-3) are calibrated using method of compensation with AR rod of known efficiency (rods swap method).

Simulation of calibration procedure was carried out for the beginning of the cycle #100 to determine calculated reactivity worth of control rods. States with experimental critical position of the shim rods during calibration by the compensation method were calculated. Positions of AR rod herewith were averaged. Table 2 shows the results of calculation of states during calibration of CR-1 rods (analogous calculations were made for CR-2 and CR-3 rods). The measured and the calculated reactivity worth of the calibrated parts of CR-1 (Δρ) is presented. Calculated differential reactivity worth was defined as reactivity (ρ) difference between the state №15 and №14, №13 and №12, etc.

Table 2. Simulation results of calibration of CR-1 rods (standard deviation 0.00015)

	Position of control rods, mm				ρ, %Δk/k	Δρ, %Δk/k	
№	AR	CR-1	CR-2	CR-3		Calculated	Experimental
1	355	600	333	0	1.06		
2	355	600	333	97	0.71		
3	355	473	333	97	0.96	0.26	0.394
4	355	473	333	169	0.47		
5	355	408	333	169	0.91	0.44	0.394
6	355	408	333	224	0.45		
7	355	348	333	224	0.98	0.53	0.394
8	355	348	333	269	0.51		
9	355	280	333	269	1.09	0.58	0.394
10	355	280	333	308	0.68		
11	355	198	333	308	1.26	0.58	0.394
12	355	198	333	346	0.82		
13	355	56	333	346	1.51	0.68	0.394
14	145	56	333	385	1.33		
15	145	0	333	385	1.44	0.11	0.037

According to Table 2, the largest deviation of calculated differential worth (Δρ) from experimental one takes place at the insertion depth of CR-1 ranging from zero to ~200 mm.

Calibration curves of CR-1, CR-2 and CR-3 were obtained by using the results from Table 2 and analogous calculations for the remaining control rods. Figure 4 shows experimental and calculated curves of integral worth of CR-1, CR-2 and CR-3 rods, normalized to unity.

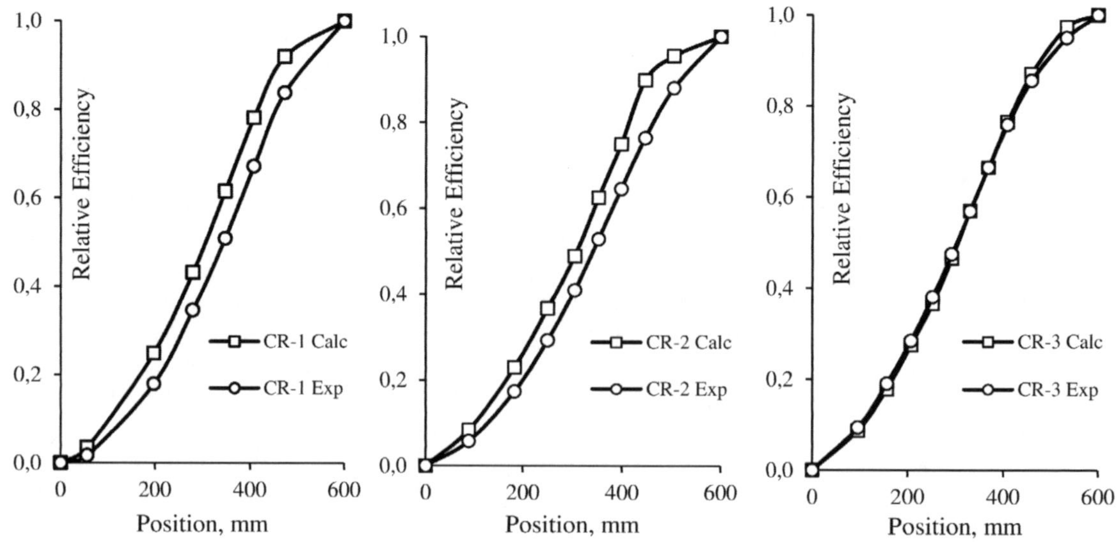

Figure 4. Calculated and experimental calibration curves of CR-1, CR-2 and CR-3 rods.

The shape of the relative calibration curves allows assumption that CR-1 and CR-2 rods are situated 4-5 cm higher than defined in the geometric model used in the calculation, while the CR-3 position over the height in the geometric model is approximately the same as actual CR-3 position over the height. In the initial geometric model at zero insertion depth of control rod, the lower end of the absorber is leveled to the top edge of the active part of the fuel assembly. The geometric model was corrected by using the results of comparison of calculated and experimental relative calibration curves. Specifically, CR-1 and CR-2 rods were shifted upwards by 5 cm and 4 cm respectively. Simulation of calibration was carried out again using the corrected geometric model. The relative calibration curves of CR-1, CR-2 and CR-3 rods obtained by using this model are shown in Figure 5.

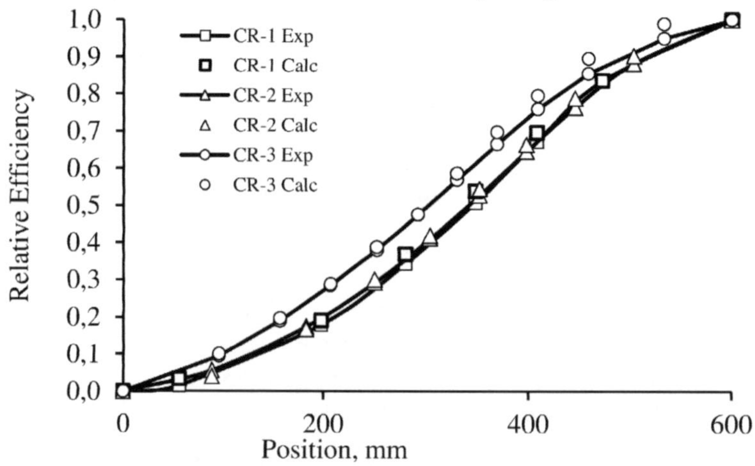

Figure 5. Calculated and experimental calibration curves of CR-1, CR-2 and CR-3 rods (corrected geometrical model)

The selected shift made it possible to achieve satisfactory agreement between the calculated and experimental calibration curves of CR-1, CR-2 and CR-3 rods.

Table 3 shows calculated and experimental integral reactivity worth of IRT-T reactor control rods.

Table 3. Comparison of calculated and experimental integral reactivity worth of control rods.

	Reactivity worth, %Δk/k			
	AR	CR-1	CR-2	CR-3
Calculated	0.530	2.96	4.04	4.45
Experimental	0.394	2.40	3.69	4.15

The discrepancy between the calculated and measured integral reactivity worth of CR-2 and CR-3 is less than 10% (relative). For the group of CR-1 rods the discrepancy is 23%. To reveal the reasons of CR-1 integral reactivity worth overestimation in the simulation additional investigations are required. It should be noted that the absorber depletion cannot be such reason. The estimation of the absorber depletion for the IRT-T was made on the basis of the irradiation history of the rods [5, 6]. The reactivity worth loss because of the absorber depletion was found to be ~3% for the CR-1 and CR-2 rods and ~0% for CR-3 rods. Thus, estimated CR-1 and CR-2 worth loss connected with the depletion is not enough to compensate the overestimation. Also it should be noted that CR-1 and CR-2 rods are used similar during the reactor operation while CR-3 group is usually withdrawn, hence the worth overestimation of CR-1 and CR-2 rods because of the absorber depletion would be the same, while the integral worth of CR-1 rods only is overestimated in the simulation. In addition, the uncertainty of location of horizontal experimental channels (HEC) relative to the housing of the reactor core may influence the value of efficiency of CR-1. According to operating data, the gap between bottoms of HECs and housing of the reactor core ranges from 0.5 cm to 2.5 cm. In the calculation model this gap is 2.5 cm for all HECs. Moving all HECs closely to the housing of the reactor core leads to CR-1 efficiency decreasing by 2.3% (relative). Thus, the error of calculated integral reactivity worth of CR-1 rods equals ~20 %.

5. Conclusion

Simulation of the operating history of IRT-T reactor enabled to obtain the extensive material for verification and validation of MCU-PTR code. The results of the presented research were used for development of verification report for MCU-PTR code [6]. The verification report was approved by Rostekhnadzor.

Further studies are planned to identify and eliminate the causes of differences between calculated and experimental data on criticality and reactivity worth of control rods, which would make the developed calculation model more precise.

References

[1] Alekseev, N. I., et al. "MCU-PTR program for high-precision calculations of pool and tank type research reactors." Atomic energy 109.3 (2011): 149-156.

[2] Alekseev, N. I., et al. "The status of MCU-5." Physics of Atomic Nuclei 75.14 (2012): 1634-1646.

[3] Shchurovskaya, M. V., and Alferov, V. P. "Calculation and experiment to determine the operational characteristics of a research reactor." Atomic Energy 101.4 (2006): 706-713.

[4] Naymushin, A., et al. "Thermal analysis of IRT-T reactor fuel elements." IOP Conference Series: Materials Science and Engineering. Vol. 93. No. 1. IOP Publishing, 2015.

[5] Lebedev I. I., et al. "Calculations of IRT-T control rods resource". Physical-technical problems of nuclear science, energy and industry. 2014. p.64

[6] Naymushin A. G., Shchurovskaya M. V., Alferov V. P., Chertkov Yu. B., Anikin M. N., Lebedev I. I., Verification of MCU-PTR code with MDBPT50 evaluated nuclear data for calculation of IRT-T and IRT MEPhI research reactors neutronics characteristics.TPU report. 2015. Tomsk.

IOP Conf. Series: Materials Science and Engineering **135** (2016) 012033 doi:10.1088/1757-899X/135/1/012033

The application of computed tomography and magnetic resonance imaging at diagnostics of the human maxillofacial system

V Nikitin[1], E Karavaeva[2], Yu Cherepennikov[3] and I Miloichikova[3]

[1] Perm National Research Polytechnic University, 29 Komsomolskiy prospect, Perm, 614990, Russia
[2] Perm State Medical University named after Academician E.A. Wagner, 26 Peter and Paul Street, Perm, 614990, Russia
[3] National Research Tomsk Polytechnic University, 30 Lenin Avenue, Tomsk, 634050, Russia
E-mail: sortior@mail.ru

Abstract. The application of computed tomography and magnetic resonance imaging has entered into wide practice at diagnosis of the maxillofacial system. Computed tomography allows us to obtain information about only bone structures. Magnetic resonance imaging gives information about bone and soft tissue structures of the maxillofacial system. The sagittal and coronal projections should make for complete diagnosis of the temporomandibular joint, because the articular disc is very mobile structure. We suggest that the temporomandibular joint can influences the internal carotid artery at medial displacement of the articular disc. As a result of analysis of the literature and our own studies concluded that changes TMJ affect the internal carotid artery.

1. Introduction

The application of computed tomography and magnetic resonance imaging has entered into wide practice now at diagnosis of the human maxillofacial system. Computed tomography allows us to give information about bone structures: the mandible and the skull. Significant disadvantage of computed tomography is impossibility of evaluation of the condition of soft tissue structures of the maxillofacial system (muscles, ligaments, articular disc) [1, 2, 4].

One of basic elements of the maxillofacial system is the temporomandibular joint disc. It consists of the condyle of the mandible, glenoid fossa and tubercle of the temporal bone and articular disc between the mandible and the temporal bone [1]. The temporomandibular joint disc is are incongruent, i.e. these articular surfaces do not coincide in shape. The disc takes particular importance in incongruent joints because it smooths out geometric differences of the articular surfaces for realizations of various movements [2, 3, 4].

Many diseases of the maxillofacial system are associated with soft tissues. For example, of the temporomandibular joint dysfunction, when the temporomandibular joint disc does not take normal place [1]. At the same time, computed tomography can not visualize of the disc and its position relatively articular surfaces. This dysfunction can be caused by incorrect functioning of muscles. In this case, the muscles can have values of efforts close to limit values (maximal values, which can be created). Also in this case, computed tomography can not visualize of the muscles. Also in this case, computed tomography can not visualize of the muscles. Besides, computed tomography causes additional radiation exposure on the human, reduction of which is highly relevant issue [5, 6].

In this paper, authors present necessity of application of magnetic resonance imaging at diagnostics of the human maxillofacial system for development of method of individual treatment of diseases of the maxillofacial system. This method allows us to include position of the temporomandibular joint disc and directions of muscles.

2. Materials and methods

The temporomandibular joint is one of the most interesting joints, because the disc makes significant displacements allowing perform different movements [2, 3, 4].

One of important characteristics defining condition of the temporomandibular joint is the articular disc (Fig. 1). The position of the disc in norm and pathology is shown in Fig. 1, when the jaws are closed [1].The position of the temporomandibular joint disc is very important parameter, because it determined the point of application of the temporomandibular joint reaction. This point belongs both surfaces of the disc and the mandible. It is determined as the shortest distance between the articular surfaces of the temporomandibular joint (Fig. 2) [2].

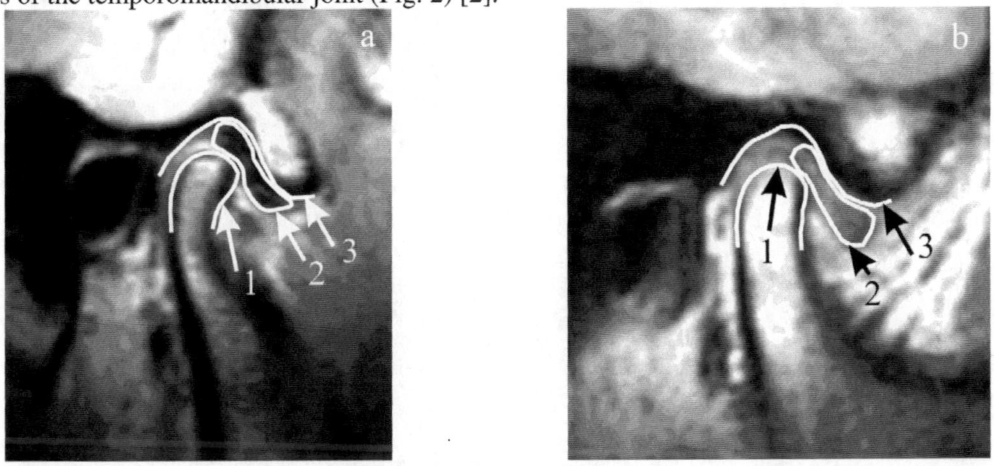

Figure 1. Position of the temporomandibular joint disc: a - is norm; b - is anterior displacement (pathology) [1] (1 is condyle; 2 is disk; 3 are glenoid fossa and tubercle of the temporal bone)

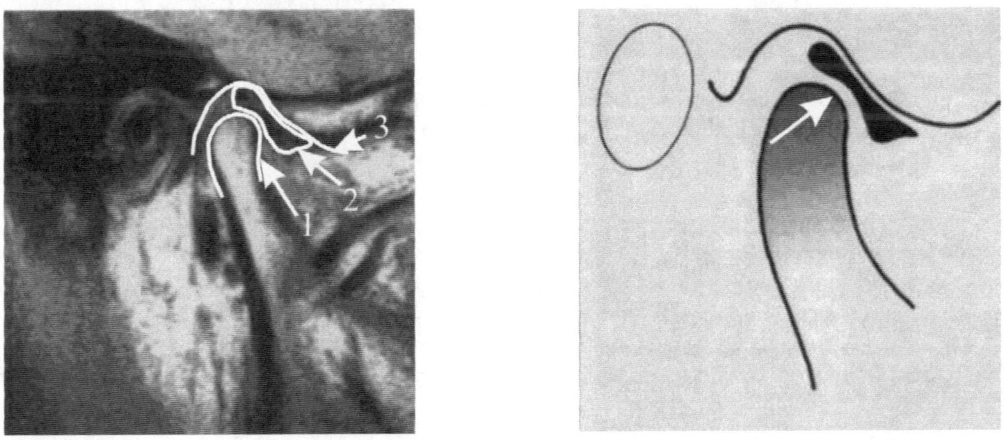

Figure 2. Magnetic resonance imaging of the temporomandibular joint in oblique sagittal projection (projection performed under angle to the sagittal projection): 1 is condyle; 2 is disk; 3 are glenoid fossa and tubercle. Interposition of articular structures in central occlusion is normal. Arrow shows the central zone of the disk and vector of chewing load in the diagram [2]

The direction of chewing load in norm is shown at the normal position of the disc in Fig. 2. It can be seen that the load is applied in the central zone of the disc and action line of chewing load passes through shortest distance between the articular surfaces [2]. Accordingly, the direction of the temporomandibular joint reaction is directed to the opposite side.

But, only sagittal projection of the temporomandibular joint is not allows us to characterize of position of the disc relatively the articular surfaces. The coronal projection of the temporomandibular joint should be used (Fig. 3) [7].

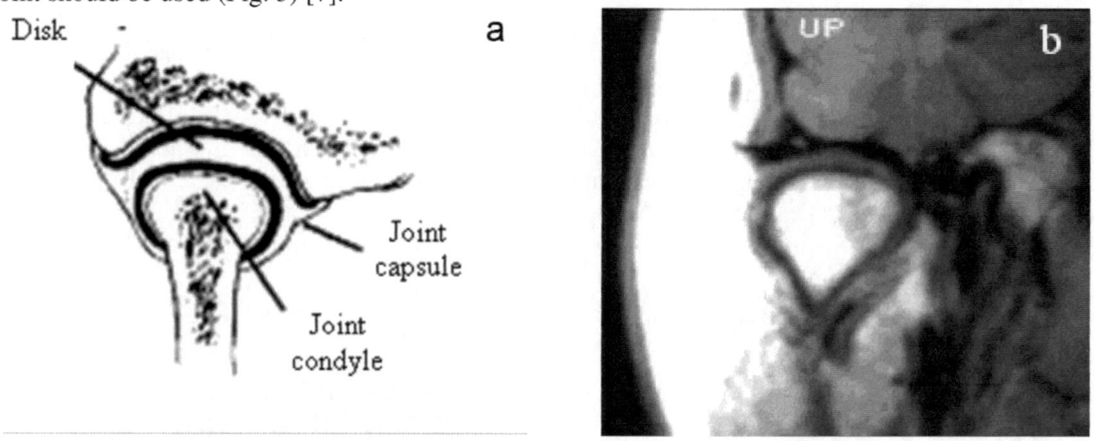

Figure 3. Articular disk in norm as the first target of the damage at redistribution of biomechanical masticatory pressure: a - is schematic figure; b - is magnetic resonance imaging (MRI) [7]

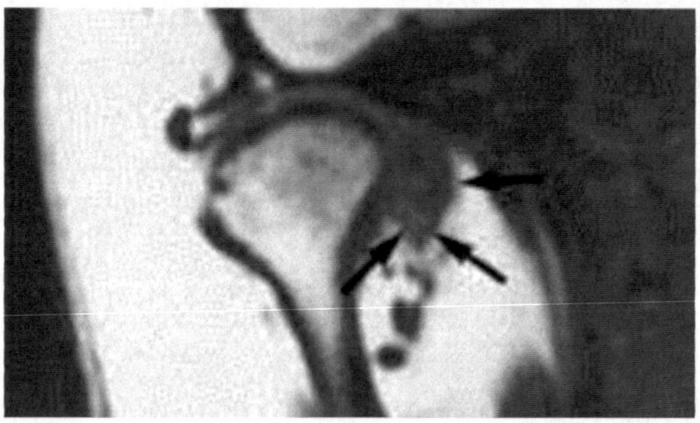

Figure 4. Magnetic resonance imaging of dislocation of the disk on medial surface of the temporomandibular joint condyle (black arrows show represent border of the disk shifted medially) [8]

The sagittal and coronal projections of the temporomandibular joint allow us to estimation of possibility of development of pathological processes into the disc [9]. Its position influences position of determine the point of application of the temporomandibular joint reaction (see Fig. 2).

The information about coordinates of the point of application of the temporomandibular joint reaction allows us to determine values of efforts of muscles and value and direction of the reaction [10].

3. Results and discussion

Magnetic resonance imaging allows us to obtain the attachment points and sizes of muscles and ligaments of the human maxillofacial system. These parameters characterize individual features of the maxillofacial system. Thus, they influence efforts of the human maxillofacial system.

But, magnetic resonance imaging is used less frequently than computed tomography in the practice of dentists. Thus, computed tomography is not includes diagnosis of major structures of the maxillofacial system. According to the authors, magnetic resonance imaging should be used together this computed tomography for individual estimation of condition of the maxillofacial system and development of method of individual treatment of diseases of the maxillofacial system.

Changes of the maxillofacial system influence different processes and systems of the human organism [11-19]. For example, cardiovascular system includes in these system. The internal carotid artery is one of arteries involved in blood supply to the brain. Internal carotid artery is located close to the temporomandibular joint.

Coronal projections of temporomandibular joints and internal carotid arteries are shown in Fig. 5. This case illustrates possible influence of condition of the temporomandibular joint on the internal carotid artery.

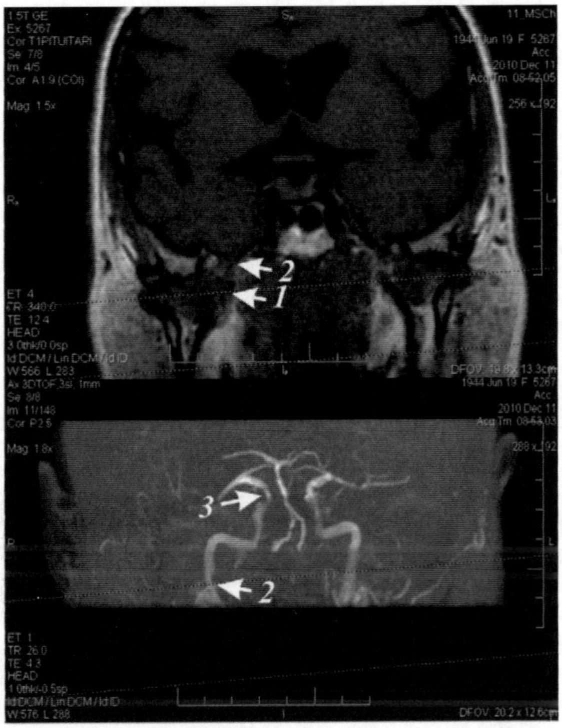

Figure 5. Magnetic resonance imaging of the patient A 68 years with infringement of the support zone (absence of right molars) of the right temporomandibular joint in general mode (upper part) (1 is asymmetric enlargement of the right temporomandibular joint due to structures of the medial pole; 2 are fragments of the right internal carotid artery); in the vascular mode (lower part) (tandem stenosis of the right internal carotid artery at the level of the right temporomandibular joint (2) and in the area of the bifurcation of the vessel on the brain artery (3). Pathology of the left internal carotid artery have been identified [8]

4. Conclusion

In this paper, authors present sagittal and coronal projects of magnetic resonance imaging, which show condition of the temporomandibular joint. Magnetic resonance imaging allows us to estimation of position of the temporomandibular joint disc, development of pathological processes, which is not indicated at using of computed tomograghy.

Abnormal position of the disc entails to changing of the efforts of the maxillofacial system and stresses into the disc. Large values of stresses compared to limit stresses entail to development of processes into the disc and also the bone tissue of the condyle [2, 4].

Acknowledgments
This work was partly supported by the Competitiveness Enhancement Program of Tomsk Polytechnic University.

References
[1] Mello C F Jr, Saito O C and Guimarães Filho H A 2011 Sonographic evaluation of temporomandibular joint internal disorders *Radiol. Bras.* **44 6** 355-359.
[2] Khvatova V A 2005 *Clinical gnatology* (Moscow: Meditsina) (in Russian).
[3] Ingawalé S M, Goswami T 2012 *Biomechanics of the temporomandibular joint. Human Musculoskeletal Biomechanics* ed. T Goswami (Croatia)
[4] Nordin M, Franke V H 2001 *Basic biomechanics of the musculoskeletal system* (Philadelphia: Lippincott Williams & Wilkins)
[5] Stuchebrov S, Batranin A, Verigin D and Wagner A 2014 Setups for Tomographic Imaging with Submillimeter Spatial Resolution *J. of Phys.: Conf. Ser.* **517** Article number 012046
[6] Stuchebrov S, Batranin A, Verigin D, Lukyanenko Y, Siniagina M and Wagner A 2014 Estimation of radiation doses in X-ray visualization of biological objects *Adv. Mat. Res.* **880** 53-56
[7] Information on http:// www.tmjsurgeon.com/tag/tmj-surgery
[8] Information on http:// www.imaging.consult.com
[9] Nikitin V N, Tverier V M, Nyashin Y I and Oborin L.Ph. 2014 The temporomandibular joint reaction and masticatory muscle tensions *Russian Journal of Biomechanics* **18 2** 167-179
[10] Norer V, Pomaroli A and Dize O 1989 Evaluation of degerativnye process in the articular disc of the temporomandibular joint *Diseases of the temporomandibular joint* (Moscow) pp 105-112 (in Russian)
[11] Desvarieux M, Demmer R T, Rundek T, Boden-Albala B, Jacobs D R J, Papapanou P N and Sacco R L 2003 Relationship between periodontal disease, tooth loss, and carotid artery plaque: the oral infections and vascular disease epidemiology study (INVEST) *Stroke* **34** 2120–2125
[12] Joshipura K J, Douglass C W and Willett W C 1998 Possible explanations for the tooth loss and cardiovascular disease relationship *Ann. Periodontol* **3** 175–183
[13] Joshipura K J, Hung H-C, Rimm E B, Willett W C and Ascherio A 2003 Periodontal disease, tooth loss, and incidence of ischemic stroke *Stroke* **34** 47–52
[14] Joshipura K J, Rimm E B, Douglass C W, Trihopoulos D, Ascherio A and Willett W C 1996 Poor oral hearth and coronary hearth disease *J. Dent. Res.* **75 9** 1631–1636
[15] Oborin L F, Patlusova E S 2009 Interaction of biomechanical and hemodynamic factors under congenital and acquired injuries of the temporomandibular joint Russian Journal of Biomechanics **13 4** 88–99
[16] Oborin L F, Shmurak M I 2010 About mechanism of influence of the collateral circulation of the brain on development of the atrophic, pain, and others syndromes of the dentofacial system *Russian Journal of Biomechanics* **14 1** 64–71
[17] Paunio K, Impivaara O, Tiekso J and Maki J 1993 Missing teeth and ischaemic heart disease in men aged 45-64 years *Eur. Heart J.* **14** 54–56
[18] Schillinger Th, Kluger W, Exner M, Mlekusch W, Sabeti S, Amighi J, Wagner O, Minar E and Schillinger M 2006 Dental and periodontal status and risk of progression of carotid atherosclerosis: the inflammation and carotid artery risk for atherosclerosis study dental substudy *Stroke* **37** 2271–2276
[19] Simanovskaya E Y, Elovikova A N, Tverier V M and Nyashin Y I 2004 Biomechanical description of the functional features of human masticatory apparatus in norm and in various pathological processes *Russian Journal of Biomechanics* **8 4** 15–26

The Organization of the Internal Irradiation Monitoring System in Conditions of Nonstandard Radionuclide Intakes

[1] **A V Ovchinnikov**, [1] **K M Izmestyev**, [2] **D G Demyanyuk**, [1] **D D Krivoshein**, [1] **S Yu Poluektov**

[1]Siberian Group of Chemical Enterprises, Seversk, Russia
[2]National Research Tomsk Polytechnic University, Tomsk, Russia

E-mail: alviov@ro.ru

Abstract. Scientific knowledge presently available in the area of monitoring the internal radiation due to nonstandard radionuclide intakes gives no way of identifying the location and nature of intakes in a reasonably accurate and expeditious manner. Both theoretical models and practical methods of personnel internal radiation screening exhibit the lack of research. To this end, the present paper deals with the experience gained by the SGChE in the monitoring of the nonstandard radionuclide penetration into internals and tissues of the personnel. It provides recommendations for the organization and implementation of such monitoring procedures, and describes the practical method for the vulnerary intake containment.

1. Introduction.

Joint-Stock Company "Siberian Group of Chemical Enterprises" (SGChE) incorporates four plants for nuclear materials handling: Isotope Separation Plant, Sublimate Plant, Radiochemical Plant and Chemical and Metallurgical Plant. SGChE plants integrate high radiation and nuclear hazard industries and facilities, engaged in the development, production, operation, storage, transportation, utilization of nuclear materials and goods. The Company is integrated into "TVEL" Fuel Company of "Rosatom" State Corporation.

The process technology employed at the SGChE as one of "Rosatom" State Corporation nuclear fuel cycle and weapons-related enterprises involves external and internal irradiation of personnel. Internal irradiation results from the penetration of radionuclides into the worker body. Radioactive substances have the capability to accumulate and travel in human tissues, and, as this takes place, they can reach the health-hazardous levels. There are different paths of radionuclide penetration into the human body to include: ingestion, inhalation, or intakes through intact or damaged skin. The path by which radioactive substances penetrate into the body significantly dictates the behavior and course of radiologic pathology.

The most detailed description in research works [1, 2] dedicated to the issue of human internal exposure is given to the chronic inhalation intake of radionuclides. However, instances of the acute radionuclides intake crop up frequently in the manufacturing environment. The "acute intake" is hereinafter considered to mean a vulnerary or critical single-time radionuclides intake by the worker body. By the example of the analysis done on of the SGChE plutonium production workers examined at the Biophysical Lab No. 81 of the State Center for Sanitation and Epidemiological Supervision, research work [3] shows that acute intakes of actinides entail larger radiation exposer of individual workers. It primarily pertains to plutonium and americium intakes through the damaged skin.

2. Problem Definition

The initial stages of the nuclear industry development showed poor radiation safety culture both inside and outside of the RF. There occurred a large number of occupational injuries involving radionuclides intake. Research [4] reports on 1,250 incidents of radionuclides intake through the damaged skin registered in respect of the Windscale personnel during the period 1979 to 1989. Research work [5] cites data on 137 wounds potentially contaminated with transuranic elements at the Los-Alamos National Laboratory over the period 1960 to 1972. More than 80% cases describe wounds localized in the area of hands or fingers. These data correlate well with the observations taken at the JSC "SGChE" Chemical and Metallurgical Plant site.

Over the period of the JSC "SGChE" Chemical and Metallurgical Plant plutonium facilities operation, the total amount of recorded skin injuries ran to about 200, of which 100 involved radionuclide contamination (the graph in Picture 1 shows the yearly distribution of radionuclides intakes as a result of skin injuries). Like those at similar plutonium production facilities, most of injuries were accounted for stabs and cuts of hands primarily attributed to imperfections in process equipment used in the plutonium production.

Due to the implementation of the action plan meant to prevent abnormal intakes, the number of nonstandard radionuclide penetration events for the JSC "SGChE" workers has been considerably reduced since 2000.

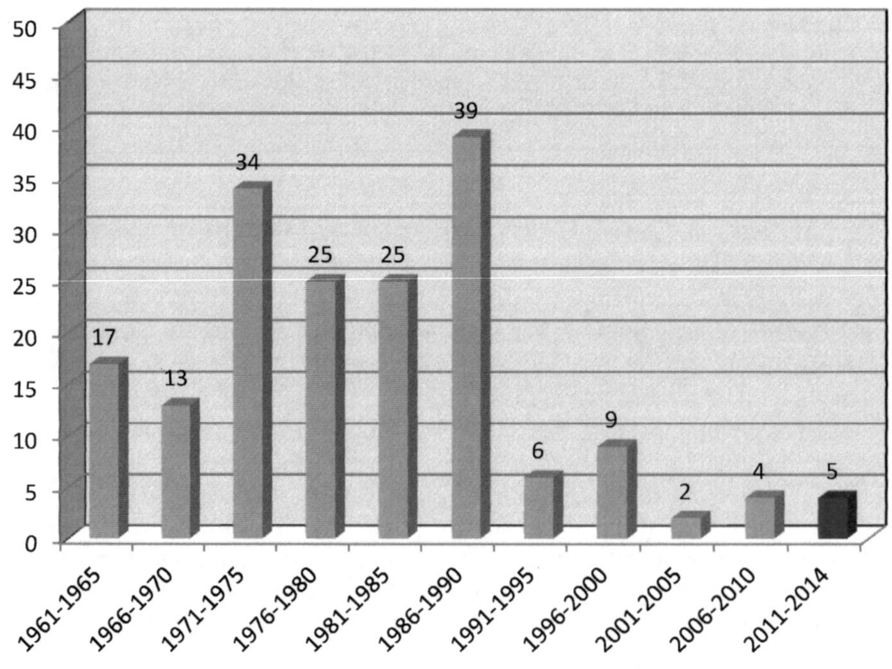

Figure 1. Incidents of Nonstandard Radionuclide Intakes by the JSC "SGChE" Employees during 1961 to 2014

For the time being, however, the issue of improvements in the internal irradiation monitoring system for acute intakes continues to be relevant for the nuclear industry. The reason is that the existing working environment does not make it practicable to completely rule out such intakes.

There are frequent occasions when the vulnerary intake either goes unnoticed or is deliberately kept back by the worker until it shows up in the course of examination by the spectrometer of human irradiation (SHI) or at the biophysical laboratory. In such events, because of long time between the intake and the adoption of measures to minimize health hazards, the worker receives a considerable radiation burden.

Accordingly, it is of current concern nowadays to explore the issue of improvements in the acute intake detection and containment methods.

3. Empirical Research

During 2013 to 2014, the JSC "SGChE" carried out empirical research into the enhancement of methods used to detect and localize acute intakes of radioactive substances in the worker body.

The group under study included 115 workers of critical trades (molders, equipment operators) engaged in the operations at the JSC "SGChE" Chemical and Metallurgical Plant. The periodicity of the SHI examination for the said group was scaled down from annual to quarterly.

The research used the "Lung SHI" detection units. Advantages offered by such detection units as affected spot location finders include the maximum detection area and the minimum time needed to conduct one examination. As part of the measurement, two detectors were fixed at the chest level, and the other two were attached to the examinee's hands.

The potential acute inhalation intake was detected with the aid of measurement procedure "^{241}Am Detection in Lungs" [6]. The ^{241}Am presence in hands of the worker was determined qualitatively through the comparison of the measured spectra and the background. In the event of increased readings, measures were taken to exclude any surface contamination (by means of measuring the density of the α-emitting flux using the DKS-96 radiation-measuring dosimeter). If no surface contamination was identified, ^{241}Am and ^{239}Pu intakes were assessed quantitatively using the "Wound SHI" detection unit. The final stage was to find the exact location of the vulnerary intake with the aid of the lead collimator installed onto the "Wound SHI" detection unit.

To speed up the detection of radionuclides present in other parts of the body in the event of abnormal intake, it was suggested to use hand-operated dose-metering gamma-monitor IR-560. The IR-560 enables the qualitative analysis of the worker skin for presence/absence of radionuclides.

The drawback to the appliance of the IR-560 is that it can be used only in the low-background camera, the lower limit of detection being not less than 500 Bq, by an experimental estimate; in addition, radionuclides location error will amount to a minimum of 10 cm (experimental evaluation).

Once the location of radionuclides has been roughly determined, the "Wound SHI" detection unit [7] was used to obtain more precise location data. In this event the localization is confined to the area of the "Wound SHI" detector measuring window that is 40 mm in diameter.

After that, the lead collimator is mounted onto the measuring window of the "Wound SHI" detection unit. The collimator represents a lead plate, 3 to 4 mm in thickness, fitted with an opening, the opening diameter varying in the range from 100 mm to 3mm depending on the expected area of the radionuclides location. The lead collimator is mounted in such a way that the measuring window of the "Wound SHI" detection unit is limited to the diameter of the collimator opening. The above measures make it possible to localize radionuclides to an accuracy of 1-2 mm followed by the applicable actions to confine/resect soft tissues in the affected area.

4. Conclusion

Owing to the enhanced radiation monitoring procedures, in 2013 the relevant JSC "SGChE" experts detected two cases of the abnormal radionuclide intake (local contamination of tissues in the knee-cap and forefinger areas) by workers engaged in the JSC "SGChE" plutonium production operations.

In 2014, the systematic quarterly examination of the SGChE Chemical and Metallurgical Plant critical group resulted in the detection of one case of vulnerary ^{241}Am and ^{239}Pu penetration into the examinee's body.

The use of the contaminated tissue confinement method developed at the JSC "SGChE" permitted taking expeditious and comprehensive measures to confine contamination and resect the affected skin that enabled a significant decrease in the radiation exposure of workers.

The described method for the vulnerary intake confinement makes it possible to considerably expedite the localization of radionuclides in the skin. The work done demonstrated the following advantages:

- Possibility of scheduled "critical group" examinations for the presence of radionuclides in soft tissues of hands;
- Detection of unrecorded vulnerary or abnormal radionuclide intakes that is of particular importance in case that the worker is unaware of the radionuclide penetration in his/her body;
- Reduced examination interval that expedites responsiveness to unrecorded vulnerary or acute inhalation intakes, and hence considerably reduces the radiation burden of the worker; and
- Localization of radionuclides with an accuracy of 1-2 mm with a view to taking further measures to resect soft issues in the affected region.

Based on the research findings, the JSC "SGChE" has been conducting scheduled health surveys of workers engaged in critical occupations at the Chemical and Metallurgical Plant since 2014.

ACKNOWLEDGMENT
This work was performed on the unique scientific IRT-T equipment and financially supported by Government represented by the Ministry of Education and Science of the Russian Federation (RFMEFI59114X0001).

References

[1] Golutvina M.M., Sadikova N.M. Radioactive Body Burden Control. – M.: Atomizdat, 1979, 135 pages.
[2] Antipin E.B., Potsupyan N.P., Hohryakov V.V. Enhancement of the Personnel Internal Irradiation Individual Dosimetry Control System. – M.: "Anri" Periodical, 2011
[3] Maslyuk A.I., Bogdanov I.M., Simonenko P.D. Specifics of Internal Irradiation Exposure of the Siberian Group of Chemical Enterprises Plutonium Production Personnel. – Siberian Medicine Bulletin, 2, 2005
[4] Schofield G.B. Absorption and Measurement of Radionuclides in Wound and Abrasions. – Clin. Rad., 1963, v. 15, p. 50-54
[5] Johnson L.J., Lawrence J.N.P. Plutonium Contaminated Wound Experience and Assay Techniques at the Los Alamos Scientific Laboratory. Health Phys., 1974, v. 27, p.55-59.
[6] Pronina E.A., Romantsov V.P., Isakov A.P. ^{241}Am Determination in Human Lungs. Measurement Methodology. – Obninsk: Research and Production Company "Radiation Control. Instruments and Methods" Ltd., 2005 г.
[7] Pronina E.A., Romantsov V.P., Isakov A.P. Determination of Wound Soft Tissue Penetration and Penetrated Am and Pu Isotopes Activity. – Obninsk: RPC "Radiation Control. Instruments and Methods" Ltd., 2005 г.
[8] Pronina E.A., Romantsov V.P., Isakov A.P. ^{241}Am Determination in Human Knee Joint. Measurement Methodology. – Obninsk: RPC "Radiation Control. Instruments and Methods" Ltd., 2005 г.
[9] Golutvina M.M., Abramov Yu.V. Monitoring of Human Body Radionuclide Intake and Concentration. – M.: Energoatomizdat, 1989. 168 pages.
[10] OSPORB 99/2010. Basic Sanitation Rules for Ensuring Radiation Safety. SP 2.6.1.2612-10

IOP Conf. Series: Materials Science and Engineering **135** (2016) 012034 doi:10.1088/1757-899X/135/1/012034

[11] NRB-99/2009. Radiation Safety Standards. SanPiN (Sanitary Regulations and Norms) 2.6.1.2523 - 09

[12] Molokanov A.A. Contemporary Methodology for the Personnel Internal Irradiation Dose Control. – M.: "Anri" Periodical, 2011

IOP Conf. Series: Materials Science and Engineering **135** (2016) 012036 doi:10.1088/1757-899X/135/1/012036

Investigation of seasonal dynamics of *β*- and *γ*-radiation fields vertical profile in the surface atmospheric layer

K S Ryabkina[1], A G Kondratyeva[1], P M Nagorskiy[2], V S Yakovleva[1]
[1]National Research Tomsk Polytechnic University, 30 Lenin Ave., Tomsk, 634050, Russia
[2]Institute of Monitoring of Climatic and Ecological Systems Siberian Branch of the Russian Academy of Sciences (IMCES SB RAS), 10/3 Academichesky avenue, Tomsk, 634055, Russia

E-mail: vsyakovleva@tpu.ru

Abstract. This paper is devoted to the seasonal dynamics of β- and γ-radiation field vertical profile in the surface atmospheric layer study. The findings revealed the following: a) significant seasonal variation in vertical distribution characteristics fields β- and γ-radiation; b) reduced β-radiation flux density and γ-radiation dose with increasing height is maintained in the range of spring and fall, namely, starting with snow melt and finishing its setting; c) distortion vertical profile in the winter season, an inverse relationship β-flux density and dose rate of γ-radiation from the heights above the earth's surface was registered.

1. Introduction

Remarkable indicator properties of radionuclides and ionizing radiation are known and actively used since ancient times to obtain new knowledge of dynamic processes in the atmosphere and the lithosphere, as well as to forecast natural and man-made hazards. Models of gases and aerosols transport have been improving based on that knowledge.

Ionizing radiation (IR) and natural radioactivity are all-important in radioecology and radiobiology, radiological protection, construction and geophysics. Atmospheric field of IR are of particular interest in the areas of radioecology and radiobiology for evaluation small (background) doses of the population and environmental exposure. In this regard, research groups and governmental structures carry out radiation monitoring of the ground atmosphere.

Interpretation of the data derived from observations of the atmosphere ionizing radiation background is a complex task [1–5]. This task solution requires the use of IR transfer modeling results, as well as knowledge of atmospheric ionizing radiation sources function.

The scientific team [5] has developed a comprehensive approach to radiation monitoring. The study of the vertical profile of ionizing radiation field characteristics in the system soil-atmosphere is the main feature of this technology approach.

The aim of this study was to investigate the dynamics of the vertical profile of β- and γ-radiation fields in the surface layer of the atmosphere on the annual and above synoptic time scale.

2. Materials and Methods

Radiation monitoring is implemented since the end of 2008 in Tomsk Observatory of Radioactivity and Ionizing Radiation (TORIR). Its technology is constantly being improved. At present TORIR includes Dosimetry Laboratory (Bldg. 10, TPU) and two experimental platforms, which are located on the territory adjacent to the IMCES SB RAS building and its mezzanine. Radiation monitoring in the surface atmosphere includes synchronous continuous automated high sampling frequency (1–10 min.) measuring the characteristics of α-, β-, and γ- radiation fields. Radiation monitoring scheme is given in figure 1.

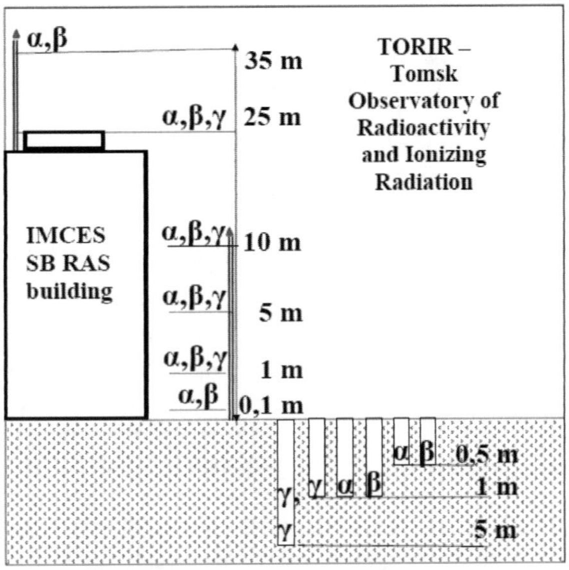

Figure 1. Radiation monitoring scheme

Instrumental equipment included: highly sensitive scintillation α-, β- and γ-radiation detectors, respectively, BDPA-01 BDPB-01, BDKG-03 (ATOMTEX, Republic of Belarus); gas-discharge γ-and β-radiation counters of STS-6 and SBM-19 types (RF); radon isotopes and their DPs radiometers RTM 2200 (SARAD, Germany), RRA-01-03 (RF), Ramon-01 (Kazakhstan); self-engineered devices for measurements of radon flux densities from the ground surface [5]. Ionizing radiation detectors or counters were installed in twos or in threes at series of heights as shown in figure 1.

BDPB-01 and BDPA-01 detectors were installed at heights of 10 cm, 1, 5, 10, 25 and 35 m to measure the flux densities α- and β-radiation in the surface atmosphere, and at depths 10, 20, 50 cm and 1 m – to measure soil radon concentration activity.

BDKG-03 detectors and STS-6, SBM-19 counters were installed at heights of 1, 5, 10 and 25 m to measure the flux density and ambient dose equivalent rate of γ-radiation in the surface atmosphere.

These detectors have been chosen based on the requirements of a wide range of operating conditions: temperatures range from - 30 to + 50 ° C and relative humidity – up to 98%.

Since the intensity of cosmic rays influences on the formation of the background ionizing radiation, the data of neutron monitor of Geophysical Observatory "Klyuchi" of Altay-Sayan Branch of Geophysical Survey SB RAS (Novosibirsk, RF) were used in this work. The data have been registered with 1 minute tact.

In order to determine the degree of external factors influence, the monitoring of meteorological (air temperature, pressure, air relative humidity, wind speed and direction, surface temperature and soil temperature at series of depths), actinometrical and atmospheric-electrical values was performed via automated information measuring system of IMCES. Incoming solar radiation is measured by pyranometer Kipp & Zonen CM-11 and the photometer NILU-UV-6T. The precipitation and snow thickness data were taken from http://rp5.ru/8218/ru.

3. Results and discussion

Analysis of radiation monitoring data revealed a strong seasonal variability of the vertical distribution β- fields and γ-radiation characteristics. An example of the dynamics of the vertical profile of β-radiation flux density and dose rate of γ-radiation is shown in figure 2. This figure also shows the dynamics of the neutron component of cosmic radiation (CR) and variations of the meteorological values. The data was smoothed rectangular filter duration of 10 days.

Since the beginning of April until October and November the vertical profiles of β-radiation flux density and the dose rate of γ-radiation agrees well with model representations [6]. The results of simulation [6] using the Monte Carlo method shows that the dependence of the dose rate of γ-radiation on the height above the Earth's surface is well described by an exponential law, and decreases with increasing altitude. The dependence of β-radiation from a height can be represented as a sum of two exponents [6].

The transition to the winter season is accompanied by a synchronous decrease in the γ-background at all altitudes. The β-background dynamics is substantially similar to, but with small differences. The appearance of snow cover leads to the following: a) first, β- and γ-background ceases to depend on the height; b) then, with increasing thickness snow cover, begins to grow with height. Should be noted, the appearance of snow cover leads to a change the sign of the vertical gradient of β- and γ-background. The above is illustrated in figure 2 as follows: a two-year dynamic of relationships of β-radiation flux density at altitudes of 25 m and 1 m (β(25m)/ β(1m)) and the dose rate of γ-radiation ((γ(25m) / γ(1m) is shown in the third top panel. These data directly point the existence of the annual cycle and a high consistency of year to year variations.

The variations β- and γ-background have weakly expressed maximum in the summer the annual cycle. For β-background the annual cycle changes from year to year, in contrast to the γ-background, for which there is a maximum in November and a minimum in February and March.

Between the height snow cover and atmospheric levels of β-, γ-background there is an obvious dependence. The increase in snow cover from 0.5 m to 1 m leads to the formation of "failure" in the annual cycle of β- and γ-background. This failure clearly expressed in figure 2. Since the "failure" depth (figures 2, 3) depends on the thickness of snow cover, then we construct the dependence of the level of atmospheric β-, γ-background from snow depth (figure 3). The same relationship is shown in figure 3 for the neutron component of the secondary cosmic radiation.

Straight lines in this figure are the regression lines. When carrying out the regression analysis data, satisfying the following conditions: a thickness of snow cover > 1 cm and an ambient temperature <-1 °C, were selected.

Comparing the regression analysis results (figure 3) and snow depth data has shown that in periods of the snow cover growth the levels of β-, γ-background decrease practically linearly and at the same time the dispersion of the experimental data is small.

The dispersion of levels β-, γ- background increases significantly during periods of stabilization of snow depth. If the change in depth of snow cover affects slightly on the level of variations of the neutron component of cosmic rays, the atmospheric pressure change leads to a consistent pressure and neutron component level change. This is due to the fact that the optical thickness of the atmosphere for primary cosmic radiation is much greater than 1 (figure 4, right panel).

The dependence of levels of β-, γ-background on the atmospheric pressure are much less expressed. It is caused by the fact that the pressure is at a maximum in the winter months of the annual cycle, when there is snow cover.

In the process of temperature change during the annual cycle the level of β- background practically unchanged, and the γ-background levels and CR grow poorly (figure 5). To analyze the levels of β-, γ-background and CR dependency on pressure and temperature (figures 4 and 5), all data in the annual cycle, and not only the data period of snow cover have been used as in figure 3.

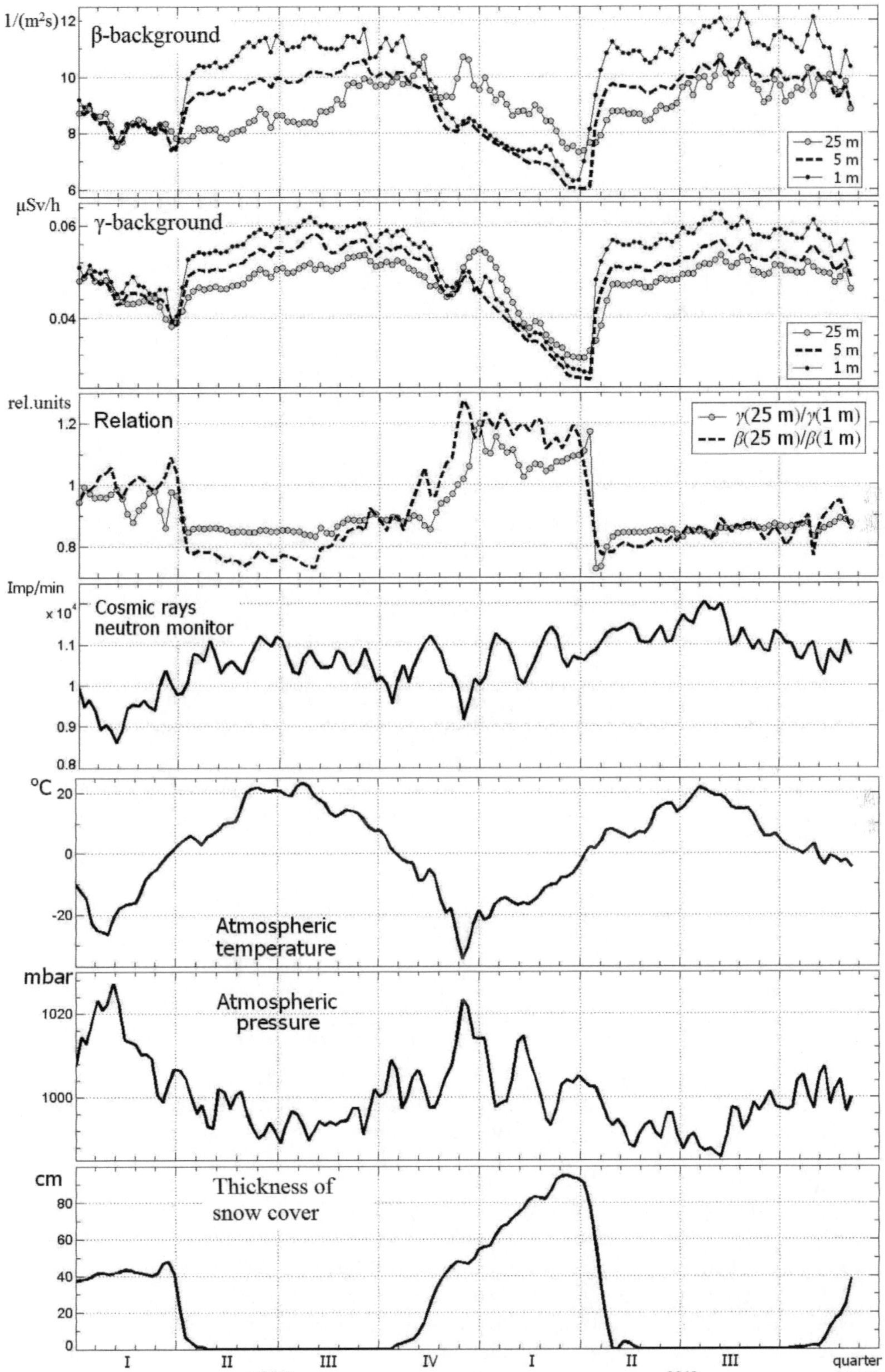

Figure 2. Dynamics of radiation and meteorological values

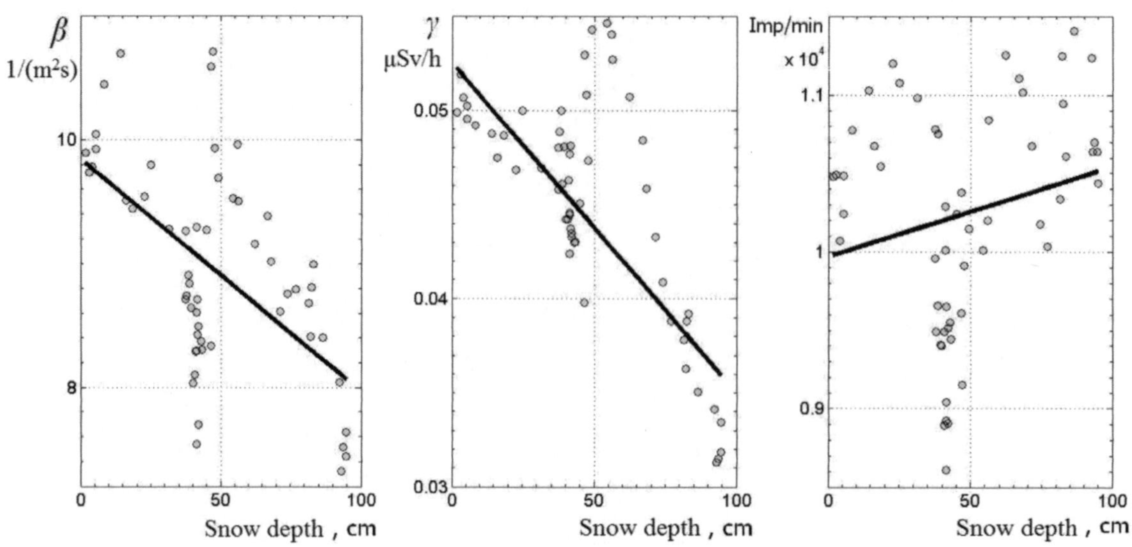

Figure 3. The level β-, γ- backgrounds and CR neutron component depending on the height of the snow cover (left - right)

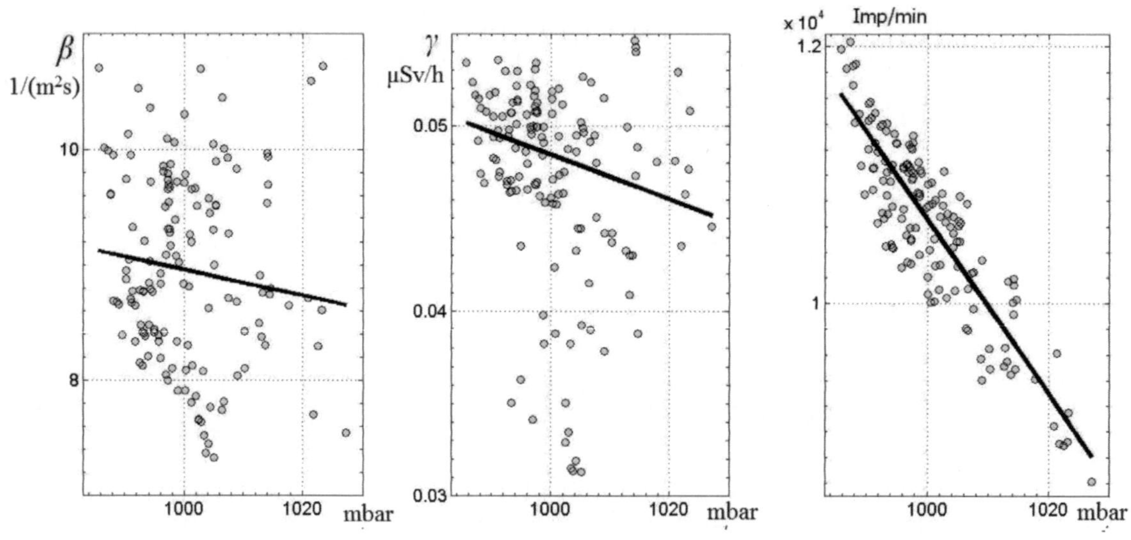

Figure 4. The level β-, γ- backgrounds and CR neutron component depending on the atmospheric pressure (left - right)

5. Conclusion

Analysis of the years' research results revealed a significant seasonal change vertical field profile β- and γ-radiation.

It was found that in the absence of snow cover there is a decrease of β-radiation flux density and γ-radiation dose rate occurring with increasing distance from the earth's surface.

It has been established that in the presence of snow and freezing temperatures in the winter, there is an inverse relationship, i.e., increase of β-radiation flux density and γ-radiation dose rate occurring with increasing distance from the earth's surface.

Furthermore, a linear dependence between snow depth and registered atmospheric β- and γ-background levels in periods of the snow cover growth was revealed.

It is shown that in Western Siberia the radiation dose of people who are outdoors in winter is reduced. This is pressing issue in terms of radiobiology.

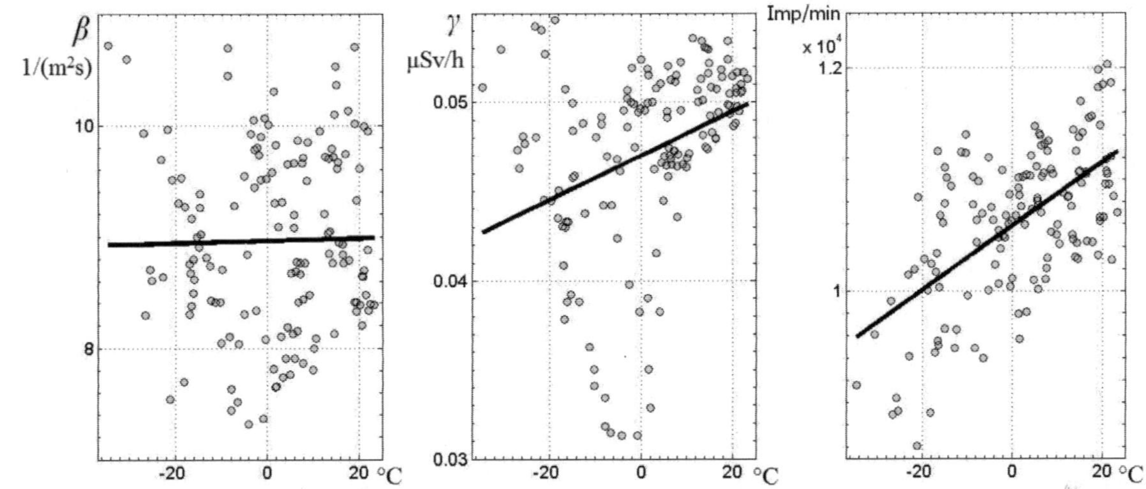

Figure 5. The level β-, γ- backgrounds and CR depending on the atmospheric temperature (left - right)

Acknowledgments

This work is funded by Russian Federal Target Program № 14.575.21.0105 (RFMEFI57514X0105) of Ministry of Education and Science of Russian Federation.

References

[1] Mercier J F, Tracy B L, d'Amours, Chagnon R F, Hoffman I, Korpach E P, Johnson S and Ungar R K 2009 *Journal of Environmental Radioactivity* **100** 527-533

[2] Sobolev, A.I., Tikhomirov, V.A., Verbova, L.F., Mitronova, Yu.N., Zhunov, I.K., 2005 Fundamental research 6, 74-75

[3] Sedunova, Yu.S., 1991. Atmosphere. Handbook. Gidrometeoizdat, Leningrad, 511 p

[4] Burnett J L Croudace I W and Warwick P E 2010 *Journal of Radiological Protection* Volume 30 N.**3.** 525-**532**

[5] Yakovleva V S and Nagorsky P M 2015 *Bulletin of KRAUNTS. Physical and mathematical sciences* **1** 65

[6] Yakovleva V S, Karataev V D and Zukau V V 2011 *Bulletin of KRAUNTS. Physical and mathematical sciences* **1** 65

IOP Conf. Series: Materials Science and Engineering **135** (2016) 012037 doi:10.1088/1757-899X/135/1/012037

Digitally focused array ultrasonic testing technique for carbon fiber composite structures

Y Salchak, V Zhvyrblya, D Sednev and A Lider

National Research Tomsk Polytechnic University, 634050 Tomsk, Russia

yasalchak@gmail.com

Abstract. Composite fiber reinforced polymers are highly promising structures. At present, they are widely used in different areas such as aeronautics and nuclear industries. There is a great number of advantages of composite structures such as design flexibility, low cost per cubic inch, resistance to corrosion, lower material costs, lighter weight and improved productivity. However, composites degradation may be caused by different mechanisms such as overload, impact, overheating, creep and fatigue. Comparing to inspection of other materials some unique consideration is required for testing and analysis. Ultrasound testing is the most common method for inspection of composite structures. Digitally Focused Array Technology is considered as novel approach which enables fast and effective quantitative automatic testing. In this study new methodology of quality assurance of composite structure components based on DFA is performed.

1. Introduction

Application of low quality equipment in manufacturing of composite structures that does not provide required accuracy can cause occurrence of defects that decrease physical and mechanical properties of the material. Furthermore, defects can appear because of manufacturing process failures such as impurity of the material surface or other.

The manufacturing process of composite carbon fiber reinforced polymers is very complicated and longterm procedure. Any inconsistencies can affect the integrity and reliability of the component. Performance of a composite is mostly affected by structural discontinuities, especially if they are exploited under special conditions such as high-temperature loading.

There are several methods of quality assurance that can be applied for composite structures. There are different visual, eddy-current, ultrasonic and radiographic techniques. Composites susceptible to manufacturing defects that compromise structural integrity and generally are not visible to the naked eye. Two factors contribute to increase of the need for better method of evaluating the quality. On the one hand, composite materials are usually used in structurally critical areas, such as military, nuclear and aeronautic. On the other hand, application of composites becomes more and more popular and requires faster inspection as well as ability to detect smaller defects [1].

Ultrasonic inspection has become the most widely used technique for detection of internal defects in composites laminates. It is relatively low-cost, highly sensitive and does not cause any additional exposure as radiographic methods. Moreover, radiography often cannot be applied because the material sufficiently absorbs X-Rays.

2. Principles of Digitally Focused Array Technique

Nowadays, ultrasonic techniques are usually based on application of phased array transducers. These types of sensors consist of several elements that can generate an acoustic beam. The main principle is to get interference of acoustic waves. It is possible to change delay laws for each element in order to form particular wave front with higher amplitude on given distance from the surface. Thus, there is a focus point within the highest amplitude that enables to detect smaller flaws and flaws with uncertain boundaries. However, for determination of another focus point the operator will have to generate new delay laws which is time-consuming. Also, it will take a lot of separate measurements to build an image of the whole volume of the component.

In this study we propose the methodology of application the state-of-the-art ultrasonic called Digital Focused Array (DFA) [2-3]. The main principle is to generate a separate impulse by each element of the array, one after another (see Figure. 1).

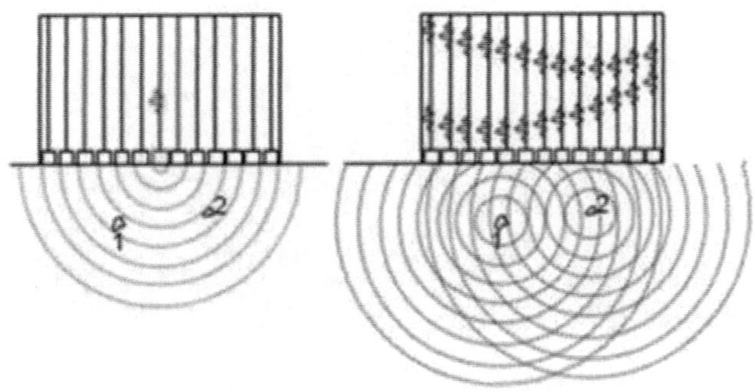

Figure 1. Principles of Digitally Focused Array Technique.

Taking into account that an element has a broad beam pattern, generated waves are propagating in all directions. Further, the defects that occur in the material will reflect the wave and will become new sources of secondary waves. Secondary waves are received by all of the elements and form an electrical signal. These signals contain the time of propagation and amplitudes of reflected waves. The achieved signals are then amplified, digitized and filtered in order to get higher signal-noise ratio.

This algorithm is implemented for each element of the phased array, which has a scanning path along the profile in both directions with a given step. The achieved results are used for creating 2D and 3D images of the controlled object.

The resolution depends on the duration of the pulse, dimensions of the array and the velocity of ultrasound. The maximum resolution is half of the ultrasound wavelength.

DFA technique has a number of noticeable advantages such as:

1) Simultaneous scanning on each angle of incident;
2) Possibility to increase resolution by decreasing inspection time;
3) Obtaining reliable results by determining the position of each flaw;
4) The results are generated as 3D images that can significantly simplify interpretation procedure.

As follows, in this study the methodology of quality assurance of composite structures based on DFA algorithm is proposed.

3. Experimental Procedure

There are high accuracy and reliability requirements to modern ultrasonic systems, especially to the instruments that are used for testing complex shaped objects. The ultrasonic equipment used for the experiment is demonstrated in Figure 2. The system consists of three main parts: instrumental cabinet unit within the PC (1), immersion tank (2) and robotic manipulator (3).

The scanner performs positioning of the transducer on the given distance from the controlled object.

Correct choice of the transducer is to determine the reliability of testing objects made of different materials and with different geometries.

As it was mentioned earlier, the maximum resolution depends not on the methodology but on the wavelength. Within increasing the transducer's frequency, the resolution is also increasing, however the control depth gets smaller. The transducer used for the experiment is manufactured by Doppler Electronic Technologies Co., its main characteristics are shown in Table 1.

Figure 2. Experimental setup.

Table 1. The parameters of the transducer.

Frequency, MHz	5
Number of elements	128
Element width, mm	1
Gap between elements, mm	0.2
Element length, мм	8

The loss of the amplitude of a signal can be caused by scattering or divergence. This entails the necessity to provide accurate setting for the control system.

While testing well-known materials such as steel, for example, the calibration of instrument sensitivity is made by DGS diagrams. DGS diagram consists of curves representing the character of amplitude attenuation within the increase of the defect occurrence depth. These curves demonstrate the amplitude behavior for defects of different dimensions (Figure 5).

In Figure 5 the vertical axis corresponds to amplitude decrease and the horizontal axis represents relative depth measured in the number of transducer near fields.

However, the manufacturing process requires constant change of composite layers' material that will unpreventably lead to the changes of acoustical properties. That is why for the inspection of composites it is proposed to produce a reference sample according to the given acoustical properties and manufacturing parameters.

There is one major requirement for the reference sample: it is to represent all necessary defects at all possible depth of occurrence. Also, it is preferable for the sample to imitate different geometrical parameters of the controlled object. The first thing that should be taken into account before designing the sample is multiple reflections of the signal from the specimen surface (see Figure 3).

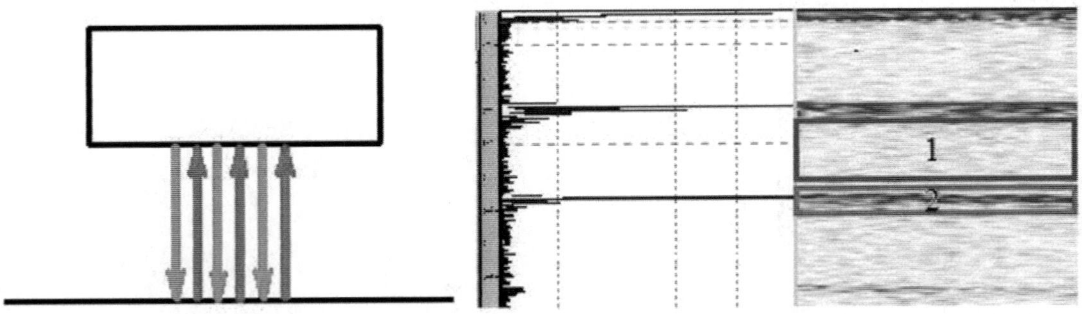

Figure 3. Multiple reflections of the signal between transducer and the surface.

This situation may lead to the interference of multiply reflected signal from the surface and signal from the defect that can occur in wrong interpretation. Figure 3 demonstrates that if, for example, defect occurs in zone 1 it will be easily detected, however if the impurity is in zone 2 the operator will miss it.

In order to avoid this problem, the distance between the controlled object and the transducer is to be chosen considering that the time of flight of ultrasound in the coupling material is longer than inside the component. The sound velocity in water is 1500 m/s and in carbon fiber structure it is about 2900 m/s. Thus, the maximum time of flight of ultrasound in the object is:

$$t_o = \frac{2 \cdot h}{2900}, \qquad (1)$$

where h is the width of the component.

For generating the second multiple reflected echo behind the back-wall echo, the distance between the surface and transducer is to be calculated according to the following equitation:

$$l > t_o * c = \frac{2 \cdot h}{2900} \cdot 1500 \approx 1.035 \cdot h, \qquad (2)$$

where l – distance between the surface of the object and the transducer.

4. Results and Discussion

The proposed methodology of composites quality assurance is adjusted on the specified reference sample for system calibration and control samples for the validation procedure.

Delamination is the type of defects that can often be found in composite structures. Delamination is caused by improper surface preparation, inclusion of foreign matter or impact damage during handling [4]. The experimental sample that was used for inspection of delamination is shown in Figure 3 and Figure 4.

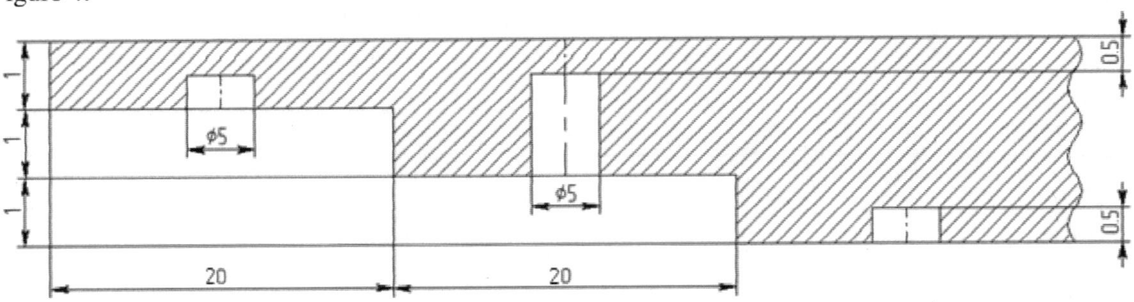

Figure 3. Reference sample technical drawing.

The reference sample is made as stepped structure with flat-bottom holes. The step is 1 mm, the length of a step is 20 mm, the diameter of each hole is 5 mm. In order to simulate different defect occurrence depth, the holes are made near the surface or at the bottom. The depth of each flat-bottom hole interchanges in one step. This construction allows to simulate components of different depth and with different defect occurrence depth.

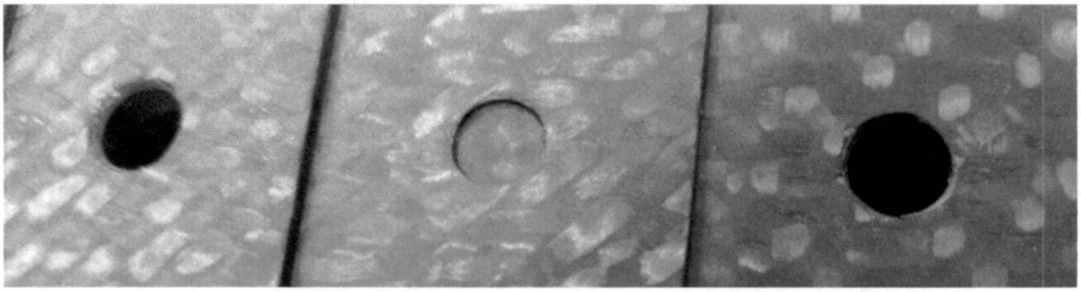

Figure 4. Flat-bottom holes on the reference sample.

The calibration of the ultrasonic system requires DGS diagram of the controlled object. For this purpose, it is necessary to determine attenuation of the back-wall echo for different values of depth. In this study the attenuation was evaluated according to the following algorithm:

1) The ultrasound transducer is placed above the first step that corresponds to the thickness of 1 mm;

2) The gain is adjusted in order to achieve 80% of the screen for the back-wall echo (16 dB);

3) The ultrasound transducer is placed above the following step and the procedure is repeated.

The results are shown in Table 2 that contains gain values that are to be chosen for signal attenuation compensation (Table 2).

Table 2. Gain values.

Defect depth, mm	1	2	3	4	5	6	7	8	9	10	11
Gain, dB	16	19	23	27	30	35	36	39	42	44	46

Taking into account that DGS diagram in the x-point equals to zero we can build DGS diagram (see Figure 5).

For the implementation of this relation time corrected gain was applied. Unfortunately, the software does not allow application of nonlinear time corrected gain. Thus, time corrected gain with constant tilt was applied. It can be shown as a linear correlation graph with the tilt of 3 dB per 1 mm. The measurement results after the correction are demonstrated in Figure 6 below.

IOP Conf. Series: Materials Science and Engineering **135** (2016) 012037 doi:10.1088/1757-899X/135/1/012037

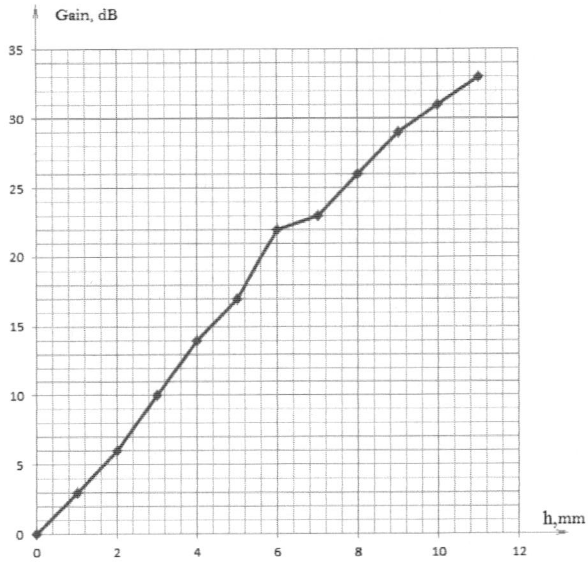

Figure 5. DGS diagram of composite component.

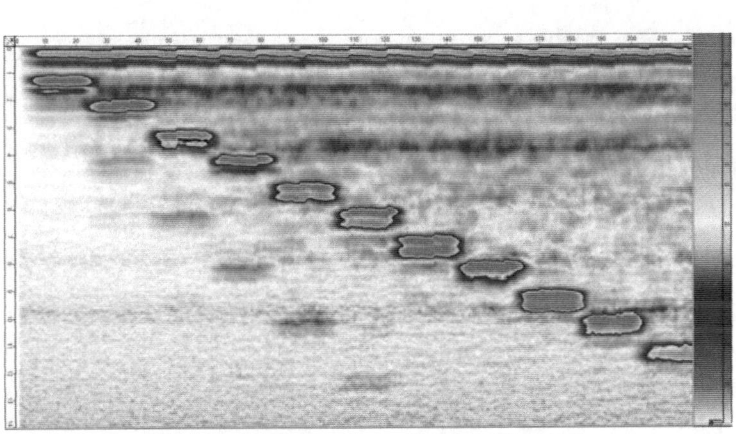

Figure 6. The back-wall echo (B-scan).

The inspection of flat-bottom holes is made after the calibration procedure due to the fact, that the back-wall echo is much stronger than the echo from the bottom. Thus the gain is preliminary increased by 3 dB (see Figure 7). The holes near the surface (1), bottom holes and shadows are clearly shown (3).

In order to validate the proposed technique, the experiment of testing the control sample was conducted. The control sample was made with artificial delamination. The results of the experiment are demonstrated in Figure 8 a that represents C-scan and Figure 8 b that represents tomography results. It is quite evident that there are 7 indications that correspond to delamination of the structure.

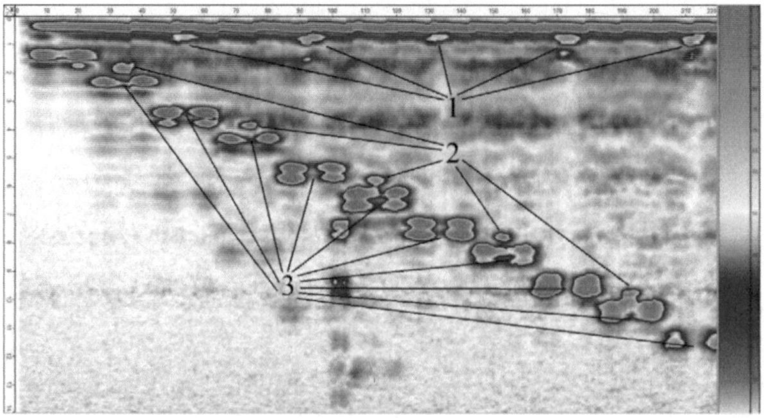

Figure 7. Flat-bottom holes' echoes (B-scan).

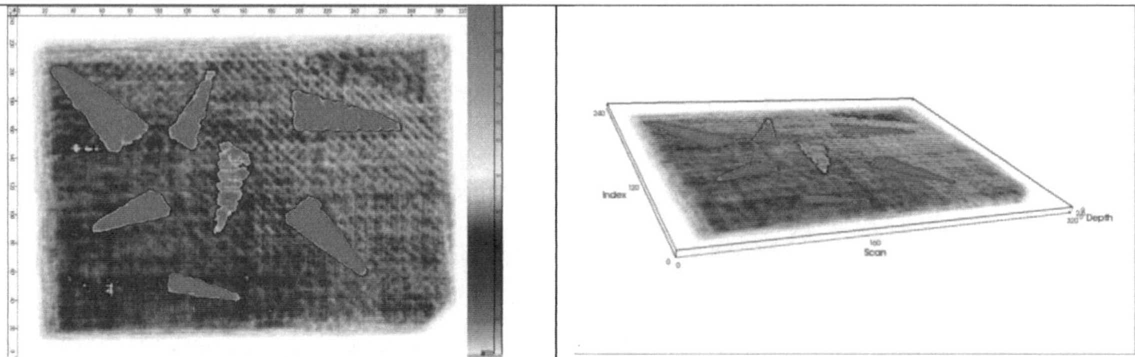

a) C-scan of experimental results

b) Tomography results of the control sample testing

Figure 8. Experimental results.

Conclusion

In this study the methodology for composite quality assurance is proposed. It is based on ultrasonic DFA technique. The methodology was validated according to experimental results. In further work it is planned to use the suggested technique for testing composite materials produced by additive manufacturing method. Moreover, within the development of high-quality automatic evaluation method it will be possible to create an innovative system for testing complex-shaped objects.

Acknowledgments

Research has been conducted with financial support of the Government of the Russian Federation (The Ministry of Education and Science) agreement № 02. G25.31.0176.

References

[1] Kapadia A, *Best Practice Guide*, 2007 48.
[2] Bernus L, Bulavinov A, Dalichow M *In: Insight*. **9**, September 2006, pp 545-549,
[3] Bulavinov A, Kröning M, Pudovikov S, Oster R, Hanke R, Hegemann U, Reddy K and Venkat R 2008 *International Symposium on NDT in Aerospace* p 21
[4] Smith R *Materials Science and Engineering* 2009 **3** pp 103-143
[5] Bulavinov, A, Pinchuk, R, Pudovikov S *European Conference for Non-destructive Testing*, Moscow 2010 (pp. 1-9).
[6] Mallick P *CRC Press Taylor & Francis Group* 2007 **3**

Preparation and structural stability of ordered nanocomposites: opal matrix – lead titanates

M. I. Samoylovich [1,4], A.R. Mkrtchyan[2,3], A. F. Belyanin[1,4], S.A. Bagdasaryan[4], A.A. Kiziridi[3]

[1] Central Research Technological Institute "Technomash", Russia, Moscow
[2] Institute of Applied Problems of Physics of NAS, Republic of Armenia, Yerevan
[3] National Research Tomsk Polytechnic University, Tomsk, Russia
[4] Joint Stock Company "Scientific Engineering Center "Technological developments of telecommunication and radio frequency identification", Russia, Moscow

E-mail: malpic@sci.am

Abstract. The conditions for the formation of nanocomposites based on the basis of lattice packings of SiO_2 nanospheres (opal matrices) with included crystallites of lead titanates ($PbTiO_3$ and $PbTi_3O_7$) in interspherical nanospacing are considered. For the formation of nanocomposites are used sample opal matrices with dimensions of single-domain regions ≥0,1 mm.[3] The diameter of SiO_2 nanospheres was ~260 nm. Obtained nanocomposites volume >2 cm^3 in filling >20% of interspherical nanospacing $PbTiO_3$, $PbTi_3O_7$ crystallites were size of 16–36 nm. Using X-ray diffraction and Raman spectroscopy are studied composition and structural stability when heated nanocomposites to 550°C, which allowed the identification of a local phase transition with change of the space group. The dependence of the composition of synthesized materials on the conditions of their preparation is submitted.

Keywords: nanocomposites, opal matrix, lead titanates, X-ray diffractometry, Raman spectroscopy.

1. Introduction

Recently there is growing interest to multiferroics - perspective materials for the production of elements of devices for reciprocal conversion of magnetic and electric signals. This property is related to a big magnetoelectric effect observed in series of compounds, for example, in manganese and titanium oxides - the titanates of rare earth elements (REE). In these materials the antiferromagnetic, magnetic and ferroelectric properties coexist at low temperature and they can be used as magnetic carriers and detectors [1-3]. There are reasons to suppose the emergence of construction of devices with tunable phase velocities in the optical, microwave and terahertz range with the use of such materials [4,5].

For massive materials, the frequency restrictions often are associated with the dispersion of dielectric permittivity and magnetic permeability, therefore, it is promising the use of multiferroic materials in the form of ordered 3D nano-lattices of crystals (clusters) of the size 15-50 nm, the use of which will allow to extend the frequency range of their application and to design high operation speed devices controllable by electric and magnetic fields. The ordered 3D nano-lattices of multiferroic crystals can be created with the use of opal matrices, by filling their between spherical cavities with multiferroics [6,7]. The opal matrices are 3-demensional lattices of close packing of nanospheres of X-ray-amorphous silica (SiO_2) with the nearly same diameter (Δd<5%), the sizes of which, depending on the production process conditions, may vary within given limits of 200-350 nm. The perspective materials for filling in the between-spherical cavities of the opal matrices are multiferroics such as

REE titanates or metals with the high melting temperature and having segnetoelectric properties. The antiferromagnetic properties are detected for dititanates of REE (frustrated materials – $Er_2Ti_2O_7$, $Dy_2Ti_2O_7$, $Yb_2Ti_2O_7$ и $Ho_2Ti_2O_7$) [8].

The composition and the structure of the materials in the synthesis in a closed nanosize volume (the process of formation or recrystallization under the conditions of limited geometry of the crystallizer) in the role of which the between spherical cavities of the opal matrices are used, are significantly different from the results of the synthesis of massive samples. The synthesis in a nanosize volume of between-spherical cavities of opal matrices contributes to the stabilization at low temperatures of the high-temperature phase and to the decrease of the formation temperature of the synthesized material [6,9]. Earlier, the construction, the magnetic and dielectric properties of nano-composites on the basis of opal matrices have been obtained and studied, in the between-spherical cavities of which the crystals of REE titanates —oxo-titanates with general formula R_2TiO_5 and de-titanates $R_2Ti_2O_7$ (Er_2TiO_5, $Dy_2Ti_2O_7$, $Gd_2Ti_2O_7$, La_2TiO_5, $Pr_2Ti_2O_7$, $Tb_2Ti_2O_7$ и $Yb_2Ti_2O_7$) were synthesized [9].

In the present paper, we investigate the influence of the formation conditions on the composition and the structural stability of the nano-composites on the base of the opal matrices in the between-spherical cavities of which the crystals of lead titanates ($PbTiO_3$ и $PbTi_3O_7$) are obtained.

2. Obtaining and the structure of the opal matrices and nano-composites on their basis

The opal matrices were obtained by the hydrolysis reaction of tetra-ether of orthosilicic acid ($Si(OC_2H_5)_4$) with the solution of ethanol (C_2H_5OH) in the presence of ammonium hydroxide (NH_4OH). The samples of 3D ordered packings of SiO_2 nano-spheres were manufactured with the diameters of ~260 nm, with the volume of 2–3 cm^3 and with the size of mono-domain areas >0,1 mm^3 and $\Delta d \approx 5\%$. In Fig. 1a, we demonstrate two layers (1,2) of the densest cubic lattice of SiO_2 nano-sphere packing (3). Each nano-sphere A with the densest planar layer is surrounded by 6 triangular gaps of different orientations (B and C). The upper layer of the nano-spheres is relatively oriented with respect to the lower one in the positions B or C. Under the conditions, used in the paper, a three-layer (cubic) structure .../ABC/... has been formed. The densest packing of the nano-spheres forms tetrahedron and octahedron between-spherical cavities, the volume of which equals 25,95% of the opal matrix volume [10]. By connecting the centers of the six and four nano-spheres, forming the cavities, we obtain octahedrons (Ok) and tetrahedrons of different orientations (T1 and T2), respectively (Fig. 1,a). The octahedrons and tetrahedrons entirely fill the space (Fig 1,b).

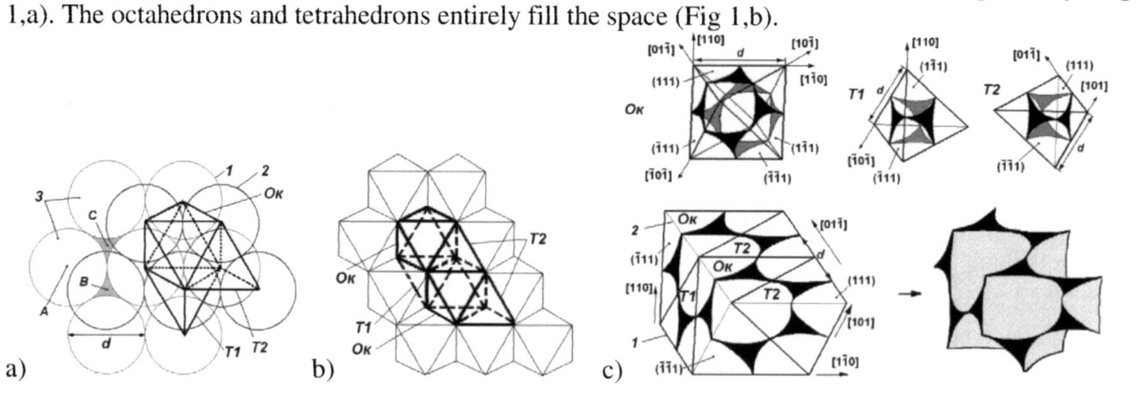

Fig. 1. a) The structure of two densely packed layers of SiO_2 nano-spheres of diameter d (projection on the plane {111}); the centers of SiO_2 nano-spheres, forming face-centered cubic lattice, form octahedron and tetrahedron cavities. b) The presentation of the opal matrix structure in the form of octahedrons (Ok) and tetrahedrons (T1 and T2) filling the space. c) Octahedrons (Ok) and tetrahedrons (T1 and T2) formed by SiO_2 nano-spheres and the volume model of the material, filling the between-spherical cavities highlighted in b of two octahedrons and three tetrahedrons.

In Fig 1, c , on the faces of tetrahedrons and octahedrons, the X-cuts of the between-spherical cavities in the region of the contacts of tetrahedrons and octahedrons are shown. The between-spherical tetrahedron and octahedron cavities can be filled with different materials. In Fig. 1,c, we show the volume model of the material, filling five between-spherical cavities, existing in two-layer packing (two octahedron (Ok) and three tetrahedron (one-T1 and two T2)).

The nano-composites on the base of the opal matrices, the between-spherical cavities of which are filled by the crystallites $PbTiO_3$, $PbTi_3O_7$, were obtained by the impregnation method including the filling the cavities with the solution of salts of different chemical compound of lead or terbium followed by the thermal treatment, during which the required material is formed. The successful application of the impregnation method is conditioned by the rather wide channels of between-spherical cavities of the opal matrices, which have maximal size ~0,37*d* on the joint of tetrahedrons and octahedrons (shown in Fig. 1, c).

For obtaining lead titanates, in the between-spherical cavities of the opal matrices, we have introduced by impregnating the lead nitrate, dissolved in the 15% solution of trichloride of titan ($TiCl_3$) in a weak concentration solution of hydrochloric acid (HCl). The impregnation was conducted up to 8 times with the following low-temperature thermal treatment in the range of tempertures 300–500°C along with a gradual filling the between-spherical cavities of the opal matrices. The concentration of the dissolved lead nitrate was calculated according to the ratio of the components of the synthesized material. The different phase composition and the phase structure of the synthesized material depends on the composition of the impregnation solutions and on the conditions of the thermal treatment of the nano-composites. The synthesized materials filled >20% of the volume of between-spherical cavities of the opal matrices.

3. The Phase Composition of Nano-Composites

X-ray diffraction

For the identification of crystal phases synthesized in the between-spherical cavities of the opal matrices, the diffractiometer *ARL* X'tra (Cu *k*α radiation, λ=0,154178 nm, an energy dispersion solid detector with Pelette cooler, step size 0,01°, continuous mode 1°/min) was used. The X-ray-grams were analyzed by using the automated database *ICDD PDF*-2.

The X-ray diffractograms of the opal matrix samples, the between-spherical cavities of which are filled by lead titanates, are presented in Fig.2. The X-ray-grams are characterized by different sets and relations of the intensities of reflections (I). The X-ray-phase analysis allowed to confirm the presence of the following crystal phases: $PbTiO_3$ (tetragonal crystal system, space group *P4mm*) (Fig. 2, curve I); $PbTi_3O_7$ (monoclinic crystal system, $P2_1$) (Fig. 2, curve 2). At temperatures higher than 800°C when titan containing salts were inserted, as a rule, polycrystal particles of different titan oxides were formed: TiO_2 (rutile, tetragonal crystal system $P4_2/mnm$); $Ti_{0.72}O_2$ (synthetic anatase, tetragonal crystal system, $I4_1/amd$).

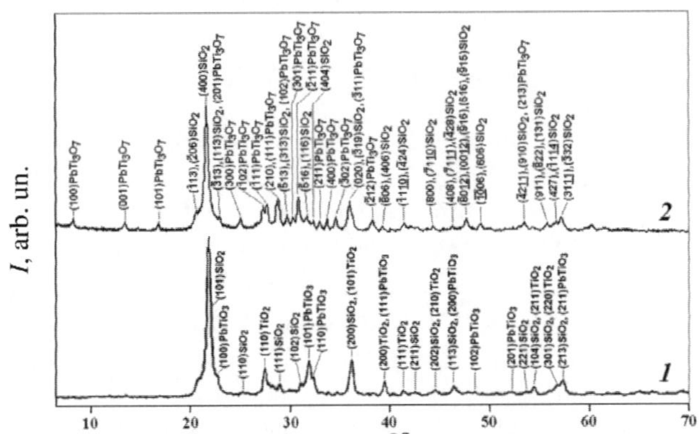

Fig. 2. The X-ray diffractograms of the opal matrix samples, the between-spherical cavities of which are partially filled with $PbTiO_3$ (*1*) and $PbTi_3O_7$ (*2*) crystals.

The high temperature thermal treatment in the synthesis of lead titanates is accompanied by the formation of SiO_2 crystal phases (re-crystallization of X-ray-amorphous silica) (Fig 2): SiO_2 (crystobalite, tetragonal crystal system, $P4_12_12$); SiO_2 (tridymite, crystal system, $P6_3/mmc$); SiO_2 (monoclinic crystal system, *I2/a*). At temperatures ≤700°C, SiO_2

crystal phases are not formed. At temperatures ≥1000°C the interaction takes place between the silicon forming materials synthesized in the between-spherical cavities, containing Pb, and silica, during which SiO_2 nano-spheres are deformed. Under the conditions, used used in this paper, the synthesized titanates of Pb don't interact with SiO_2. The crystal size (the areas of coherent scattering of X-rays, L_{OKP}) of the crystal phases $PbTiO_3$, $PbTi_3O_7$ is determined by the widening of the diffraction maxima on the X-ray diffractograms by the Selyakow-Scherrer formula. The synthesized lead titanates had L_{OKP} in the range of 16-36 nm. The synthesized materials present equiaxed crystals. It was established that L_{OKP} for $PbTiO_3$, $PbTi_3O_7$ doesn't depend on the degree of crystallinity.

Spectroscopy of Raman scattering of light

Besides crystal phases, the studied samples contain X-ray-amorphous phases. The analysis of different materials, including X-ray-amorphous ones, is possible with Raman scattering (RS) spectroscopy of light. The RS spectra were registered by using the RS micro-Raman spectrometer LabRAM *HR*800 (HORIBA Jobin-Yvon) (line 632,8 nm of a He-Ne laser; the laser power <300 mWt; diameter of the beam spot ~4 mcm^2; the depth of the analyzed layer ~3 mcm), equipped with a console *THMS*600/720 (Linkam Scientific Instruments) for sample heating. The RS spectra for the nanocomposites, containing $PbTiO_3$ and $PbTi_3O_7$, are presented in Fig.3,4. The position (RS shift Δv) and the intensity (*I*) of the stripes on the RS spectra depend on the conditions of their obtaining.

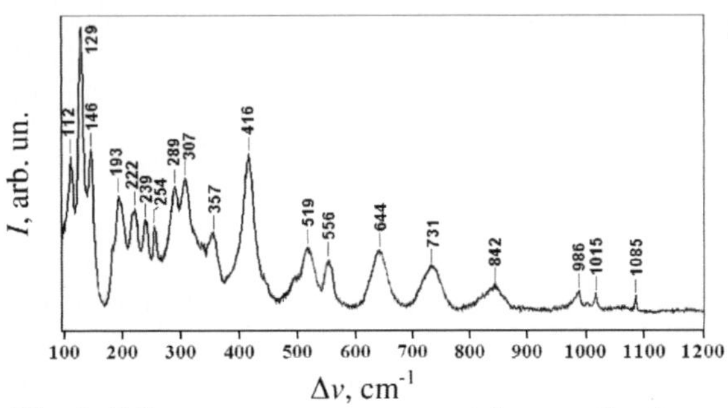

Fig. 3. RS spectra of the opal matrix sample, the between-spherical cavities of which are partially filled with $PbTi_3O_7$ crystals.

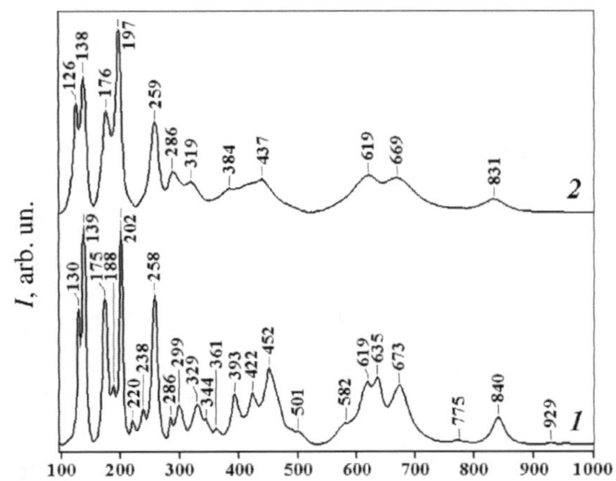

Fig. 4. RS spectra of the opal matrix sample, the between-spherical cavities of which are partially filled with $PbTiO_3$ crystallites, the initial *(1)* and heated to the temperature 550°C (*2*).

For the nanocomposites containing $PbTiO_3$ we have obtained the RS spectra during the sample heating process (Fig. 4). At room temperatures, $PbTiO_3$ has tetragonal crystal system of perovskite type (*P*4*mm*). When heated to the Curie temperature, equal to 493°C, a segnetoelectric phase transition takes place from the tetragonal crystal system to the cubic (*Pm*3*m*) one. The influence of the temperature and other factors leads to the change of phase composition (including the X-ray amorphous phases) of the titanates. The presence and the ratio of the stripe intensities on the RS spectra depend on the conditions of heat treatment. The exposure at high temperatures does not change the features of the RS spectra.

On the RS spectra of the nano-composites containing $PbTiO_3$, $PbTi_3O_7$ and $TbMn_2O_5$ the stripes of different modes of SiO_2 are seen. In the presented RS spectra of the nano-composites containing $PbTiO_3$ and $PbTi_3O_7$ (Fig. 3,4) the stripes are seen for Δv equal to 220–222, 286–289, 299 and 775 sm^{-1}. The derivation of the spatial groups of crystal symmetry is based on the requirement of the translation invariance of its lattice, and also on the finiteness of discrete (point) subgroups of the orthogonal group. The information obtained by the RS spectroscopy, as a rule, in the case of crystals, expresses the peculiarities of the mentioned subgroups (crystal class). Taking into account that such subgroups are not the crystal movement groups, the RS studies enable to obtain the data on the local phase transformation.

4. Conclusion

In the present we have investigated the influence of the formation process on the composition of the nanocomposites on the basis of opal matrices in the between-spheric cavities of which the crystal titanates of lead ($PbTiO_3$, $PbTi_3O_7$) are synthesized. In obtaining the nanocomposites, the samples of opal matrices (3D lattice packings of nano-spheres of X-ray amorphous silica (SiO_2) with the diameter of ~260 НМ) having the volume >2 cm^3 and the mono domain area size >0,1 mm^3 were used. The nanocomposites on the basis of the opal matrices, the between-spherical cavities of which are filled with $PbTiO_3$, $PbTi_3O_7$, were obtained by the method of impregnation, including the filling up the cavities with the solutions of different chemical compounds with the following thermal treatment. In the synthesis of the lead titanates, the solution of Pb nitrate in the 15% solution of $TiCl_3$ in HCl was introduced in the between-spherical cavities. For the crystallization of materials with a given composition, a high-temperature treatment was carried out at temperatures 700–1000°C. The synthesized materials filled >20% of the volume of the between-spherical cavities of the opal matrices. The different phase composition and the structure of phases of the synthesized materials depend on the composition of the impregnating solutions and on the conditions of thermal treatment of the nanocomposites.

We have studied the peculiarities of crystallization and phase transitions (when heated to 550°C) of the compounds ($PbTiO_3$), synthesized in the between-spherical cavities of the opal matrices. The composition of nanocomposites is studied by the methods of X-ray diffraction and RS spectroscopy. It is confirmed that the materials synthesized in the between-spherical cavities are made of crystal (in the form of crystallites) and X-ray-amorphous phases of titanates with the presence of oxides of metals. $PbTiO_3$ and $PbTi_3O_7$ synthesized in the between-spherical cavities had the size of crystallites in the range 16-36 nm. The obtained data are results of the synthesis under the limitations with external walls of the between-spherical cavities of crystallite sizes (the process of the formation and re-crystallization under the conditions of limited geometry of a crystallizer).

The understanding of the peculiarities of phase transitions in the between-spherical cavities of opal matrices, in obtaining given materials, will allow to create metamaterials with controllable improved properties of the functional properties and operation characteristics, which is very promising for the future applications of devices. It is assumed that the approach connected with the use of such nanocomposites will be most efficient in the development of different solid state microwave devices: filters, delay lines, phase shifters and other controlling elements.

Acknowledgement

The work is partially supported by RFFS (Grant 12-07-00529A) and by the Competitiveness Enhancement Program of Tomsk Polytechnic University.

References

[1] Kimura T., Goto T., Shintani H. *at al* 2003 *Nature*. **426** 55
[2] Hill N.A. 2000 *The journal of physical chemistry B*. **104** 6694
[3] Inomata A., Kohn K. 1996 *Journal of physics condensed matter*. **8** 2673
[4] Sarychev A.K., Shalaev V.M. *World scientific and Imperial college press*, 2007. 200 p.
[5] Kong J.A. 2002 *Progress in electromagnetics research, PIER*. **35** 1
[6] Samoylovich M.I., Belyanin A.P., Bagdasaryan A.S. *at al* 2015 *Naukoemkie tekhnologii*. **16** 75
[7] Samoylovich M.I., Belyanin A.P., Bagdasaryan A.S. *at al* 2015 *Naukoemkie tekhnologii*. **16** 42
[8] Bramwell S.T., Field M.N., Harris M.J. *at al* 1999 *Journal of physics condensed matter*. **12** 483
[9] Samoylovich M.I., Belyanin A.P., Bagdasaryan A.S. *at al* 2015 *Nanoinzheneriya*. **11** 12

IOP Conf. Series: Materials Science and Engineering **135** (2016) 012039 doi:10.1088/1757-899X/135/1/012039

Ultrasonic fingerprinting by phased array transducer

D Sednev, O Kataeva, V Abramets, P Pushenko and T Tverdokhlebova

National Research Tomsk Polytechnic University, 634050 Tomsk, Russia

sednev@tpu.ru

Abstract. Increasing quantity of spent nuclear fuel that must be under national and international control requires a novel approach to safeguard techniques and equipment. One of the proposed approaches is utilize intrinsic features of casks with spent fuel. In this article an application of a phased array ultrasonic method is considered. This study describes an experimental results on ultrasonic fingerprinting of austenitic steel seam weld.

1. Introduction

Modern ultrasonic testing (UT) procedure more often starts to use phased array transducers for both manual and automated inspections. However, reconstruction of inspection results possible only with automated testing, because it is necessary to provide a precise data on position of a transducer. Ultrasonic fingerprinting procedure is requiring a precise position data also. Moreover, the maximum efficiency will be achieved when reconstructed ultrasonic images is using as fingerprint. Manual fingerprinting technique approved a feasibility of proposed concept [1]. Automatization of the procedure should increase a matching rate. However, a coupling issue still stays vital for performing a UT, especially for investigation weak reflectors like grains, that is in use for fingerprinting. The best option for coupling is immersion contact. It could be realized by local contact or immersion tank. Unfortunately, there is no commercially available equipment for local immersion contact for complex shape parts, because usually it is customer-tailored product.

This paper describes a series of experiments on immersion ultrasonic fingerprinting procedure.

2. Experimental samples and installation

Experiments were conduct on 4 samples made from austenitic steel (Russian nomenclature is 12X18H10T). Special characteristics of this steel allow to use it in different industrial applications, e.g. nuclear industry. Russian-type casks for spent nuclear fuel (SNF) made from this material [2]. In order to model SNF cask samples were made from two plates jointed by butt weld.

Edge preparing mechanically before welding. Edge and the adjacent surface are stripped from the scale, paint, oil and other contaminants on the surface width of 20 mm outer and 10 mm inner side. The surface is degreased with pure alcohol. The welded joint is performed by an automatic TIG welding consumable electrode. The centering device is applied during welding. The deviation from linearity does not exceed 0.8 mm per 100 mm. The following materials are used during welding: welding wire St 04H19N11M3 (Ø 1,2 mm), tungsten electrode EVL (Ø 2,4 mm), argon as a protective gas. Additional heat treatment is not carried out [2].

Weld metal characterized by cellular-shaped dendritic crystallization. This form is an intrinsic feature of the material, providing a set of scattering centers, that allows to generate the fingerprint.

IOP Conf. Series: Materials Science and Engineering **135** (2016) 012039 doi:10.1088/1757-899X/135/1/012039

The experiments were conducted on laboratory multichannel ultrasound system. The block diagram is shown in Figure 1.

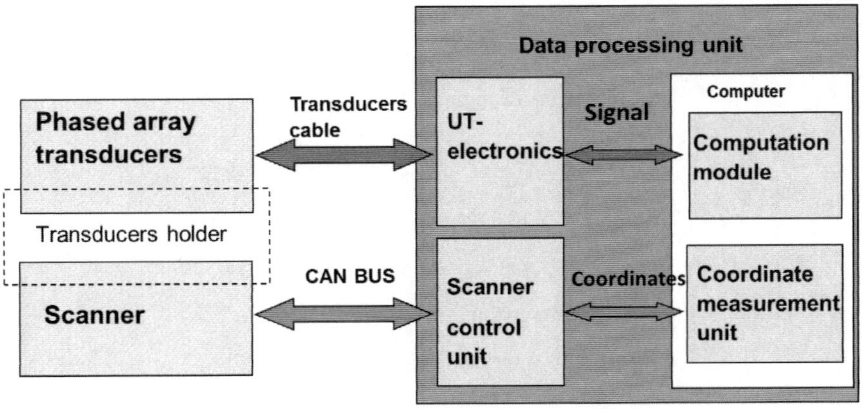

Figure 1. Ultrasound multichannel system block diagram.

The basis for the system is a unit of ultrasonic OPTUS electronics, manufactured by I-Deal Technologies GmbH. The unit has a transmitter\receiver 128 channels module that could be used for both single- and multichannel measurements. Thus, this unit can be used for a wide range of automated ultrasonic inspection tasks, e.g. the ultrasonic fingerprinting.

As a phased array were used linear probe 5L16-1.0·16-D89-P1 produced by Doppler Electronic Technologies Co. Ltd. The array contains 16 elements with 1 mm pitch. Frequency is 5 MHz. The transducer were placed on plastic wedge with 50° angle.

The samples were scanned by the three-axis manipulator that is shown on figure 2. Acoustic data is written in pair with position data of transducer, the coordinates of probe are monitored by a measurement unit.

Figure 2. Immersion tank with linear array on manipulator.

2.1. Control procedure

In order to set the optimum amount of data for the volume reconstruction a meander scan path were selected. Path is characterized by two main parameters. Ss - the distance between the measuring positions, Sp - the distance between the measuring lines (fig. 3).

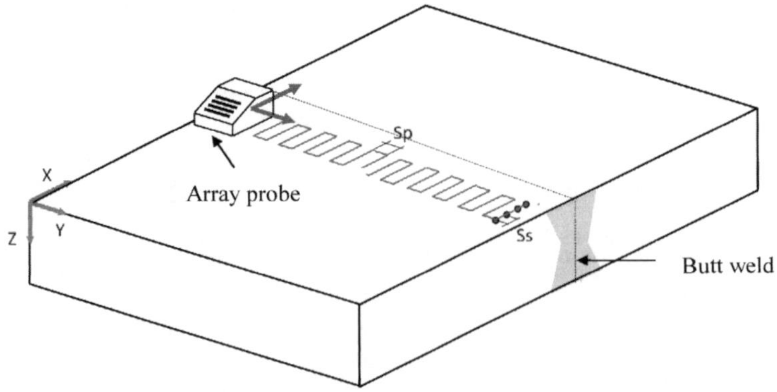

Figure 3. Testing measurement scheme.

The Ss value is 1 mm and the measuring the distance is 2 mm. The linear phased array is installed across the joint. The parameters are due to the requirements of the method of Digital Focus Array (DFA) reconstruction [3, 4], which is implemented in the software that controls OPTUS unit. When it is not possible to cut the reinforcing bead from testing welds, it is necessary to select the scanning area so that the scan path length is at least twice the wall thickness of the sample and the path ends just before the reinforcing bead edge.

Measured data is used for two-dimensional reconstruction of the sector scans in real time. Reconstructed sector scan incrementally fills a three-dimensional matrix, thus creating a three-dimensional model of the weld structure.

Testing procedure of each sample is carried out 3 times for each side of the weld. The reference fingerprint is formed as a result of this procedure. Additional measurements were made in order to determine the procedure positioning errors: offset on axes X, Y, Z and rotation.

3. Results and discussion

The measurement procedure carried out 38 times. As a result were formed 14 fingerprints for further mathcing. Example of the fingerprint is shown in Figure 4.

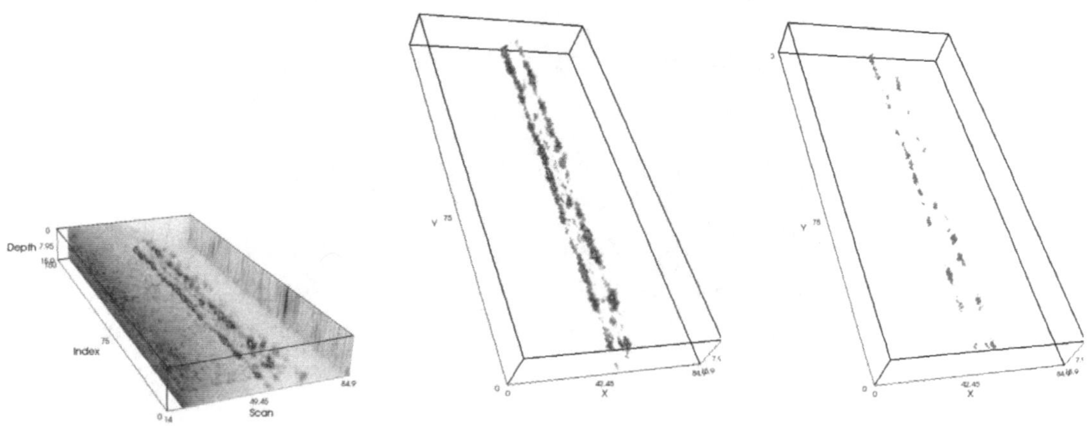

Figure 4. The process of fingerprint formation with three-dimensional visualization: a) a three-dimensional ultrasound image made with DFA method; b) the results of a segmentation process; c) the reconstructed three-dimensional fingerprint after binarization.

The fingerprint matching rate analysis were carried out. Arithmetic mean and standard deviation were calculated. Results are presented in Tables 1-5. A letter in the first row of each table is used below as a reference. Here and below, "1" - correlation, "2" - intersection, "3" - hi-square, "4" - Hellinger distance.

Table 1. The matching rate for the same position on the one sample.

a, %						
sample/method	1	2	3	4	mean	deviation
1	94.2	96.7	93.2	95.1	94.8	1.5
2	93.5	97.0	94.1	96.2	95.2	1.7
3	94.4	96.1	95.0	97.1	95.7	1.2
4	94.1	95.4	97.3	95.4	95.6	1.3

Table 2. The matching rate for the different position on the one sample.

b, %						
sample/method	1	2	3	4	mean	deviation
1	71.1	73.1	77.3	70.2	72.9	3.2
2	68.2	78.3	73.1	70.7	72.6	4.3
3	72.4	77.2	80.1	79.1	77.2	3.4
4	74.2	79.3	76.1	72.1	75.4	3.1

Table 3. The matching rate on the different samples.

c, %						
sample/method	1	2	3	4	mean	deviation
1	20.3	22.6	12.4	22.4	19.4	4.8
2	21.3	25.9	10.5	22.1	20.0	6.6
3	25.1	22.8	10.4	21.3	19.9	6.5
4	21.8	23.9	8.1	26.6	20.1	8.2

Table 4. The matching rate for the same position on the one sample with transducer rotation.

d, %						
sample/method	1	2	3	4	mean	deviation
1	49.3	57.5	47.1	51.4	51.3	4.5
2	46.3	44.4	52.7	53.4	49.2	4.5
3	48.4	46.1	56.3	47.3	49.5	4.6
4	52.9	45.8	45.4	47.9	48.0	3.4

Table 5. The matching rate for the same position on the one sample with transducer offset.

	e, %					
sample/method	1	2	3	4	mean	deviation
1	80.1	81.9	76.3	74.9	78.3	3.3
2	80.8	78.8	75.8	76.3	77.9	2.3
3	81.3	74.9	77.9	80.8	78.7	3.0
4	82.3	78.0	80.4	81.0	80.4	1.8

The results of experiment "a" show the matching rate of 95.4 ± 1.3%, which is 17% different from the correlation level obtained for one-element transducer [1]. More important from a practical point of view is the result of the experiment "b" where the matching rate is 75.3 ± 3.3%. This means that one weld inspected from different sides of the joint, is authenticated as the same weld. Thus the difference between different welds is still evident. The result of the matching rate for the experiment "c" shown the lowest value with high error rate – 19.3 ± 7.5%.

A special attention should be paid to the experiment "e" results, where the values exceeds a selected threshold of 15% [5], which confirms the possibility of compensation the transducer offset within 3 mm by using a linear phased array . The array rotation influence (experiment "e") is still significant. These values are on average 49.8 ± 4.1%.

The analysis of each of the four correlation methods were carried out. The results are shown in Table 6, the average value was 95.4 %. The best results obtained by using the intersection method, where the matching rate was 96.2 ± 0.6%.

Table 6. Results of matching rate for different correlation methods

Method	1	2	3	4
Mean	94.9	96.2	94.7	95.8
Average	95.4			
Standard deviation	0.34	0.60	0.52	0.26

4. Conclusion

Comparing the results obtained by a single-element transducer and linear phased array, should be noted a significant increase in quality of the results (17.1%). Along with this application of linear array allows authenticate the same weld with different positions and to avoid errors connected with the transducer offset on X or Y axis. Industrial applications of this technology is limited to the need to implement an immersion contact between the controlled object and the transducer, and measurement sensitivity to the array rotation. The latter factor is due to the difference in the longitudinal and transverse resolution of an array, since the resolution in the lateral direction depends on the frequency of the selected transducer, and longitudinally from the distance between adjacent array elements.

Acknowledgements
Research has been conducted with financial support of the Government of the Russian Federation (The Ministry of Education and Science) agreement № 02.G25.31.0176.

References

[1] Krening M, Demyanyuk D, Sednev D, Lider A 2012 *IFOST - 2012: Proceedings* **2** 190
[2] Bolotina I, Bulavinov A, Lider A, Sednev D, Shtaynbreher A. 2015 *Materials Science and Engineering* **81** 173

[3] Filippov G, Zhvyrblya V, Sharavina S, Salchak Y 2016 *MATEC Web. Conf.* (to be published)
[4] Bulavinov A, Kröning M, Pudovikov S, Oster R, Hanke R, Hegemann U, Reddy K M and Venkat R S 2008 *International Symposium on NDT in Aerospace* p 21
[5] Demyanuk D, Kroening M, Lider A, Chumak D, Sednev D 2013 *ESARDA Bulletin* **50** 79
[6] Larionov VV, Lider A M, Garanin G V 2015 Adv. Mater. Res. **1085** 335

IOP Conf. Series: Materials Science and Engineering **135** (2016) 012040 doi:10.1088/1757-899X/135/1/012040

Study of the Materials Microstructure using Topological Properties of Complex Networks

Mikhail Semenov, Kira Lelushkina

National research Tomsk polytechnic university, Lenin Avenue, 30, Tomsk, Russia, 634050

E-mail: sme@tpu.ru

Abstract. A method for mapping a two-dimensional color image of the microstructure of the material to a complex network is proposed. Each image elements is assigned to node network. A weighted combination of distance metrics – the Euclidean distance and the Manhattan distance – defines whether there is or not an edge between corresponding nodes. The first metric is used to calculate the spatial distance between the picture elements (pixels), the second metric takes into account the contrast between the brightness of pixels in the gray scale. On the basis of the topological properties of the constructed network the edge pixels were detected that allows us to identify the border areas in the microstructure of materials. The proposed method can be used in automated systems of materialographic analysis.

1. Introduction

There are many problems from different areas can be solved by the application of algorithmic graph theory. Basic and applied research in material science, engineering and medicine can use a network model as a prototype of a real network. Many studies are conducted on network models in which the number of nodes as well as the number of edges incident to them can be changing in time [1] or at a constant number of nodes of the graph the number of edges changes [2], that affects the network characteristics.

We note a number of studies addressed to image processing in the context of networks with constant topology. In the paper [3] a visualization approach of complex three-dimensional structure of osteocyte networks based on the voxel skeleton was proposed. This network topology includes three entries: nodes, links, and cells. A cell is a connected region with a defined distance and a size restrictions. A node is a cluster of voxels with more than two neighbors each, and a link or branch (if connected to one node only) is a group of connected voxels with two neighbors each. In the paper [4] a model for mapping a grayscale image into weighted undirected network is proposed. The conversion of images into complex networks concentrates mainly on the weight of corresponding edge, and edges in that network depend on both the Euclidean distance and the gray-level difference. This model is based on the weighted combination of the Euclidean distance and gray-level similarity between pixels. In the paper [5] proposed a graph based algorithm of grayscale image processing. Every pixel is represented by one vertex in weighted graph, the weight of each edges is defined by function in which absolute difference of intensities and distance between a pixels pair are taken account.

The aim of this paper is to research the materials microstructure based on complex networks and their topological properties. To achieve the identified aim it is necessary to solve the following problems: a) to map a color image into a complex network, b) to calculate topological properties of complex network, c) to detect pixels on object boundaries.

2. Network Construction

Suppose we have a color image, characterized by the size of N pixels in width and M pixels in height. Each pixel of a color image is described by a position (p_x, p_y) and a triple of intensities: red (r), green (g), and blue (b). In order to design a model, we will use a spatial distance as well as the contrast between the brightness of picture elements (pixels).

First, in the proposed method we need to convert an image from color to grayscale. There are various ways to convert color images to grayscale image, in our study the weighted average luminosity method was used. We have applied the Rec. 709 coefficients [6] to mapping a color image into grayscale. Thus, the linear luminance Y, which is given by:

$$Y = 0.2126 \cdot r + 0.7152 \cdot g + 0.0722 \cdot b.$$

Each pixel of the original image corresponds to 8 bits (1 byte) of information that allows us to transfer 256 gray levels. The result of the conversion of color images saves in the bitmap, N is the number of rows and M is the number of columns. Elements of bitmap take values from the interval $[0, 255]$, where the left boundary of the interval 0 corresponds to the black color, and the value 255 – to the white color.

Second, each element of the bitmap is assigned one node of the network. For given image with pixel set $P = \{p_i : i = 1, 2, \ldots, n = N \cdot M\}$, a network can generally be represented as a set $G = (V, E)$, where the node set is given by $V = \{v_i : i = 1, 2, \ldots, n\}$, n is the total number of nodes; the set of edges is given by $E \subseteq V \times V$. To define the set E we use distance metrics: the Manhattan distance, d_1, and the Euclidean distance, d_2. To account for the brightness contrast between a pair of pixels $(p, q) \in P$ in gray scale we use the Manhattan distance:

$$d_1(p, q) = |p_g - q_g|, \tag{1}$$

where p_g and q_g are gray level of pixels respectively. The value of Euclidean distance, d_2, of the two pixels defines whether there is or not an edge between corresponding nodes $(p, q) \in V$:

$$d_2(p, q) = \sqrt{(p_x - q_x)^2 + (p_y - q_y)^2}, \tag{2}$$

where (p_x, p_y), and (q_x, q_y) are coordinates of pixels $p, q \in P$ respectively. Then the set E can be determined as follows: $E = \{e_i = (p, q) : 0 < d_2 \leq R, p, q \in V, i = 1, 2, \ldots\}$, where $R \in [0, \max(N, M)/2]$ is a model parameter that is called the radius. If we choose $R = 0$, then zero-graph will be generated, all of whose nodes are isolated. If we set $R = 1$, we will obtain the two-dimensional model of crystal with square lattice.As one should expect, the number of edges m exponentially increases with the radius R. In the limiting case we will deal with the complete graph, for which the total number of edges is equal to $m = |E| = n(n-1)/2$. In practical cases we used $1 < R < \max(N, M)/2$. We note that for the Manhattan distance, d_1, Eq. (1) the impact of an outlier is reduced as compared with the Euclidean distance, d_2, Eq. (2).

Finally, the weighted undirected network G is denoted by a symmetric weight $n \times n$ matrix, $W = [w_{pq}]$, where $w_{pq} = \sqrt{(\alpha \cdot d_1(p, q))^2 + (\beta \cdot d_2(p, q))^2}$, if there is an edge between nodes p and q and $w_{pq} = 0$ otherwise, where $\alpha, \beta > 0$ are the scaling parameters of gray-level difference and spatial distance respectively.

Thus, based on the symmetric weight matrix W we constructed the weighted undirected network $G = (V, E)$, with one connectivity component.

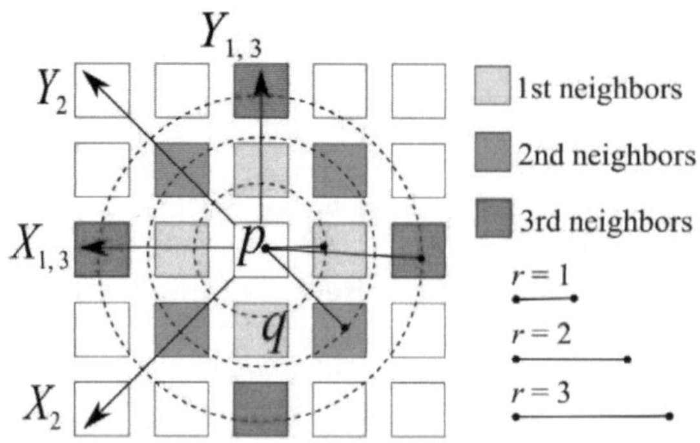

Figure 1. A fragment of the square lattice with the coordinate systems for the nearest, second and third neighbors. Edges are omitted.

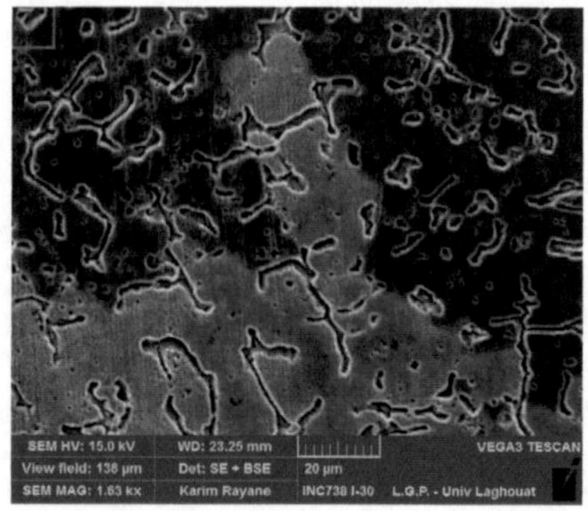

Figure 2. Original color microscopy image [9] with red borders (top left corner).

3. Calculation of Local Node Dimension

Authors [4] using the measure of dimensionality of individual nodes to the image of edge detection. Following the study [7] the local node dimension $D_i(r)$ can be obtained by the angular coefficient (slope) of the $B_i(r)$ curve on the interval $[0, R]$, where $B_i(r)$ is number of nodes at different topological distances $r \in [0, R]$ from a focal node p, R is radius in the proposed model. For defining the local node dimension $D_i(r)$ the $B_i(r)$ curve was approximated with a linear regression. It is necessary to note, for a small number of points the role of the chosen regression method becomes important to the resulting values. Currently, authors [7] addressed this problem by using the least squares method defined directly for the logarithmic scales.

In figure 1 there are shown three circles ($r = 1, 2, 3$) of the fragment of a two-dimensional model with square lattice. For any pair of nodes $(p, q) \in V$, we shall use the (X, Y) coordinate system whose axis are run through the corresponding nodes that are denoted by squares. In the case of the nearest and third neighbors in the coordinate system the axes are respectively denoted by $X_{1,3} \times Y_{1,3}$, and for second neighbors – $X_2 \times Y_2$ [8].

In order to observe the distribution of node dimension, the set of node dimension $D(R) = \{d(p_i) : i = 1, 2, \ldots, n\}$ can be mapped to a grayscale image. Gray level in the grayscale image responds to the value of the node dimension [4]. We applied a contrast enhancement technique to obtain the grayscale image. A linear scaling transform has been used in order to improve the pixel contrast for output image: $g = a \cdot f + b$, where a and b are the transform parameters which can be determined based on the required change of the dynamic range. As a result of the image processing we get the scale $[g_{min} = 0, g_{max} = 255]$, then

$$\begin{cases} g_{min} = a \cdot f_{min} + b, \\ g_{max} = a \cdot f_{max} + b. \end{cases} \Rightarrow \begin{cases} a = (g_{max} - g_{min})/(f_{max} - f_{min}), \\ b = (g_{min} \cdot f_{max} - g_{max} \cdot f_{min})/(f_{max} - f_{min}). \end{cases}$$

Here f_{min} and f_{max} are extreme values of node dimension interval.

4. Detection of Pixels on Object Boundaries

In the paper [4] the authors carried out the computational experiments and found that pixels on object boundaries have a lower node dimension. The node dimension interval directly depends on the original image. To determine the pixels on object boundaries, we suggest using a p-quantile of the node dimension distribution. A pixel of output image, p, can be obtained by values of node dimension interval $[f_{min}, f_{max}]$ as follows:

$$p \text{ is } \begin{cases} \text{edge pixel, } d(p) < p\text{-level quantile,} \\ \text{non-edge pixel, } d(p) \geq p\text{-level quantile.} \end{cases}$$

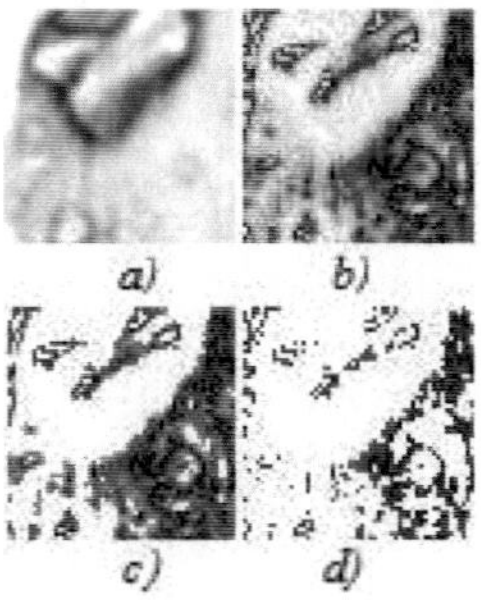

Figure 3. a) Original image and images with different thresholds of edge pixels: b) 0.25-quantile, c) mean, d) 0.75-quantile.

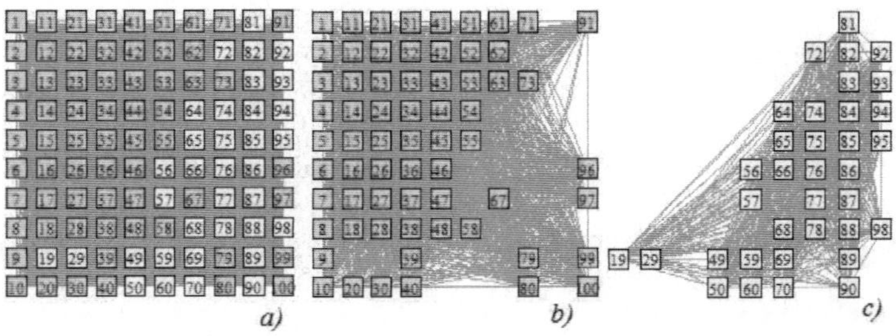

Figure 4. a) Original network, b) network of non-edge nodes, c) network of edge nodes. Nodes corresponding to edge pixels are denoted by yellow nodes, 0.25-quantile is the threshold of edge pixels.

5. Computational Experiments

To conduct the computational experiment, we chosen the picture from the related paper [9], which is shown in figure 2. Square 55 × 55 pixels part has been selected from the picture, in figure 2 borders of chosen picture are marked with red line (top left corner).

In order to design the complex network G we used the following values of the model parameters: $\alpha = \beta = 1$, and $R = 2$. Then we calculated the local node dimension, the first, second and third 4-quantiles are $\{0.005, 0.013, 0.0269\}$ respectively, and the extreme values are $\{0, 0.0421\}$. To determine the boundary pixels, we varied the value of p-quantile as thresholds. The original image and images with different thresholds of edge pixel are shown in figure 3, colors were inverted. The part of original network G, containing 10 × 10 numbered nodes, is shown in figure 4. In our experiment, based on the identified borders the original network G (figure 4a) was divided into two networks G_1 and G_2 (figure 4b, c). The density, average path length and degree distribution were calculated for each network G, G_1, and G_2: graph density are $\{0.92, 0.82, 1.00\}$, average path length are $\{1.07, 1.17, 1.00\}$, and average degree are $\{91.64, 52.67, 34.0\}$ respectively.

6. Conclusion

To study the material microstructure a model of a complex network is proposed. On the basis of the topological properties of the constructed network the local node dimension was calculated, and pixels on object boundaries were defined. The proposed approach allows us to apply the algorithmic graph theory to the image processing. The outcome of the designed network is therefore that basic characteristics of a network can be calculated. Thus our approach can be used in automated systems materialographic analysis.

References

[1] Barabasi A-L and Albert R 1999 *Science* **286** 509–512
[2] Watts D and Strogatz S 1998 *Science* **393** 440–442
[3] Kerschnitzki M et al. 2013 *Journal of Bone and Mineral Research* **28** 1837–1845
[4] Wu Z, Lu X and Deng Y 2015 *Physica A: Statistical Mechanics and its Applications* **440** 9–18
[5] Jieqi Kang J, Lu S, Gong W, Kelly P 2014 *Proceedings of IEEE International Conference on Image Processing - ICIP'2014* 2051–2055
[6] Recommendation BT.709-6 2015 URL `https://www.itu.int/rec/R-REC-BT.709/en`
[7] Silva F N and Costay L F 2013 URL `http://arxiv.org/abs/1209.2476`
[8] Rodrigues C G, Rabelo J N T and Zubov V I 2006 *Brazilian Journal of Physics* **36** 592 – 602
[9] Boumerzoug Z, Hamza S and Ji V 2016 *Journal of Thermal Engineering* (accepted for publishing)

IOP Conf. Series: Materials Science and Engineering **135** (2016) 012041 doi:10.1088/1757-899X/135/1/012041

Research of the thorium purification at monazite refinement processes

V V Shagalov, V I Sobolev, M V Turinskaya, A V Malin

National Research Tomsk Polytechnic University, 634034, Tomsk, Russia

vsobolev1989@tpu.ru

Abstract. This paper is aimed to the research of the thorium purification processes at monazite refinement processes. We have investigated different solution containing thorium with different mix of rare-earth elements. It was found that the application of cation resin is well-recommended if we want to reach the highest yields of thorium purification process.

1. Introduction

One of the principle problems of nuclear energetics is the guarantee of safe handling with long-life components of nuclear fuel; those components produce a hard radioactivity in exhausted nuclear fuel after its decomposition. Application of thorium reduces the income of long-life actinides. There is a concept of new VVER-T reactor and the concept of high-temperature gas-cooled thorium reactor of low power with the realization of closed fuel cycle based on ^{232}Th as breeding material [1-5]. Monazite sand is one of the most important sources of thorium and it contains monazite, ilmenite and zircon. Monazite concentrate could be enriched on thorium using different techniques using various methods of physical enrichment. Also there are thorium deposits available which were obtained during the rare metals ores refinement. Such deposits could not be used for rare metals extraction because of high radioactivity and that is why they are stored special places [5-7]. Thorium exists in insoluble form in both cases, especially in phosphate form [1].

We made various investigations in the field of ion exchange separation of thorium and rare-earth metals in phosphate solutions and compared the effectiveness of anion resin and cation resin resins using for thorium and rare-earth metals separation. Those solutions were obtained by dissolving of monazite concentrate in phosphoric(V) acid.

2. Experimental

The initial monazite concentrate was pre-purified from zircon impurities using electromagnetic separation and form ilmenite impurities using electrostatic separation. Purified monazite concentrate was dissolved in phosphoric(V) acid solution at 400 °C within 2 hours with vigorous stirring. To prevent the thorium and rare-earth metals hydrolysis side process the 5M phosphoric(V) acid solution concentration was used. The qualitative and quantitative determination of thorium and rare-earth metals was performed using EDXRF spectrometry and ICP-AES spectrometry [2].

EDXRF analysis was performed using ThermoFischer Scientific ARL QUANT'X spectrometer using fundamental parameters methodic.

ICP-AES analysis was done using ThermoFischer Scientific iCAP 6300 DUO ICP-AES spectrometer.

It was found that thorium forms stable anion-based complex and does not sorb on cation-based resins. There is another situation with rare-earth metals set. Rare-earth metals except yttrium don't form anion-based complexes and well-sorb on cation-exchange resins. Yttrium forms various unstable complexes - anion-based and cation-based and could be extracted from the solutions using both types of resins.

3. Results and discussion

Results means that purification of thorium form rare-earth metals can be easily done by anion-exchange resins with the resulting extraction of thorium and using cation-exchange resins with the resulting extraction of the rare-earth metals set. We have investigated which variant is the most effective one at the static conditions with the 1-hour exposure time. Both resins were places in the phosphoric(V) acid solution of thorium and rare-earth metals. The results are shown at the table 1.

Table 1. Molar ratio of thorium ions and rare-earth metals ions on various anion resins.

Ions	Initial solution (molar %)	Anion resin after sorption (molar %)	Cation resin after sorption (molar %)
La	11.5	0.0	22.6
Ce	22.9	0.0	36.4
Nd	18.0	4.9	27.0
Y	26.5	27.5	13.4
Th	21.2	67.6	0.6

As we can see from the table 1 anion-exchange resin does not provide desirable purification from yttrium and contains considerable amount of neodymium. Cation-exchange resin sorbs well all rare-earth metals set and does not sorb thorium. It means that cation-exchange resins are the best ones for purification of thorium from the set of rare-earth metals [8].

There is significant separation of thorium and rare-earth metals on cation resin, but thorium is hold in the solution in this case and could be easily precipitated in the form of its peroxides and any other insoluble compound; ions of rare-earth metals are absorbed in the cation resin in this case [9].

As the result of ionic exchange in dynamic conditions the rare-earth metals sorption queue on cation resin was determined in comparison with its ionic radius. The precipitation queue is shown at the table 2. The unsaturated cation resin was analyzed using EDXRF spectrometry. The cation resin was flushed using 10 ml 5M phosphoric(V) acid solution [10-13].

Table 2. Molar ratios of thorium ions and rare-earth metals in various parts of anion resin from top to bottom.

Metal	Top	Middle	Bottom
	Metal molar ratio (mole parts)		
La	0.23	0.12	0.08
Ce	0.36	0.31	0.26
Nd	0.27	0.25	0.21
Y	0.14	0.32	0.45
Th	0.00	0.00	0.00

As we can see from the table 2 lanthanum has the highest sorption ability because of its largest ionic radius. Yttrium is worth absorbed and it moves directly to the column bottom. Thorium is not absorbed at all and easily flushed by phosphoric(V) acid solution. Thorium and rare-earth metals concentrations at the after-sorption solution are shown at figure 1 and figure 2.

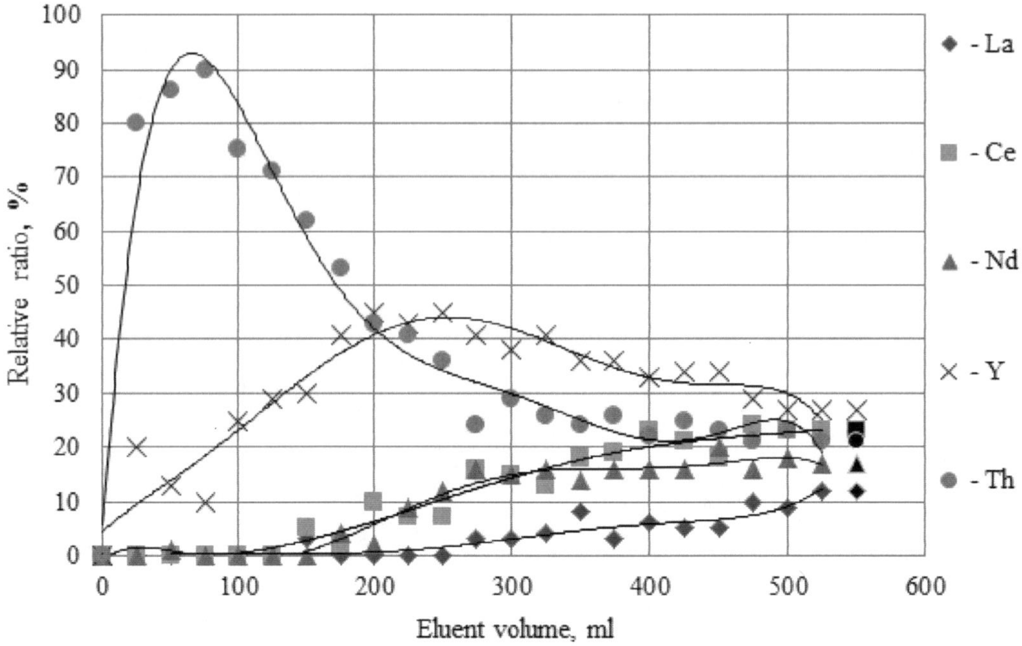

Figure 1. Thorium and rare-earth metals concentrations at the after-sorption solution.

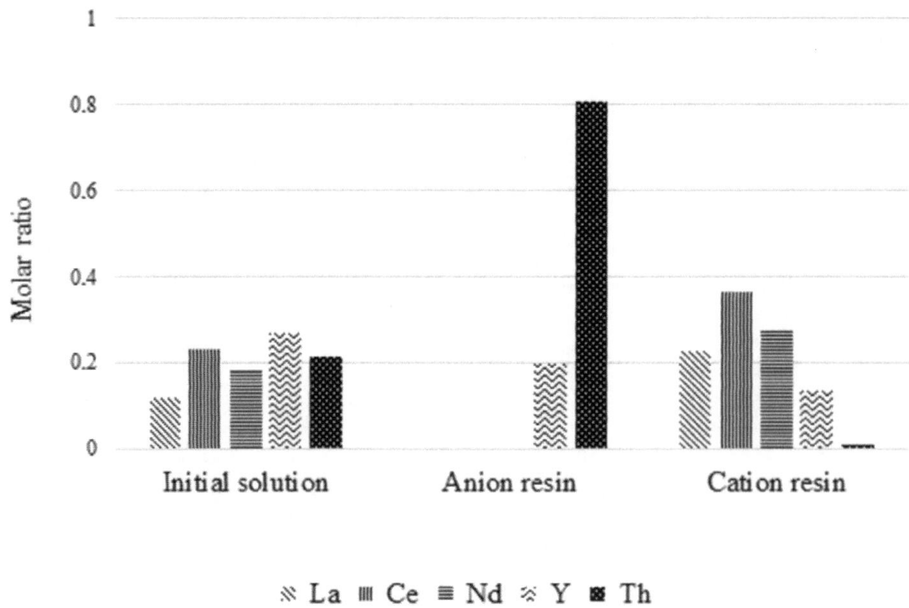

Figure 2. Thorium and rare-earth metals concentrations at the after-sorption solution.

As we can see from the figure 1 and figure 2 the relative concentration of thorium is decreased while eluent is moving through the column. It can be explained by cation resin saturation by rare-earth metals ions and further ion slippage to eluate. In order to get the purest thorium, it was necessary to carry out the ionic exchange process with the large excess of cation resin. In this case all rare-earth metals ions will be absorbed on cation resin and thorium will be hold in the solution. Thorium could be precipitated in the form of its peroxide and with further heating we can easily get thorium(IV) oxide. The metallic thorium could be obtained using simple metal thermic procedures [14-15].

Rare-earth metals can be extracted from resin using strong mineral and organic acids, and also using various complexation compounds like EDTA. Then the concentration of rare-earth metals occurs [16-17].

4. Conclusion

Thorium can be obtained from the monazite concentrate and moved to phosphate solution. I this case rare-earth metal also moved to the solution. Anyway thorium purification procedure is necessary in case of its usage in modern energetics.

We have researched the separation of soluble thorium and rare-earth metals phosphate compounds by ionic exchange method. It was shown that thorium ions are mostly absorbed on anion resin with the slight sorption of yttrium ions. It means that it is not necessary of anion resins use for the soluble thorium phosphates separation.

Ions of rare-earth metals are well-absorbed on cation resin and thorium phosphates are hold in the solution. This method provides more effective separation of elements. Thorium can be precipitated in the form of peroxide for further refinement. Rare-earth metals can be extracted using strong mineral or organic acids.

References

[1] Frybort J 2014 *Radiation Physics and Chemistry* **104** 408
[2] Waris A, Pramuditya S, Perkasa Y S, Ȧrif I 2014 *International Journal of Nuclear Energy Science and Technology* **8** 1
[3] Liu Y L, Yan Y D, Han W, Zhang M L, Yuan, L Y, Lin R S, Ye G A, He, H.c, Chai, Z.-F.a, Shi, W.-Q.a 2013 *Electrochimica Acta* **114** 180
[4] Vu T M, Fujii T, Wada K, Kojima T, Kitada T, Takaki N, Yamaguchi A, Watanabe H, Unesaki H 2013 *Annals of Nuclear Energy* **57** 173
[5] Dekusar V M, Kalashnikov A G, Kapranova E N, Kolesnikova M S, Korobytsin V E 2012 *Atomic Energy* **112** 311
[6] Breza J, Dařílek P, Nečas V 2010 *Annals of Nuclear Energy* **37** 685
[7] Raitses G, Todosow M, Galperin A 2009 *International Conference on Nuclear Engineering, Proceedings* **2** 897
[8] Glebov A P, Klushin A V 2009 *Atomic Energy* **106** 305
[9] Frybort J *Proceedings of the 2014 15th International Scientific Conference on Electric Power Engineering* 693
[10] Sobolev V, Ivlev S, Shagalov V, Ostvald R, Zherin I 2016 *Key Engineering Materials* **683** 269
[11] Sobolev V, Radchenko V, Ostvald R, Filimonov V, Zherin I 2014 *Advanced Materials Research* **1040** 337
[12] Ivlev S, Sobolev V, Hoelzel M, Karttunen A J, Müller T, Gerin I, Ostvald R, Kraus F 2014 *European Journal of Inorganic Chemistry* **36** 6261
[13] Ivlev S, Woidy P, Sobolev V, Gerin I, Ostvald R, Kraus F 2013 *Zeitschrift fur Anorganische und Allgemeine Chemie* **639** 2846
[14] Sobolev V, Zherin I, Ostvald R, Ivlev S, Filimonov V, Postnikov P 2012 *Proceedings - 2012 7th International Forum on Strategic Technology* 6357502
[15] Woidy P, Bühl M, Kraus F 2015 *Dalton Transactions* **44** 7332

IOP Conf. Series: Materials Science and Engineering **135** (2016) 012041 doi:10.1088/1757-899X/135/1/012041

[16] Woidy P, Kraus F 2014 *Acta Crystallographica Section E: Structure Reports Online* **70** 98
[17] Woidy P, Karttunen A J, Rudel S S, Kraus F 2015 *Chemical Communications* **51** 11826

Studying radiation hardness of a cadmium tungstate crystal based radiation detector

M M Shtein[1], L F Smekalin[1], S A Stepanov[1], I A Zatonov[1], T V Tkacheva[2], E Yu Usachev[2]
[1]National research Tomsk polytechnic university, Lenin Ave. 30, 634050, Tomsk, Russia
[2]Moscow technological university (MIREA), Vernadskogo Ave. 78, 119454, Moscow, Russia

e-mail: shtein38@gmail.com

Abstract. The given article considers radiation hardness of an X-ray detector used in production of non-destructive testing instruments and inspection systems. In the course of research, experiments were carried out to estimate radiation hardness of a detector based on cadmium tungstate crystal and its structural components individually. The article describes a layout of an experimental facility that was used for measurements of radiation hardness. The radiation dose dependence of the photodiode current is presented, when it is excited by a light flux of a scintillator or by an external light source. Experiments were carried out to estimate radiation hardness of two types of optical glue used in detector production; they are based on silicon rubber and epoxy. With the help of a spectrophotometer and cobalt gun, each of the glue samples was measured for a relative light transmission factor with different wavelengths, depending on the radiation dose. The obtained data are presented in a comprehensive analysis of the results. It was determined, which of the glue samples is most suitable for production of detectors working under exposure to strong radiation.

1. Introduction

Instrument facilities of customs inspection play an important role in solving tasks of cargo tracking. These facilities normally include systems of X-ray diagnostics, the main part of which is X-ray detectors [1]. One of the main requirements to detectors is stable sensitivity in the process of accumulation of the absorbed radiation dose. That is why studying radiation hardness of detector components and finding the most robust ones is of a particular practical interest. One of the common demands to the detectors is their stable sensitivity, so as sensitivity to time. Temporary instability can be connected with its change of dosage sensitivity during the process of absorbed radiation dose accumulation. That is why the study of the radiation resistance of the detectors components and the selection of more persistent ones have a particular practical interest.

Studies on the radiation resistance of the 16-channel detector based on silicon p-i-n- photodiode and scintillator CsI (Tl)[2,6], in which the radiation resistance of the detector was determined at the level of $5x(10^5$-$10^6)$ Rad were conducted earlier. However, the contribution of the detectors'(assembly)individual components in the radiation resistance was not considered.

2. Methodology

2.1. Detector structure

We carried out experiments on evaluation of radiation hardness of a CWO based crystal (fig. 1) both in assembly (crystal-glue-photodiode) and as a single photodiode. The photodiode has a p-i-n structure and made of high-resistivity silicon.

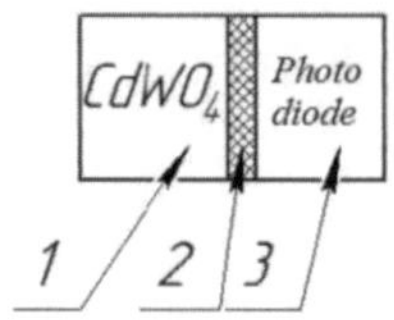

Figure 1. Detector structure, where 1 – cadmium tungstate crystal; 2 – glue; 3 – photodiode.

In the studied detector, cadmium tungstate acts as a scintillator. Optical glue is a bonding element between the scintillator and the photodiode, thus reducing light reflection where they are not tight enough.

2.2. General layout of experiment

Cobalt gun "Radian-2" with radioactive nuclide $^{60}_{27}Co$ (fig. 2) was used as a radiation source in experiments, the measuring circuit was taken outside the beam coverage area. The samples under study (assembled detector first, then photodiode separately) were placed at the output window of the gun collimator.

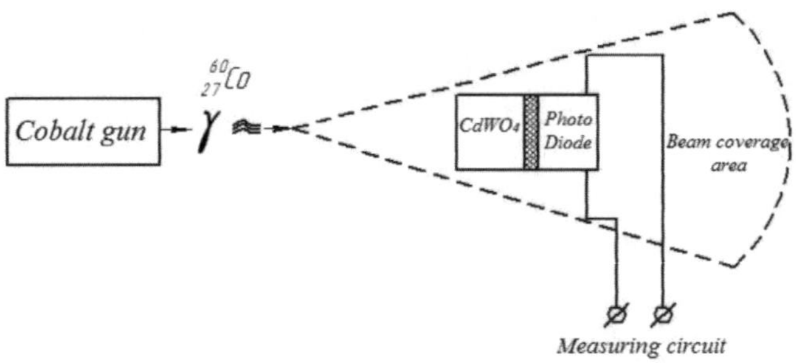

Figure 2. Experiment layout

Measuring circuit is shown in fig. 3. Photodiode current registered by a DC ampere meter Keithley 6485 was measured during experiments. When photodiode only was examined, it was exposed to a light flux from a 24V incandescent lamp, whose voltage was provided by a stabilized power source.

The photodiode efficient current was about three orders more than the dark current when exposed to both scintillator, and lamp. To provide a linearity of the light flux measurement, about 4V of reverse voltage was applied to the photodiode. All data of the photo-electric current measurements are given in relative units reduced to the initial point, where absorbed dose may be considered as zero.

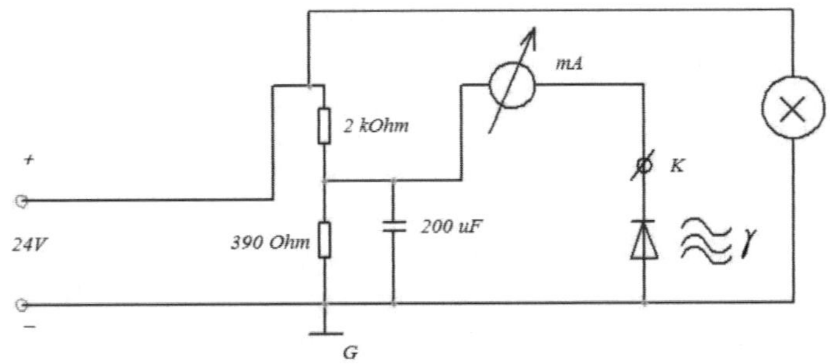

Figure 3. Measuring circuit.

3. Results

When scintillation detector entirely and single photodiode were exposed, the dose rate level was the same and made 5.5 rad/sec. The dose rate was monitored by DKS-101 dosimeter (s/n #71) equipped with ionization chamber BMK-06 (s/n #0871).

3.1. Finding radiation dose dependence of photodiode currents

Fig. 4 shows measurement results of the photodiode both assembled and separately depending on the absorbed radiation dose.

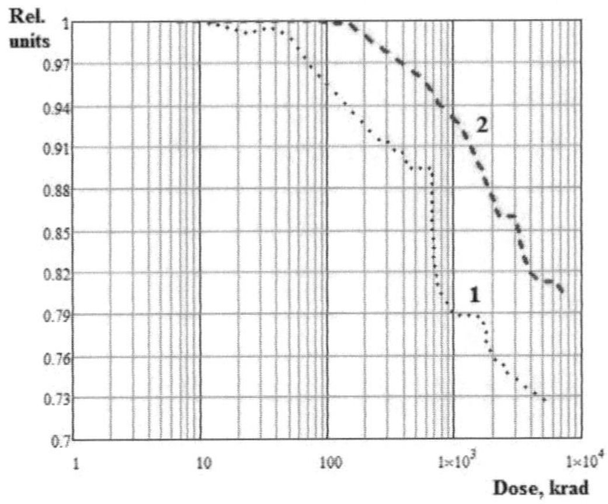

Figure 4. Dose dependence of photodiode currents, where 1 – assembly (scintillator-glue-photodiode); 2 – single photodiode.

The obtained results show that reduction of photo-electric current by 20% in the assembly (scintillator-glue-photodiode) takes place when $8\text{-}9 \cdot 10^5$ rad dose is accumulated, whereas the single photodiode shows the same current reduction at $7\text{-}8 \cdot 10^6$ rad, i.e. is more radiation-proof by about an order.

It is known [3] that oxide crystal based scintillation sensors (including cadmium tungstate) are highly radiation-proof. Together with the above-given experimental results, this fact tells us that the possible reason of the early failure of the assembly (scintillator-glue-photodiode) is low radiation hardness of the used glue.

3.2. Comparative tests for radiation hardness of two types of optical glues

To verify this conclusion we performed comparison tests for radiation hardness of two types of glues: optical silicone rubber based glue (further – SR-glue) and epoxy based glue. Applied on a polyethylene substrate, the SR-glue freezes as a thin uniform layer and is a clear bubble-free gel coat film, which can easily be cut by a scalpel. Epoxy glue freezes unequally, there is a small droplike thickening in the center. Multiple little air bubbles can be seen in it, the sample itself is quite tough and quite difficult to cut by a scalpel.

The core of the tests was as follows. Glues frozen on the polyethylene substrate were used to make samples 5-6 mm in diameter, and were put in special containers integrated in spectrophotometer SF-256 UVI. Then each sample was exposed to gamma-radiation on "Radian-2", the absorbed dose value being 5.5 rad/sec. A tin-lead plate (90% Sn and 10% Pb) 4.5 mm thick was placed in front of the sample. Content and thickness of the plate allowed modelling absorption of the primary and output of the secondary radiation in scintillation crystal of cadmium tungstate. In certain periods of exposure time, the samples in containers were put into spectrophotometer SF-256 UVI where light transmission factor in the range of wavelength from 190 to 1100 nm was measured. The layout and method of measuremetns are described in works [3, 4].

Fig. 5 shows the change of the relative transmission factor as the absorbed dose for epoxy glue increases, for two extreme lines in the visible range of light: curve 1 corresponds to the wavelength of 600 nm ("red"), curve 2 – to 430 nm ("blue"). The results show that for this glue, a significant radiation dose dependence of light transmission (almost linear) is typical, and the dependence becomes more obvious as the wavelength decreases.

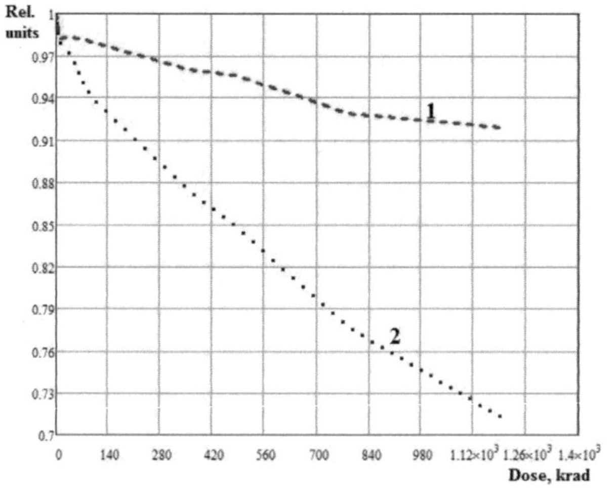

Figure 5. Change of the relative transmission factor as the absorbed dose increases (for epoxy based glue). 1 – 600 nm ("red")' 2 – 430 nm ("blue").

Fig. 6 shows the results of measuring the transmission factor for SR-glue sample. There is a quite weak dose dependence of transmission (up to 3000 krad) for this glue. We observed minor transmission enhancement as compared to the initial one, up to 5-8% ("radiation lucency"), at the initial stage of irradiation (about 20-40 krad in the area of absorbed dose), which is not seen in fig. 6 because of the small scale.

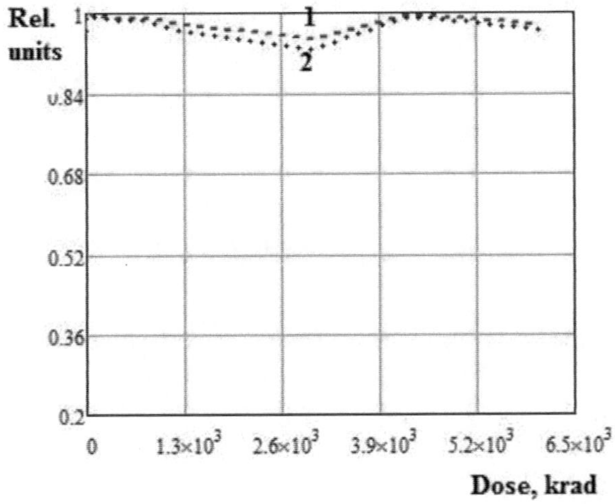

Figure 6. Change of the relative transmission factor for SR-based glue, 1 – 400 nm; 2 – 600 nm.

3.3. Analysis of results

The analysis of radiation degradation results of the assembled detector and measurement results of the transmission factor of SR- and epoxy glues gives an idea that epoxy-based glue was used in manufacture of the examined detector.

Indeed, fig. 7 shows in semi-logarithmic scale the generalized results of current measurements of the exposed assembled photodiode (data from fig. 4), and results of transmission factor measurements for epoxy-based glue (data from fig. 5). Curve 3 of photo current drop in the assembly is between the transmission factors for wavelengths 600 nm (1) and 430 nm (2). One may note that luminescence spectrum of cadmium tungstate lies in the range (380-600) nm with the peak at 480 nm. That is why, absorbed dose dependences of the photocurrent and transmission factor at 430 nm wavelength are almost the same.

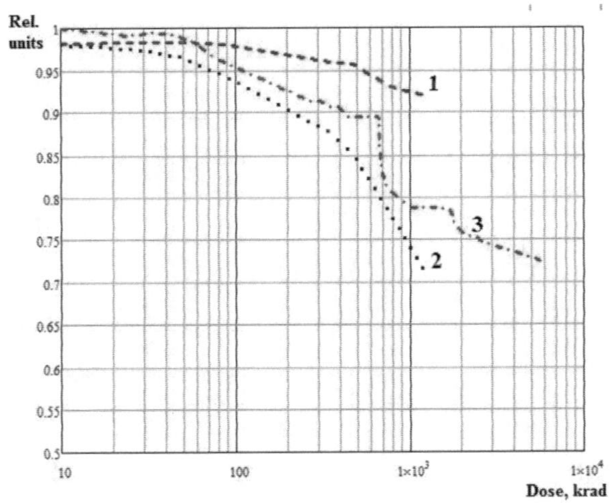

Figure 7. Change of transmission factor for epoxy-based glue (curves 1, 2) and photodiode current for assembly (curve 3).

IOP Conf. Series: Materials Science and Engineering **135** (2016) 012042 doi:10.1088/1757-899X/135/1/012042

Therefore, it is found that optical glues on the basis of silicone rubbers are much more radiation-proof (up to radiation dose values of 3000 krad) as compared to epoxy-based glues. The use of epoxy glues explains poor radiation hardness of the examined sample of the detector-assembly (cadmium tungstate – glue – photodiode).

References

[1] S. A. Schetinkin et al. 2011 Equipment for inspection of cars and cargo vehicles *XIX All-Russian scientific and practical conference for non-destructive testing and technical diagnostics*: abstracts, Samara.

[2] E.K.Gelfand.,B.S.Sychev.,T.V.Tkacheva , V.A.Rassylov, L.M. Rodionova. 1999 Radiotional resistance of multichanal detectors on the basis of "scintillation-photodiode" sistems *Theses of the International Conference «STINT-1999»*, Moscow

[3] I. M. Sol'sky 2005 Growing large-size single crystals of cadmium tungstate with high optical homogeneity *Technology and engineering in electronic equipment* **6** 47-52

[4] M. M. Gurevich 1983 *Photometry. Theory, techniques and devices* (Leningrad: Energoatomizdat)

[5] M. I. Epshtein 1990 *Measuring optical radiation in electronics* (Moscow: Energoatomizdat)

[6] Saint Gobain Crystals. Physical Properties of Common Inorganic Scintillators./ – [Electronic resource] URL: http://www.hep.caltech.edu/~zhu/papers/08_tns_crystal.pdf, free. Accessed date: November 13, 2014.

The use of $^{99m}Tc\text{-}Al_2O_3$ for detection of sentinel lymph nodes in cervical cancer patients

I G Sinilkin[1,2], V I Chernov[1,2], A Yu Lyapunov[1], A A Medvedeva[1,2], R V Zelchan[1,2], A L Chernyshova[1], L A Kolomiets[1,3]

1Tomsk Cancer Research Institute, Russia, Tomsk, Kooperativny street, 5
2Tomsk Polytechnic University, Russia, Tomsk, Lenina street, 30
3Siberian State Medical University, Russia, Tomsk, Moskovsky Trakt, 2

E-mail: Lyapunov1720.90@mail.ru

Abstract: The purpose of the study was to evaluate the feasibility of using 99mTc-Al2O3-based radiopharmaceutical, a novel molecular imaging agent for sentinel lymph node detection in patients with invasive cervical cancer. The study included 23 cervical cancer patients (T1aNxMx- T2bNxMx) treated at the Tomsk Cancer Research Institute. At 18 hours before surgery, 80 MBq of the 99mTc-Al_2O_3 were injected peritumorally, followed by single-photon emission computed tomography (SPECT) of the pelvis and intraoperative SLN identification. Twenty-seven SLNs were detected by SPECT, and 34 SLNs were identified by intraoperative gamma probe. The total number of identified SLNs per patient ranged from 1 to 3(the mean number of SLNs was 1.4 per patient). The most common site for SLN detection was the external iliac region (57.2%), followed by the internal iliac, obturator, presacral and retrosacral regions (they amounted to 14%, respectively),and the parametrial region (1%). Sensitivity in detecting SLNs was 100% for intraoperative SLN identification and 79% for SPECT image.

1. Introduction

Cervical cancer is estimated to be the 2-nd most common cancer in women worldwide, accounting for 7% of malignancies in women. Approximately 15, 000 new cases of invasive cervical cancer are diagnosed annually in Russia. The peak age at diagnosis of invasive cervical cancer is between 15 and 54 years [1].

The objective assessment of pelvic lymph nodes is an integral component of definitive surgical management for cervical cancer. Regional lymph node status reflects the metastatic potential of the tumor. Currently available non-invasive imaging methods such as ultrasound, computed tomography, magnetic resonance imaging and nuclear scintigraphy are unable to provide information about the presence or absence of lymph node metastases. Micrometastases can only be diagnosed by microscopically, using morphological or molecular biology techniques. Visualization and assessment of lymph nodes during surgical removal of the tumor in order to detect lymph node metastasis is of great importance [7].

Sentinel lymph node biopsy is becoming increasingly recognized as a standard technique for determining the status of lymphatic spread. The term "sentinel lymph node" was used for the first time by E.A. Gould in 1960 and supplemented then by R.M. Cabanas in 1977 [12, 13]. The concept of sentinel lymph node (SLN) is based on the theory of an orderly progression of tumor cells within the lymphatic system in the initial phase of the metastatic process, assuming that early lymphatic metastases are found first within the first-line tumor draining lymph node, the sentinel lymph node [13]. If the SLN is free of metastatic disease, all other lymph nodes will also be free of disease. Identification of SLNs may help some patients avoid more extensive lymph node surgery, thus minimizing postoperative morbidity and significantly improving life quality of the patients.

Clinically, radioactive nanocolloids have been used for investigations of lymphatic drainage, and recently, more specifically for identification of the sentinel node in investigations of metastatic spread of some cancers [3, 9,14].

The use of radiopharmaceuticals for SLN detection such as Nanocoll, 99mTc-nanocolloid, Nanocis, Microlite and Sulfurcolloid have several drawbacks [10, 11]. Their applications are limited by their complex manufacturing process and high cost. Furthermore, the extremely low accumulation of these radiopharmaceuticals in SLNs may lead to further redistribution of a radiotracer through the lymphatic collector and an erroneous interpretation of the results [5, 9].

In this regard, there was a need to develop a novel lymphotropic radiopharmaceutical, which would be devoid of these drawbacks.

The basic precondition for using the nanoscale Al_2O_3 gamma-alumina as a "carrier" of 99mTc label is its lower toxicity compared with antimony and rhenium compounds in combination with good absorption properties, availability, and low cost.

A specific feature of this compound is an organic coating of nanoparticles [8]. During passage through the lymphatic system, nanoparticles lose their organic coating and persistent uptake in SLN without redistribution in the body. In addition, 99mTc-Al_2O_3 has an optimum nanoparticle size from 50 to 100 nm, whereby the tracer quickly leaves the injection site and migrates through the lymphatic system [3, 9, 14].

The purpose of the study was to evaluate the feasibility of using the new radiopharmaceutical, based on technetium-99m-labeled gamma-alumina (99mTc-Al_2O_3), for the detection of sentinel lymph nodes in cervical cancer patients.

2. Materials and methods

Between 2015 and 2016, 23 patients with cervical cancer (T1aNxMx- T2bNxMx) were treated at the Gynecological Oncology Department of the Tomsk Cancer Research Institute. Non-keratinizing squamous cell carcinoma was histologically verified in all patients.

The day before surgery, four submucosal injections of 99mTc-Al_2O_3 at a dose of 20 MBq per quadrant were made in each quadrant around the cervical tumor. Patients were submitted to single photon emission computed tomography (SPECT) of the pelvis 18 hours after the injection of the radioactive isotope. Images were obtained using a double-headed gamma-camera (E.CAM 180, Siemens).

A total of 16 projection images were recorded into a 64x64 matrix (30 seconds per projection). The scan images were analyzed using the manufacturer software (e.soft, Siemens, Germany). Three-dimensional images of the abdomen, sagittal, transverse and coronal sections were obtained (fig.1). Single photon emission computed tomography scans were visually assessed.

For intraoperative detection of sentinel lymph nodes, the hand-held gamma-probe Gamma Finder II® was used, which allowed the surgeon to precisely locate gamma radiation source and obtain accurate information about the distribution of the radionuclide in tissues and organs of the patient. The intraoperative gamma probe-guidedof SLNs was performed by the surgeon after the opening of the retroperitoneal space (fig.2).The registered gamma radiation level was displayed in numerical values.

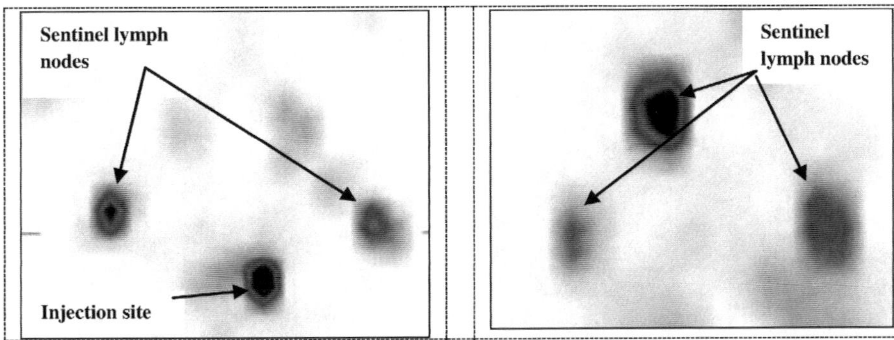

(a) Visualization of sentinel lymph nodes in the projection of external iliac lymph nodes on the right and left.

(b) Sentinel lymph nodes in the projection of the internal iliac lymph nodes on the left and the right and the right external iliac lymph nodes.

Figure 1. Sentinel lymph.

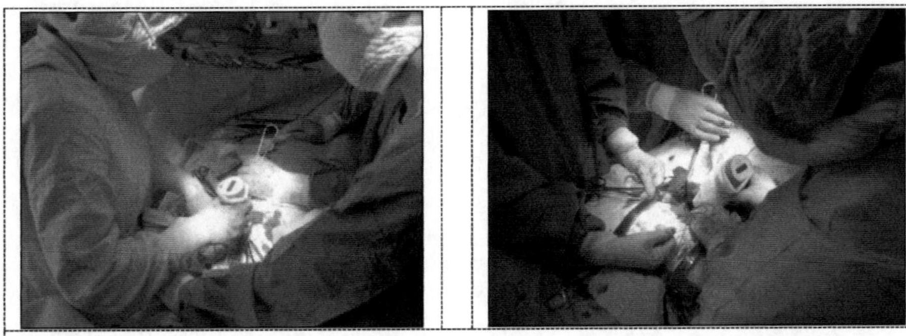

Figure 2. Intraoperative gamma probe-guided SLN identificationin the area of the iliac-pelvic lymphadenectomy using Gamma Finder II® probe.

The lymph node with radioactivity of at least three times more than the background counts was defined as a sentinel lymph node. The sentinel lymph node was marked and separately sent for express cytological examination. The lymph node was sectioned transversely into 2mm slices. Depending on the size of the node, 6–10 imprints were made from each slice by gently touching the cut surface of the node onto a glass slide. Cytological specimens were stained for 15 seconds using the kit for rapid staining of blood smears (Leucodiff 200) and examined with a Zeiss Axio Scope.
Patients were then underwent iliac-pelvic lymph node dissection. After lymphodissection, the removed lymph nodes were re-examined with gamma probe to detect SLNs missed during intraoperative examination(fig.3).

The true number of SLNs was calculated as the sum of SLNs identified removed lymph nodes. Statistical analysis was performed using the IBM SPSS statistics 20.0 package.

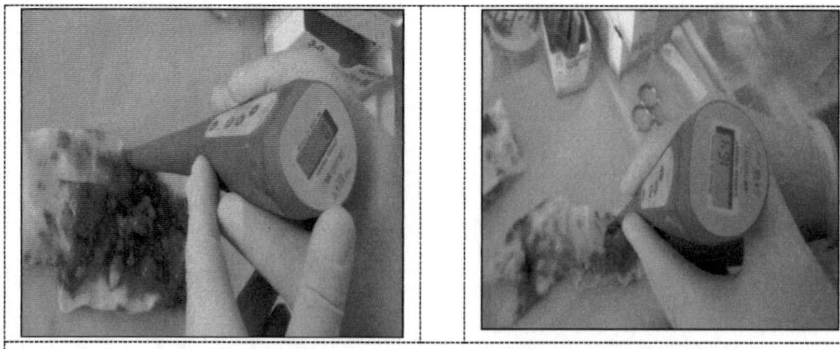

Figure 3. Direct radiometry macropreparations.

3. Results and discussion

The use of 99mTc-Al2O3 enabled SLNs to be identified by SPECT and intraoperative gamma probe procedure in all 23 patients. A total of 27 SLNs were identified by means of single photon emission computed tomography (SPECT). The proportion of radiation taken up by the SLN was 7% of the total administered dose, thereby indicating high accumulation of the radiopharmaceutical in SLNs. The intraoperative SLN assessment followed by the measurement of radioactivity in surgical specimen enabled identification of 34 SLNs. The radioactive count at the tumor injection site ranged from 1500 to 2000 cps (counts per second), whereas it ranged from 80 to 200 cps in sentinel lymph nodes. The percentage of tracer uptake was 24% of the total administered dose. High radioactivity counts were registered only in SLNs, the non-SLNs had a background count (0-3cps), indicating that there was no redistribution of the radiopharmaceutical.

The most common site for SLN detection was the external iliac region (57.1%), followed by the internal iliac, obturator, presacral and retrosacral regions (they amounted to 14%, respectively), and the parametrial region (1%). The excised SLNs were evaluated by imprint cytology with subsequent routine histological examination. All lymph nodes including SLNs and non-SLNs were metastasis-negative, thereby indicating that there was no phenomenon of "jumping" nodal metastases.

The intraoperative gamma probe procedure demonstrated 100% sensitivity for SLN detection, compared to SPECT image with sensitivity of 79%.

Various techniques of SLN identification are widely discussed in the published literature. Currently, most investigators propose that the blue dye and isotope complement each other, and the detection of SLN is more successful using combined dye- isotope technique, with sensitivity and specificity approaching 100%. In our study, the use of 99mTc-Al_2O_3, a novel molecular imaging agent has shown promising results in detecting SLNs with high sensitivity.

The frequency of lymph node metastasis in patients with stage I-II cervical cancer does not exceed 15%, which confirms the relevance of the concept of "sentinel" lymph node. The routine, widespread use of 99mTc-Al_2O_3, specially developed for SLN identification, would greatly improve the quality of life of 80-90% of patients with early-stage cervical cancer. Sentinel lymph node biopsy may be a reasonable alternative to avoid extensive pelvic lymph nodes dissection in early stage cervical cancer patients with negative SLNs, thereby significantly reducing the length of patients' hospital stay and speeding up their rehabilitation.

Conclusion

The novel radiopharmaceutical based on 99mTc-Al_2O_3 has high accumulation in sentinel lymph nodes without the evidence of further redistribution along the lymphatic collector, thus significantly facilitating scintigraphic and gamma probe-guided intraoperative SLN identification.

Single photon emission computed tomography and intraoperativegamma probe-guidedidentification performed after paratumoral injection of 99mTc-Al_2O_3 enable the detection of SLNs with 79 and 100% sensitivity, correspondently.

References

[1] Edited by A.D. Caprina, V.V. Starinskaya, G.V. Petrova. 2014 *Malignancies in Russia in 2014 (morbidity and mortality)* **250**s.
[2] VI Chernov, IG Sinilkin, ECh Choynzonov, SY Chijevskaya, AA Titskaya, RV Zelchan, OD Bragina, AY Lyapunov, VS Skuridin. 2015 *Eur J Nucl Med Mol Imaging* **42** (Suppl 1): S 704.
[3] I.Sinilkin, V.Chernov, R. Zelchan, A.A.Titskaya, V.Skuridin. 2014 *Eur J Nucl Med Mol Imaging* - **41** (Suppl 2) – S518
[4] I.Sinilkin, V.Chernov, A.Chernyshova, L.Kolomiets A.Titskaya, R.Zelchan. 2015 *Advanced Materials Research Vol.* **1084** pp 540-544
[5] S.G. Afanasiev, A.V. Avgustinovich, V.I. Chernov, I.G. Sinilkin. 2009 *Siberian Journal of Oncology* **№4** (34) .- S.27-32.
[6] V.Chernov, A.Titskaya, I.Sinilkin, R.Zelchan, N.Varlamova. 2015 *Advanced Materials Research Vol.* **1084** pp 443-446
[7] V.N. Sagaidak,L.E. Komarov. 2001 *"Magazine "Practical Oncology"* - S. **4**5-48.
[8] N.V. Varlamova, V.S. Skuridin, E.A. Nesterov, V.I. Chernov, A.A. Titskaya. 2015 *Experimental and Clinical Pharmacology*. T. **78.** № 6. S. 26-29.
[9] I.Sinilkin, V.Chernov, A.Titskaya, R.Zelchan, N.Daneikina. 2015 *Advanced Materials Research Vol.* **1084** pp 373-376
[10] C. Tax, M.M. Rovers, C. de Graaf, P.L. Zusterzeel, R.L. Bekkers. 2015 *GynecolOncol.* Dec;**139** (3):559-67
[11] R. Wuntakal, A.J.Papadopoulos, S.A.Montalto, M.Perovic, M.Coutts, O.Devaja 2015 *Int J Gynecol Cancer*. Nov;25 (9):**16**63-8.
[12] E.A. Gould, T. Winship, P.H. Philbin, H.H. Kerr. 1960 *Cancer;***13**: 77–78.
[13] R.M. Cabanas. 1977 *Cancer.***39**: 456–466.
[14] I.G. Sinilkin, V.I. Chernov, A.L. Chernyshova, L.A. Kolomiets, A.A. Titskaya, R.V. Zelchan,O.D. Bragina, A.Y. Lyapunov, V.S.Skuridin. 2015 *Eur J Nucl Med Mol Imaging* **42** (Suppl 1): S 704

IOP Conf. Series: Materials Science and Engineering **135** (2016) 012044 doi:10.1088/1757-899X/135/1/012044

Development new radiopharmaceutical based on 5-thio-d-glucose labeled technetium-99m

E S Stasyuk[1], V S Skuridin[1], E A Ilina[1], A S Rogov[1], E A Nesterov[1], V L Sadkin[1], L A Larionova[1], N V Varlamova[1], R Zelchan[2].
[1]National Research Tomsk Polytechnic University, Tomsk, Russia
[2]Tomsk Cancer Research Institute, Tomsk, Russia

E-mail: rogov@tpu.ru

Abstract. The article considers the obtaining and possibility of using 5-thio-D-glucose labeled technetium-99m for the diagnosis of malignant tumors by single photon emission computed tomography. The analysis of the level of international developments of radiopharmaceuticals based on derivatives of glucose has been carried out. Also the article provides information on of using experimental batches of lyophilisate on the basis of 5-thio-D-glucose for preliminary biomedical testing on the mice.

1. Introduction

Timely performed diagnosis and detection of malignant tumors in the early stages of development remains one of the most important problems of modern medicine. Over the past 10 years, according to the Ministry of Health and Social Development, the death rate from cancer has increased to a level of 13.87% and has become the second leading cause of mortality in Russia. Annually in our country about 480 thousand cases of cancer tumors are revealed. Unfortunately, about 60% of registered patients are detected in the late stages of disease, and this greatly reduces the possibility of treatment. Thus, in 2013 the death rate from cancer in Russia amounted to 292 thousand people (according to the Federal State Statistics Service on 05.22.2014). Therefore, the efficiency of medical care to patients with cancer depends on the level of development and introduction of modern nuclear medicine techniques into medical practice. The uniqueness of nuclear medicine techniques is that they allow to diagnose functional abnormalities of vital activity of organs at the earliest stages of the disease when a person does not feel any symptoms. This makes it easier to detect and treat a wide variety of diseases, which significantly increases the likelihood of cure.

Glucose derivatives, labeled with radioactive isotopes, are promising radiopharmaceutical (RPC) for early diagnosis of malignant neoplasms. This is due to the fact that in tumor cells there is an increased level of glucose metabolism as compared to normal cells [1]. In tumor cells the number of glucose transporters is increased, so glucose administration to the cancer cells also increases [11,12].

There are different glucose derivatives labeled with radionuclides ^{18}F, ^{11}C and others. Currently in Russia in Positron Emission Tomography (PET) a radiopharmaceutical 2-fluoro-2-deoxy-D-glucose (^{18}F-FDG) is used. This preparation comprises a positron-emitting radionuclide fluorine-18, for the diagnosis of tumors, metastases and evaluating of the anticancer therapy efficiency. Despite the high diagnostic informative value of PET with ^{18}F-FDG, the widespread use of this method is limited because of its high cost and the lack of PET centers in most regions of Russia.

The main advantage of RPC based on glucose derivatives labeled with technetium-99m, is that visualization of tumors and their use can be made by single-photon emission computed tomography (SPECT), which significantly reduces the cost of the diagnostic procedure. Now in Russia there are over 200 SPECT centers, and the most frequently used radionuclide for SPECT studies is technetium-99m. Therefore, the aim of this work is to study the possibility of obtaining technetium-99m-labeled glucose derivatives, such as 5-thio-D-glucose for the subsequent creation of a new radiopharmaceutical on its basis.

2. Methods of synthesis of radiopharmaceutical "2-fluoro, ^{18}F-2-deoxy-D-glucose"

For the first time [^{18}F] FDG was synthesized more than 35 years ago at the Brookhaven National Laboratory (USA) using the method of electrophilic fluorination [13]. But, unfortunately, it is not widely used due to the complexity of the synthesis. In 1986, a group of German scientists [14] have developed a method of obtaining a nucleophilic [^{18}F] FDG. This method proved to be more simple to perform and has a higher radiochemical yield.

Preparation of 2-fluoro, ^{18}F - 2-deoxy-D-glucose in using nucleophilic radiofluorination [15] is carried out starting from 1,3,4,6-tetra-O-acetyl-2-O-trifluoromethanesulfonyl-β-D-mannopyranose (TATM) which is synthesized by Hamacher [16] of 1,3,4,6-tetra-O-acetyl-beta-D-mannopyranose and its - from D-mannose [17]. The complete synthetic scheme is shown in Figure 1.

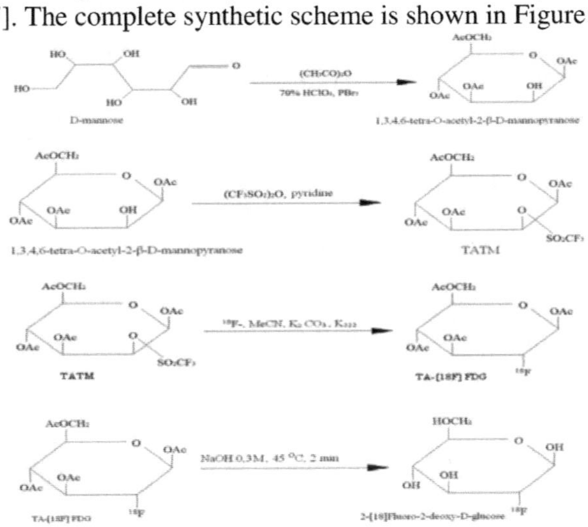

Figure 1. Scheme nucleophilic synthesis of 2-fluoro [^{18}F]-2-deoxy-D-glucose.

Before the introduction of fluoride anion ^{18}F into TATM, it is pre-separated from the target water using microcolumn filled with resin QMA Accel Waters or other anion-exchange resin with similar properties. Desorption of fluoride from the column is carried out with an eluent consisting of kryptofix, potassium carbonate, deionized water and acetonitrile. The result is a complex of fluoride-ion with a catalyst in a mixture of acetonitrile/water, which is distilled before the fluorination as an azeotropic mixture under a stream of nitrogen.

The substitution reaction of a triflate group in TATM to ion fluorine -18 is conducted in acetonitrile solution at 80°C. Its radioactive product is 1,3,4,6-tetra-O-acetyl-2-^{18}F-β-D-glucopyranose. Hydrolysis of the reaction mass is carried out at alkaline catalysis (solution 0.3 M NaOH) at 45°C for 2 min. The result is 2-fluoro-2-deoxy-D-glucose, labeled with radionuclide ^{18}F.

After neutralizing of the mixture with 0.5 M HCl solution, it is heated in the reaction vessel to 120°C and under a nitrogen stream the traces of acetonitrile are removed during 3-4 min. Reaction mixture diluted with water is transferred to purification column filled with a cation-exchange resin Supelclean LC-SCX, Supelco and neutral aluminum oxide. After purification, the isotonicity of a resulting preparation is achieved by mixing it with an aqueous solution of sodium chloride, in such a concentration that NaCl content of the finished RPC is 8-10 mg/ml. Then sterilization is performed by passing it through a sterilizing Millipore filter with a pore diameter of 220 nm. All these chemical operations can be performed by a laboratory robot RB-86 Anatech («Scanditronix», Sweden).

A synthesis scheme, shown in Fig. 1, is used in radiochemistry laboratory of the Institute of Human Brain of RAS (St. Petersburg) [18].

This method of synthesis of ^{18}F-fluorodeoxyglucose is quite complex and requires the use of special equipment, such as a cyclotron, a radiochemical module or a laboratory robot along with the PET camera costing $ 5 million. Furthermore, the reaction with the given precursor cannot be implemented with other radionuclide, such as iodine-123 or technetium-99m, convenient for conducting research using wide-spread gamma-cameras [2].

3. Methods for obtaining derivatives of glucose labeled with technetium-99m

In [3] it is shown that for tumor lyophilisate, using a gamma camera, the complexes of ^{99m}Tc with a variety of glucose derivatives can be used. Here, for the labeling by isotope ^{99m}Tc, the authors tested glucose derivatives, which contain nitrogen atoms in their composition. The preference was given to D-glucosamine, its salts and hydrates.

^{99m}Tc-labeled complexes of glucose derivatives had radiochemical purity (RCP) of 98% or more. Approximately the same parameters of RCP were achieved in study [4], when mixing 5-thio-D-glucose with sodium pertechnetate, ^{99m}Tc (1.85-3.7 GBq) in the presence of 0.01 mg of tin chloride (II) and subsequent 30-minute incubation of the mixture at room temperature. Tests of the obtained preparation in rabbits showed a level of binding with proteins of 32%, which is favorable for the detection of tumors.

Also, good results were obtained in [5], which describes the process of radiolabeling with ^{99m}Tc and preliminary biological tests of complex of dietelentriaminepentaacetic acid (DTPA) with deoxyglucose (DG) were conducted.

In the first step of this process, the reduction of ^{99m}Tc with tin chloride (II) was conducted. Thereafter , reduced technetium-99m was added to the complex of DTPA-DG, and incubation of the mixture was carried out at room temperature for 30 min. Conducted biological tests have shown that the absorption coefficient of brain tumor of the complex ^{99m}Tc-DTPA-DG in relation to the muscles was higher than that of ^{18}F-FDG. At the same time in any other organs no significant accumulation was observed. And it has been revealed that blood is rapidly cleaned through kidneys [6].

Good quality of tumor images made by the method of SPECT is also observed when using labeled glucose analog ^{99m}Tc-glukarate [7]. In another work [8] the authors studied the differences in the biological behavior of ^{99m}Tc-labeled 1-thio-beta-D-glucose 2,3,4,6-tetra acetyl analog (Tc-TG) and ^{18}F-FDG. Binding of both indicators was performed in vitro on viable tumor cells and bacteria. Both indicators were then injected into the thigh muscle of mice, which were infected with Staphylococcus aureus. Thus, it was shown that both indicators effectively bind tumor cells and provide a high accumulation ratio in infected muscle tissues. Both tracers are rapidly removed from the blood system through the kidneys, mainly accumulating in the bladder, followed by their removal from the body.

In the animal tumors one could observe a high level of accumulation during administration of technetium-99m-labeled glucose analog ^{99m}Tc-ethylene ditsestin deoxyglucose (ECDG), which was obtained from the reaction of ethylene ditsestin with glucosamine in the presence of a carbodiimide coupling agent. [9]

4. Experimental results and discussion

The most of diagnostic drugs based on technetium-99m in clinical conditions is received by mixing of eluate from the generator ^{99m}Tc with standard sets of reagents, which are intended for receiving the set radiopharmaceutical. As such standard sets to ^{99}Mo/^{99m}Tc the ^{99m}Tc-generator the lyophilisate representing the lyophilised mix of components of a dosage form of drug are used. The reducer - a dihydrate of tin chloride (II) ($SnCl_2{\cdot}2H_2O$) which in standard conditions quickly is oxidized is a part of lyophilisate, but at a freeze-drying in vacuum keeps the properties within a year that is the regulated period of validity of lyophilisate.. For development of new radiopharmaceutical based on glucose derivatives of 5-tio-D-glucose was taken as a basis. The structural formula of 5-tio-D-glucose has an appearance.

At an initial stage of researches solubility of the chosen substance in various environments of an organic or inorganic origin, for the purpose of selection of solvents and system for a chromatography was checked. As solvents tested the following solutions: normal saline solution – 0.9% NaCl solution, H_2O (a dist.), acetone, ammonia, the dehydrated alcohol, 75% alcohol solution.

For check of solubility of substance of 5-tio-D-glucose add hinge plates on 5 mg in separate bottles, and then entered the prepared solvents of 1 ml into each of bottles. Solubility of substance was defined visually with use of the specialized installation representing the light stand with the optical device. Results are provided in table 1.

Table 1. Solubility of 5-tio-D-glucose in different solvents.

Solvent	Result of process of dissolution	pH of mixes
0.9% NaCl solution	Dissolves without being heated, the solution becoming clear	5.0
H_2O (dist.)	Dissolves without being heated, the solution becoming clear	5.0
Acetone	Does not dissolve, even when heated up	6.0
The dehydrated C_2H_5OH	Slightly soluble	6.5
C_2H_5OH - 75% solution	Dissolves without being heated, the solution becoming clear	6.5
0.5 M NaOH solution	Dissolves without being heated, the solution becoming clear	12.0
0.5 M HCl solution	Dissolves without being heated, the solution becoming clear	1.5
Ammonia	Dissolves without being heated, the solution becoming clear	1.5

By results of the conducted researches the conclusion was drawn that substance 5-tio-D-glucose in dehydrated alcohol is poorly soluble, but is well soluble in 75% spirit. In alkaline condition, substance though is completely soluble, but over time collapses. Full dissolution of 5-tio-D-glucose without formation of a colloid is observed in normal saline solution, water and muriatic solution.

Now rather easy way of receiving 5-tio-D-glucose with technetium-99m is known [4], according to which the drug of ^{99m}Tc-5-tio-D-glucose prepare by mixing 10 mg of 5-tio-D-glucose from 0.074 mg of tin chloride (II) of a dihydrate with the subsequent introduction to the received mix of 2-4 ml of solution of a pertechnetate of ^{99m}Tc with activity of 50-100 mCi (1.85-3.7 GBq) and an incubation within 30 min. Thus efficiency of branding makes 98.5 ± 0.8 % and remains stable 24 hours. Authors do not consider possibility of preparation of lyophilisate therefore original stock of reagents for receiving drug cannot be stored more than 1-2 hours, owing to oxidation of tin (II) to the 4th valence state.

Approximately same technique of preparation of the marked ^{99m}Tc 5-tio-D-glucose it is offered in [19]. Here in the bottle containing 5-tio-D-glucose mix (5 mg, 0.51 mmol) and $SnCl_2{\cdot}2H_2O$ (0.080 mg), add eluate $^{99m}TcO^{4-}$ at the rate of 20 mCi/ml. After mix with in 10 min.

As experimental check showed, at preparation from these structures of lyophilisate, the exit of a marked main product in filtrates of 200 nanometers did not exceed the 40th percent. Most part of activity ^{99m}Tc, it was adsorbed on a colloid. Thus on chromatograms of the received radiopharmaceutical in Acetone it was found in addition to peak ^{99m}Tc (VII) two peaks of initial substance of 5-tio-D-glucose that gave the grounds to assume that at process of freeze-drying there is its partial decomposition. Technetium eluate for research was obtained from technetium generators "^{99m}Tc-GT-TOM" and "GT-TOM-II».

For 5-tio-D-glucose substance stabilization into composition of reaction mixture it was solved to enter as the stabilizing additive ascorbic acid in number of 0.5 mg. For its preparation, the hinge plate of ascorbic acid weighing 100 mg was dissolved in 10 ml of water (concentration of 10 mg/ml). Besides, influence of pH of environment at a size of a radiochemical exit of a main product was studied. Hydrogen ionization value of reagent before freeze-drying of mix established by introduction to its structure salt 0.05 M of solution of hydrochloric acid in quantity to 0.36 mg/ml. 5% $NaHCO_3$ solution used for increase of pH to values 5 and more.

As showed researches, optimum value rn to mix at which the maximum radiochemical exit of a main product is reached (to 98%), is in limits 3.5 – 4.0. The increase in pH to 5.0 by introduction to mix of a hydrocarbonate of $NaHCO_3$, led to noticeable decrease in an exit of a marked complex almost by 18% though radiochemical purity thus remained on rather high level (96%).

Radiochemical purity (RCP) of the received product was estimated by removal chromatogram in two mobile phases: in acetone (R_f = 0.9 $^{99m}TcO_4^-$) and in the mix C_2H_5OH : 25% of NH_4OH: H_2O ÷ 2 : 5 : 5 (R_f = 0.0 radiocolloids) with use of plates of PTLC silicagel-AF-A-UV (Sorbfil), the mm size 20×100. Putting drug on plates, in number of 5 μl, was carried out having receded from one of edges on 20 mm (the line of start). After drying of a spot on air the strip was placed in the camera for a chromatography for 10-40 min, depending on the used mix. Visualization and studying of distribution of activity ^{99m}Tc on length chromatogram carried out by means of the «Gammaskan-01A» installation. The radiochemical purity of 5-thio-D-glucose was calculated by subtracting from 100% the percents of free pertechnetate and radiocolloids.

Due to introduction to composition of reaction mixture of the ascorbic acid and strengthening of a reducer it was succeeded to increase stability of substance and as a result, on chromatogram in acetone the second peak of a product of decomposition was not found.

Finally, the following technique of preparation of drug was developed. A hinge plate of 5-tio-D-glucose weighing 15 mg entered in a bottle with a capacity of 10 ml and parted it in 1 ml of water for injections. Then consistently added to a bottle 20-25 μl freshly cooked Sn (II) solution (concentration of 7 mg/ml on $SnCl_2 \cdot 2H_2O$), 45-50 μl aqueous solution of ascorbic acid with concentration of 10 mg/ml and 200 μl, 0.05M of HCl. The received solution is passed via the sterilizing filter (pore size of 0.22 microns) in a sterile bottle with a capacity of 10 ml. After a bottle without prefreezing of mix placed in the refrigerator of a lyophilizer and carried out process of freeze-drying in the automatic mode within 24 hours.

The 4 ml of eluate ^{99m}Tc with activity of 1.5-2 GBq (40-50 mCi) entered into a bottle with reagent for the subsequent preparation of drug. The drug was ready to use incubation of mix at the room temperature in 30 min.

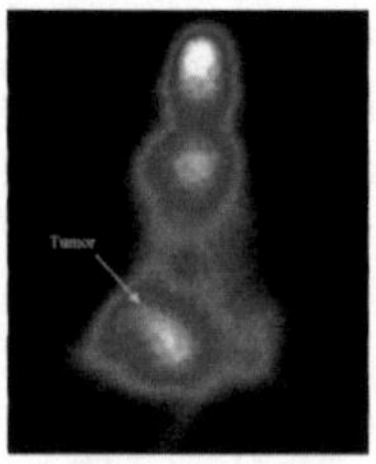

Figure 2. Scintigram of a mouse obtained in 30 min after administration of radiopharmaceutical based on 5-tio-D-glucose marked by technetium-99m.

Also studying of stability of the received radiopharmaceutical in time on an indicator radiochemical purity was carried out. Definition of RCP was carried out each 30 min within 7 hours by according to the described technique since the moment of preparation of RP. The received results showed that within 4 hours radiochemical purity to be in limits of 95-98 %, but in 4.5 hours falloff of radiochemical purity begins and by the end of the 6th hour it falls to 89 %. From what the conclusion was drawn that period of validity of a drug on an indicator of radiochemical purity makes 4 hours.

Preliminary biomedical testing of the received radiopharmaceutical ^{99m}Tc-5-tio-D-glucose, which are carried out at the Tomsk Cancer Research Institute on the mice of the C57BL/6 line with Lewis's carcinoma (LLC) graft in hip area showed functional suitability of a drug. Level of accumulation of drug in a malignant tumor made 6.5 % of the general entered activity that is enough for its reliable visualization. Scintigram of a mouse body obtained in 30 min after administration of radiopharmaceutical based on 5-tio-D-glucose labeled by technetium-99m, presented on figure 2.

5. Conclusion

As a result of experiments it has been established that:

1. Development of the composition and methods of preparation of the radiopharmaceutical formulation "^{99m}Tc-5-thio-D-glucose" in the form of lyophilisate.
2. The radiochemical purity of technetium-99m-labeled preparation, produced on the basis of lyophilizates of 5-thio-D-glucose, is in the range of 95-98 %.
3. Preparation without noticeable decrease in its radiochemical purity remains stable for 4 hours.
4. Preliminary biomedical testing the resulting radiopharmaceutical conducted at the Tomsk Cancer Research Institute in the mice C57BL line / 6 mice with tumor hip. It is confirmed its functional suitability.

Acknowledgments

This work was financially supported by Ministry of Education and Science of the Russian Federation (RFMEFI57514X0034).

References

[1] Dimitriadis G., Maratou E., Boutati E., Psarra K., Papasteriades C., Raptis S.A. (2005) Cytometry A. Mar;64(1):27-33.
[2] Skuridin V.S. Semenov A.S. (2013) Proceedings of the Universities. Physics. Vol.56. 4/2:264-267.
[3] Hee-Kyung Lee, Dae-Hyuk Moon, Jin-Sook Ryu, Jae-Seung Kim, Seung-Jun Oh (2003) Patent United States Pub. No. 2003/0120046 A1.
[4] Ozker S.K. Collier B.D. (2000) Patent United States Patent No. 60998222003 A61K 51/00.
[5] Chen Y., Huang Z.W., He L., Zheng S.L., Li JL, Qin DL. (2006) Appl. Radiat. Isot. Vol. 64. 3:342–347.
[6] Chen X., Li L., Liu F., Liu B. (2006) Bioorg. Med. Chem. Lett. Vol. 16. 21:5503–5506.
[7] Cheng D., Rusckowski M., Wang Y., Liu Y., Liu G., Liu X., Hnatowich D.A. (2011) Curr. Radiopharm. Vol. 4. 1:5–9.
[8] Welling M.M. Alberto R. (2010) Nucl. Med. Commun. Vol. 31. 3:239–248.
[9] Yang DJ, Kim CG, Schechter NR, Azhdarinia A, Yu DF, Oh CS, Bryant JL, Won JJ, Kim EE, Podoloff DA. (2003) Radiology. Vol. 226. 2:465–473.
[10] Stasyuk E.S., Nesterov E.A., Skuridin V.S., Sadkin V.L., Rogov A.S. (2012) J. Radiochemistry. Vol. 54. 4:391-394.
[11] Benard F., Turcotte E. (2005) Breast Cancer Research. 7:153-162.
[12] Sakurai H, Suzuki Y, Nonaka T, Ishikawa H, Shioya M, Kiyohara H, Katoh H, Nakayama Y, Hasegawa M, Nakano T. (2006) Gynecologic Oncology 100:601-607.
[13] T. Ido, C-N. Wan, V. Casella, J. S. Fowler, A. P. Wolf, M. Reivich D. E. Kuhl (2006) J. Labelled Compounds and Radiopharmaceuticals DOI: 10.1002/jlcr.2580140204.
[14] Hamacher K., Coenen H.H., Stoecklin G. J. (1986) Nuclear Medicine. 27:235.
[15] Hamacher K., Blessing G., Nebeling B. (1990) Applied Radiation and Isotopes 41:49.
[16] Hamacher K (1984) Carbohydrate Research. 128:291-295.
[17] Deferrari J.O (1967) Carbohydrate Research. Vol. 4 P. 432-434.
[18] Gomzina N. A., Vasil'ev D. A., Krasikova R. N. (2002) Radiochemistry. Vol. 44 4:403-409.

Electrophysical methods of separation of metal cations in the moving salts solution

V.N. Gofman[1], I.V. Tuksov[1], S.N. Timchenko[1], I.V. Shamanin[1], A.D. Poberezhnikov[1], M.A. Kazaryan[2]

[1]National Research Tomsk Polytechnic University
634050, Tomsk, Lenin Avenue, 30
[2]P.N. Lebedev Physical Institute of the Russian Academy of Sciences
117924, Moscow, GSP-1, Leninskiy Prospekt, 53

E-mail: shiva@tpu.ru

Abstract. The results of experiments on the excitation of the phenomenon of selective drift of solvated ions under the influence of an external "asymmetric" electric field to the circulating solution of calcium chloride and magnesium salts in a polar liquid dielectric – water are shown. The purpose of the experiments was to determine the influence of the field frequency and amplitude of the field strength on the excitation phenomenon, and the study of the operating characteristics of the testing apparatus - a dividing cell. The dependences of the separation efficiency of solvated cations from the frequency of the external field and the excitation threshold of the phenomenon from the field strength in the separation cell are defined.

1. State of the research

Earlier discovered phenomenon of selective drift of solvated ions under the influence of an external "asymmetric" electric field [1] for today found only one practical application in technology of the separation of cerium and yttrium cations formed during the investigation of the properties of spent nuclear fuel [2]. At the same time, the directions, in which the use of the phenomenon may bring a significant effect, sufficient clearly marked in a number of papers [3, 4, 5]. The main reason for the small number of practical uses of the observed phenomenon were two factors: a shallow study of the question of hardware implementation of technical systems that operate on the basis of the phenomenon, and the lack of experimental data on the influence of the electric field frequency on the effectiveness of separation processes solvated ions in moving (circulating) solutions.

This paper presents the results that once again confirm the importance of the observed phenomenon, its stability and reproducibility, as well as enabling to create a relatively simple experimental installation for searching of application to phenomenon of selective drift of solvated ions under the influence of an external "asymmetric" electric field. The term "asymmetric" electric field was introduced in experiments on excitation of selective drift of ions [1] and has an important meaning as the "normal" sinusoidal potential does not cause the observed phenomenon.

2. Experimental section

Block diagram of the experimental installation is shown in Figure 1. The main power source of variable frequency signal was audio-frequency generator GZ-109 1, the signal from which is fed to the input of pre-amplifier RUB2250 2 with its own power supply NES-100-12 3.

IOP Conf. Series: Materials Science and Engineering **135** (2016) 012045 doi:10.1088/1757-899X/135/1/012045

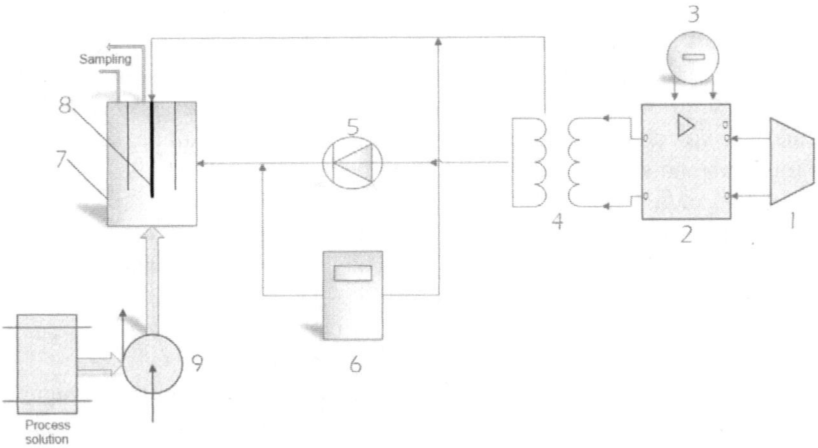

Figure 1. Block diagram of the experimental installation.

The potential, amplified by transformer 4, goes through the asymmetric waveform forming device 5 and applied to the central electrode 8 of the testing apparatus 7. Here, the second electrode is grounded. The signal was recorded with an oscilloscope 6. During the experiments the process solution from the tank was fed by a peristaltic pump 9 into a lower section of the testing apparatus. In experiments were used a dilute mixture of solutions of calcium and magnesium chlorides mark "chemically pure". A general view of the testing apparatus, which is a cylinder with arranged coaxially basic structural elements inside shown in Figure 2.

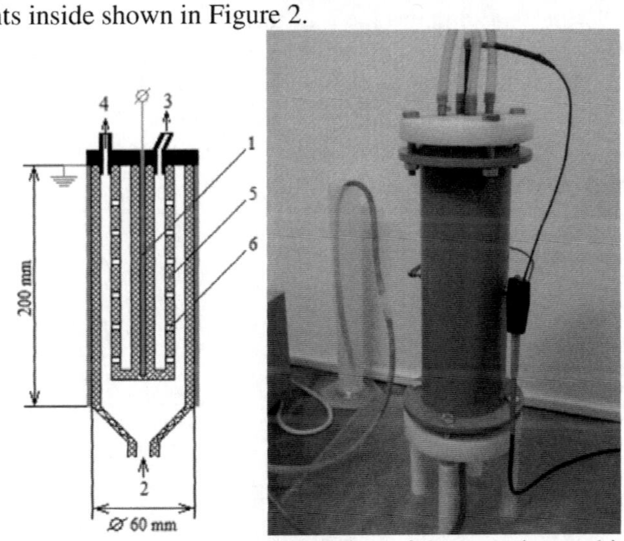

Figure 2. General view and a schematic representation of the experimental installation in the front section.

In the center is a potential electrode 1 of the steel rod, hermetically isolated from the process solution with polyvinylchloride pipe. Grounded steel electrode of cylindrical shape hermetically isolated from the process solution by the polyethylene film. The entire volume of the process solution supplied to the lower section of the installation, divided by polyethylene perforated cylinder 5 on the inner and outer zones. This cylinder performs the function of a separator between the zones of enrichment and depletion by target ions of the solution.

Perforation of cylinder 6 is formed as a vertical oblong openings arranged azimuthally symmetric along the generatrices of the cylinder along its entire surface. The openings allow moving ionic components of the process solution from one zone to another. Process solution fed through the nozzle

2 to the lower section of the installation and moving up along both zones with the same linear speed. Sampling at the top of the installation of each zone was carried out autonomously by the fittings 3 and 4 and the drain hose, equipped with special screw clamps. With these clamps carried regulating the output volume of the solution at inner and outer zones. Wherein, solution velocity in the inner zone is 4-7% higher than the velocity of the solution in the outer zone. In all experiments, the character of solution motion is corresponded to laminar liquid flow regime, and the linear velocity of the solution is not more than 4 cm / min.

During experiments solutions was sampled from both inner and outer zones at defined time intervals. Analysis of samples for magnesium and calcium ions was performed on atomic emission spectrometer with inductively coupled plasma iCAP 6300 Duo.

The result of the statistical analysis of experimental data showed that the measurement error in all of the experiments did not exceed in relative proportions 0.014 from average. It should be noted that using a device 5 (Fig. 1) was formed positive halfwave of the initial sinusoidal signal and volumes of fluid in inner and outer zones were exposed only to positive potential field of sinusoidal shape. This provided an abrupt movement of positively charged solvated cations to the electrode with zero charge 7 from the potential of a positively charged electrode 6 (Fig. 2).

The main objective was to detect a change of concentrations of calcium and magnesium cations in solutions at the inner and outer zones at certain range of frequencies. This fact would indicate that the selective transfer of one type of cation from the outer to the inner zone take place. In order to provide forced passage of cations from the outer zone into the inner, the linear velocity of the solution should be somewhat higher. Such a mode of fluid movement was mounted with clamps (Fig. 2) in each experiment. It has previously been shown [1] that the effects of changes in concentrations of ions in dilute solutions when applied to these asymmetric electric fields were observed at frequencies of hundreds of hertz. Considering this circumstance, in this paper, the research was carried out in the frequency range from 10 to 1500 Hz.

3. Discussion of experimental results

Fig. 3 shows the variation in the concentration of calcium ions in inner (curve 1) and outer (curve 2) zones in the frequency range from 20 to 180 Hz.

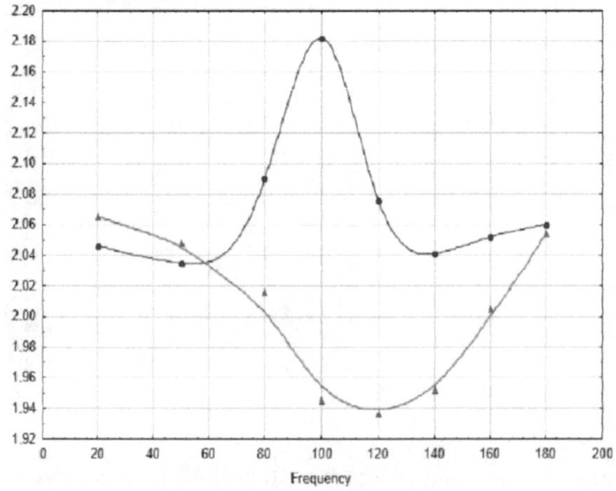

Figure 3. Influence of frequency on content of calcium ions in the separation zone at a voltage of 285 V (field strength in the cell 86.4 V/cm) and the initial concentration of 2.065 g/l. The circles marked concentration in the inner zone, triangles-concentration in the outer zone.

Earlier, in theory, it was shown [1] that the excitation of the phenomenon should be expected at frequencies of tens of Hertz, when a self-consistent field is formed in solution, and the size of the solvated ion-cluster, formed by the ion and molecules of solvent, forming solvation shell, is inversely proportional to the square root of the salt concentration values in a solvent. The values of frequencies, by-turn, is inversely proportional to the value of the moment of inertia of the cluster.

As seen, in range of 80-120 Hz is observed a distinct increase in the content of calcium ions in the inner zone of the process solution with a clearly defined maximum. Importantly, that the corresponding concentration decrease takes place in outer zone, i.e. there is a redistribution of the concentration of one ion component between zones of process solution, which is essentially the separation process. Probably, that in this frequency range there is a deformation changes in the structures of the solvation shells of calcium ions, and, possibly, partial destruction of the external solvation layers, where the binding energy of the molecules of the solvation water with the central ion has a minimum value in comparison with the water molecules that composing the primary solvation shell. This circumstance give rise to conditions for directed transfer of desolvated ions in a potential field that, in accordance with the terminology adopted earlier, determines the effect of electrically induced selective drift [1]. Since each electrolytic solution has a property of electrical neutrality, then any change in its local electric charge in local volume - positive or negative - should lead to the instantaneous compensation of this charge by moving of the corresponding sort of ions in this volume. Thus, in this case, if calcium ions being transferred from the outer to the inner separating zone, producing in the latter the excess content of positively charged ions, then in an equivalent amount of magnesium ions should be transferred from the inner zone into the outer. This phenomenon can be interpreted as a process of electromigration transport, as well, as the exchange process between ions of different sorts. Thus, the decrease in the content of magnesium ions in the inner zone, and, on the contrary, increase in concentrations of these ions in the outer zone should take place. Indeed, if we look at fig. 4, it is obvious that a change in the concentration of magnesium ions in relation to the content of calcium ions in the inner and outer zones of separation in the specified frequency range has inverse relationship.

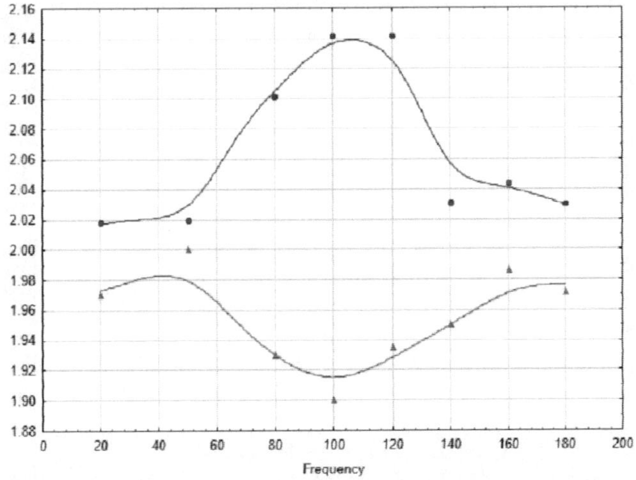

Figure 4. Variation in the concentration of magnesium ions in the separation zones at a voltage of 285 V (field strength in the cell 86.4 V/cm) and the initial content of 1.986 g/l. The circles marked concentration in the inner zone, triangles-concentration in the outer zone.

By its form, the frequency dependence of calcium ions content has a resonant character, i.e., the concentration extremes of these ions is defined in a fairly narrow range of frequencies. The same can be said about the frequency dependence of the content of magnesium ions (Fig. 5). However, the extreme values of the concentration of magnesium ions are observed in the separation zones at higher frequencies, i.e., when the external field energy is excess the threshold values for the resonance phenomena of the same calcium ions. This is due, apparently, to the difference in bond length of solvation shell with the central ion. The radius of the outer electron shell in magnesium ion is smaller than that of calcium ion, so the Coulomb interaction of the magnesium ion and the water molecules

IOP Conf. Series: Materials Science and Engineering **135** (2016) 012045 doi:10.1088/1757-899X/135/1/012045

that are part of solvation shells, forming a shorter bound that corresponds to higher eigenfrequency. Cations have only a coordinating effect on the solvent molecules in the first and second solvation spheres [6], and the number of solvent molecules in the solvation shell is determined by the shielding radius of cationic charge by cumulative charge of polarized solvent molecules [1, 7].

It has been shown in theory [1] that the directional drift of solvated ions also excited at the frequencies, corresponding to the various components of the rotational-translational motion of the system "ion - solvation shell", and at the frequencies, corresponding to the transition of the oscillatory motion into rotary.

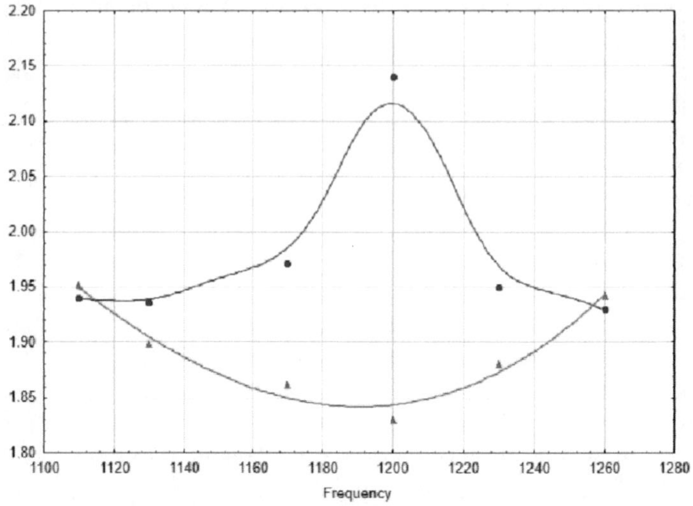

Figure 5. Frequency dependence of magnesium ions in the separation zone at a voltage of 285 V (field strength in the cell 86.4 V/cm) and the initial concentration of 1.931 g/l. The circles marked concentration in the inner zone, triangles-concentration in the outer zone.

The frequency values amounts to units of kHz. Resonance can be expected:

- at the frequency of the cooperative rotational motion of the molecules of H2O, combined in the solvation shell relatively to an axis, passing at a distance, equal to the outer radius of the solvation shell from the center of gravity;
- at the frequency of the cooperative rotational motion of the molecules of H2O, combined in the solvation shell relatively to the center of gravity of the solvated ion-cluster;
- at a frequency of rotational movement of the cluster as a whole;
- and the frequency of transition of oscillatory motion into rotational.

Formula for the calculation of the defined frequency values are obtained and given in the monograph [1].

Consequently, for the resonant deformation of solvation shells of magnesium ions will require higher energy value of an external electric field, which will increase with increasing frequency. Theoretical understanding of the electrically induced drift of solvated ions indicate that one of the factors causing this phenomenon is the value of the amplitude of the electric potential, which determines the intensity of the electric field. Experiments conducted at voltage value of 40 V (field strength in the cell 12.1 V/cm), have shown that within the errors of measurement the concentration of calcium and magnesium ions did not change and remained at their original values. Since the potential energy is the main characteristic of the electric field, apparently, there is a threshold value of electric field strength, which determines the minimum energy, which in a certain frequency range causes an effect of deformation of solvation shells of ions and consequently, their selective electrically induced drift. In this case, a threshold value of electric field strength is in the range from 12 to 100 V/cm. Depending to examine method of separation, measurements data define general separation factors of competing calcium and magnesium cations. In order to talk about single separation factor, complete information on all possible mechanisms of selective mass transfer is required. Hypothetically, electrically induced drift at the interface between the outer and inner zones is the prevailing process, leading to the exchange of ions of different sorts between these zones. Additional moving force for moving ions from the outer zone into the inner, along with the electromigration transfer, can be an excess of hydrodynamic pressure, which occurs due to the difference in the velocities of the solutions in the inner and outer zones. In this connection, the influence of the ratio of velocities of the solutions in the separation zones on separation effect is the question, requiring separate consideration. Simultaneous mass transfer of a certain quantity of competing ions for a certain period of time through the border of outer and inner zones on the entire area of their contact can apparently be interpreted as a single separation factor.

4. Conclusion

1. The conducted experiments proved the ability to use previously discovered phenomenon of electrically induced selective drift of solvated ions in salt solutions under the influence of asymmetric electric field for the organization of the technological process of solutions enrichment by the target metal.

2. It was found that the excitation of effect of the selective electrically induced drift of solvated ions may be observed in at least in two frequency ranges. Within each of the intervals are the frequencies corresponding to the Eigen frequency of the system "ion - solvation shell" as a spherical rotator (rotation of solvation shell around the central ion) or as a system of rigidly connected central ion and a hollow shell.

3. Frequency dependence of the content of calcium and magnesium ions in the mixture of their chloride solutions was studied and was showed, that for each of the competing ions there is a certain frequency range in which the speed of the directional drift of one of them increases significantly.

4. It is shown that effect of ions separation depends essentially on the value of supplied potential - the main characteristics of the external electric field.

5. The technological scheme of the process of separation of ionic mixtures by the method of electrically induced selective drift with an applied alternating electric field was developed and an experimental installation that implements this process was constructed.

References

[1] Kazarjan M.A., Lomov I.V., Shamanin I.V. Jelektrofizika strukturirovannyh rastvorov solej v zhidkih poljarnyh dijelektrikah [Electrophysics of structured salts solutions in liquid polar dielectrics]. - Moscow: Fizmatlit, 2011. – 190 p.

[2] Sposob razdelenija ionov metallov [Method for separation of metal ions]. Patent RF № 2428759 (2011) / Gavrilov P.M., Gusev A.L., Trutnev Ju.A., Kazarjan M.A., Shamanin I.V. et al.

[3] Kazaryan M.A., Shamanin I.V., Mel'nik N.N., Lomov I.V., Dolgopolov S.Yu The Structure and Radiophysical Properties of Solutions of Salts in Liquid Polar Dielectrics // Russian Journal of Physical Chemistry B, 2009, Vol. 3, No. 1, p.p. 40-45

[4] Shamanin I.V., Kazaryan M. A. Clusters Formation in Salts Solution in Polar Dielectric Liquids and Electrically-induced Separation of Solvated Ions // British Journal of Applied Science and Technology. - 2014 - Vol. 4 - №. 18, p.p. 2538-2550

[5] Shamanin I.V., Kazaryan M. A., Sidko D. F. Cluster Structure of Salt Solutions in Polar Dielectric Liquids // Advanced Materials Research. - 2015 - Vol. 1084, p.p. 97-107

[6] Puhovskij Ju.P., Saharov D.V., Safonova L.P. Issledovanie struktury sol'vatnyh obolochek ionov Na+ i K+ v N,N-dimetilformamide pri razlichnoj temperature metodom molekuljarno-dinamicheskogo modelirovanija [Investigation of the structure of ion solvation shells of Na + and K + in N, N-dimethylformamide at a varying temperature by method of molecular dynamics simulation]. - Zhurnal strukturnoj himii, Vol. 43, № 2, 2002, p.p. 307-316

[7] Burgess J. Metal ions in solutions. – Ellis Horwood: Chichester, UK., 1978. – 481 p.

Effect of the Compaction Pressure and Ni Content on the Modified Aluminum-Based Perovskite Synthesis Designed to Immobilize the Radioactive Waste in Combustion Mode

E S Tarasova, O Yu Dolmatov, D S Isachenko, A A Permikin and A O Semenov
Tomsk Polytechnic University, Tomsk, Russia

E-mail: semenov_ao@tpu.ru.ru

Abstract. The article deals with the synthesis of perovskite-like ceramics matrix material for immobilization of radioactive waste by SHS method. The dependence of the compaction pressure on the synthesis of the samples was established. Synthesis conditions for the matrix with the desired properties of the composition were determined that is acceptable for reliable isolation of radionuclides throughout the long-term storage of waste. The maximum amount of aluminum perovskite is observed when the initial mixture compaction pressure equal to 30 MPa and 25% wt. nickel.

1. Introduction

Nuclear industry is inevitably associated with the ecologically hazardous nuclear waste accumulation [1, 2]. The issue of the treatment of high-level waste (HLW), formed during the processing of irradiated nuclear fuel is considered to be far more complicated. This waste has a complex chemical composition which depends on the reactor type, exposure time, processing technology, etc.

The traditional way of HLW isolation from the biosphere is their immobilization in phosphate and borosilicate glasses of different composition. However, glass does not completely satisfy the requirements to the matrices for the preservation of long-lived radionuclides for a long period of time [3].

Along with the improvement of glasses material necessary for long-term HLW disposal in the geological environment, various types of ceramics based on the unique rock-forming minerals with high chemical, thermal and radiation stability are being developed [4]. The basic idea of this method - the inclusion of chemically bound active atoms of radionuclides and toxic waste components in the form of a solid solution into the crystal lattice of the minerals' matrix similar to the stable natural minerals, where such radionuclides exist for thousands of years with the levels of radioactivity or toxicity hundreds of times smaller than if they were in free or mechanically blocked state [5-8]. Nowadays over 20 compounds differing in capacitance depending on the elements mentioned, chemical and radioactive strength were proposed. These compounds include zircon, zirconolite, perovskite, yttrium-aluminum garnet (IAG), britholite, monazite, pyrochlore.

The analysis of the known studies in matrix immobilization showed that the main reasons hindering wide practical application of the methods, are the complexity and high cost of technologies (mixtures heated to temperatures of 1300-2000° C, the need for more sophisticated technological operations to eliminate such accompanying processes as dust and gas formation, foaming, samples compaction, etc.).

High-level waste disposal by means of a spontaneous metallothermic process with the SHS in durable synthetic mineral-like matrix is the most economical and simple way of HLW isolation from the environment. This out-of-furnace method has almost no energy costs for the refractory materials synthesis, a short time of the SHS-product formation and the absence of sophisticated equipment that facilitates the work with high-level materials in a hot cell and makes it safer [9, 10].

Thus the aim of this study was to obtain a matrix material based on a modified perovskite of ABO_3 ($A = Me^{3+}$, $B = Al$) type by SHS method. The neodymium oxide was used as a trivalent radioactive waste simulator due to the similarity in ionic radii of actinides and neodymium [11-13].

2. Methods and experiments

SHS-based materials synthesis included the following steps:
– mixing powders of initial reagents according to the stoichiometric calculation for the corresponding reaction equations;
– drying the initial mixture of reactants in a technical vacuum;
- mixing in a cubic mixer;
- compressing the initial reactants mixture into cylindrical pellets of different diameter and height with variations in material density values in samples obtained by changing the pressure;
- running the SHS process in a laboratory facility and obtaining functional materials samples.

To prepare the mixture we used industrially manufactured PA-4 aluminum powders and neodymium oxide with a chemical purity grade TU 48 – 4 – 186 – 72.

The initial mixture was prepared for the exothermic reactions (1):

$$Al + Nd_2O_3 = Al_2O_3 + Nd, \quad (1)$$
$$Al_2O_3 + Nd_2O_3 = 2NdAlO_3.$$

Initial reagents powders have been mixed in a cubic mixer AR 403 All-Purpose Equipment (Germany), which is a hollow rotating cube with three stainless steel rods, designed to enhance the mixing of the blend components.

After mixing and vacuum drying the resulting mixture was compressed into cylindrical samples 30 mm in diameter and 12-15 mm in height at different pressures – 15, 20, 25, 30, 35 and 40 MPa, to obtain samples of different density.

Pressing the initial mixture was performed using a hydraulic press PGL-12 (maximum pressure 50 MPa to 30 cm^2 surface). Metal molds were made of tool steel. Duration of pressing - 15 min.

The synthesis was carried out on the test bench for pyrometric studies of SHS laws, including a SHS reactor.

The phase compositions of the synthesized compound have been determined using a diffractometer Rigaku D / Max-2200 (CuKα radiation, a step 0.02, 1 second exposure).

3. Results and discussions

By heating the initial sample to a temperature of about 500-600 K (depending on the preparation of the initial mixture of components), the combustion wave was initiated at the ends' edges that spread over the sample volume. In this process the temperature increased rapidly, then stabilized, and burning continued almost isothermally under the temperature of 1000-1250 K. After the passage of the combustion wave along the sample volume, it cooled down to ambient temperature of the medium.

The study of the described system showed that sustainable propagation of the combustion wave is observed in the cases where the value of the system pressure equals 15 ... 30 MPa for all values of preheat temperature. However, when the initial mixture pressure is approximately 35 MPa and higher there was a significant increase in the specific energy release of the reactions occurring per unit volume of the sample; thus leading to the thermomechanical destruction of the samples samples during synthesis.

Figure 1 shows an X-ray phase analysis of the samples obtained the compaction pressure of 25 and 30 MPa.

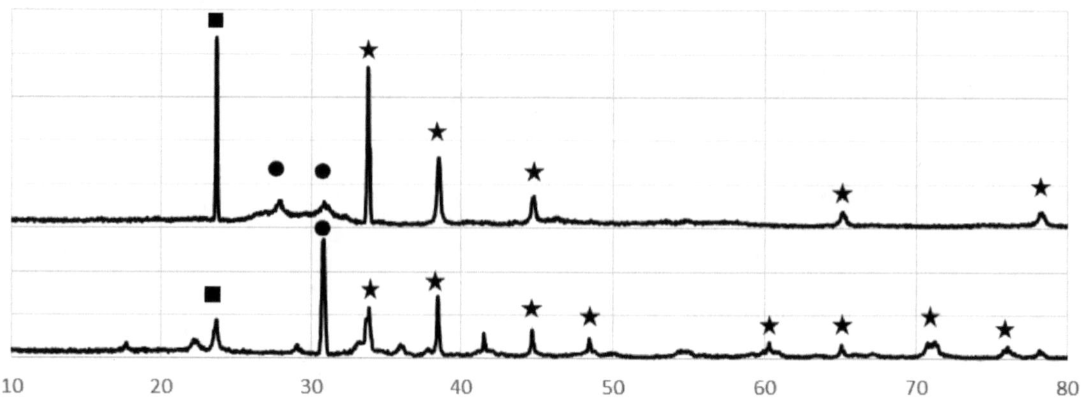

Figure 1. X-ray phase analysis of the samples obtained the compaction pressure of 25 (lower) and 30 MPa (upper), where ■ – $NdAlO_3$, ● – Nd_2O_3, ★ – Al.

Phase composition was the following:

– phase composition of the sample (% by weight) with the compaction pressure 25 MPa: 63.2% Al, 24.6% Nd_2O_3, 12.2% $NdAlO_3$;

– phase composition of the sample (% by weight) with the compaction pressure 30 MPa: 69.1% Al, 19.6% Nd_2O_3, 11.3% $NdAlO_3$.

The X-ray phase analysis shows that most part of Al left unreacted and the presence of the desired perovskite-like phase is only about 10% by weight in samples subjected to synthesis. Most likely, this phenomenon can be explained by the insufficient temperature growth in the synthesis of the final product and it is necessary to increase the total heat impact required for the reaction. This means that the process temperature must be increased so that the temperature conditions throughout the entire volume of the sample would allow to complete the SHS.

As it was previously noted, increasing the sample compaction pressure higher than 35 MPa leads to thermo-mechanical degradation, therefore, it is necessary to look for other ways to control the SHS process.

One solution to this problem could be the introduction of the stoichiometric reactive additives to the initial reactants mixture, which react exothermically with the components of the mixture during the synthesis, thereby increasing the total heat effect of SHS.

As a required additive we selected c.p. PNA OT1 Ni powder. Adding nickel involves the formation of the following phases: NiAl; Ni_2Al_3; $NiAl_3$. Ni2Al3 and $NiAl_3$ phases, in contrast to the negative component NiAl, should provide a high strength characteristics, because have irregular crystal lattice structures, which are more resistant to mechanical stress [14].

To obtain the result, which is the most appropriate to the characteristics, the samples with various contents of Ni (10 ... 30 wt.%) and compaction pressure (15 ... 30 MPa) were prepared.

Experiments have shown that the combustion temperature depends on the nickel additive amount in the initial mixture of reactants. Thus, when the content of additional components in a sample is about 25% wt. there is a significant increase in temperature (up to 2700 K) of the process; in case of 30% wt. – thermochemical degradation of the sample during synthesis for any stoichiometry considered due to significant heat release of the reactions in the mixture.

Figure 2 is shows a X-ray pattern of the samples containing 20% wt. of additives:

- phase composition of the sample (% wt.) with the compaction pressure 15 MPa: 59.3% Ni_2Al_3, 4.6% $NdAlO_3$, 9.9% NiAl, 22.5% Al;
- phase composition of the sample (% wt.) with the compaction pressure 25 MPa: 43.1% Ni_2Al_3, 10.2% $NdAlO_3$, 21.1% NiAl, 23.7% Al.

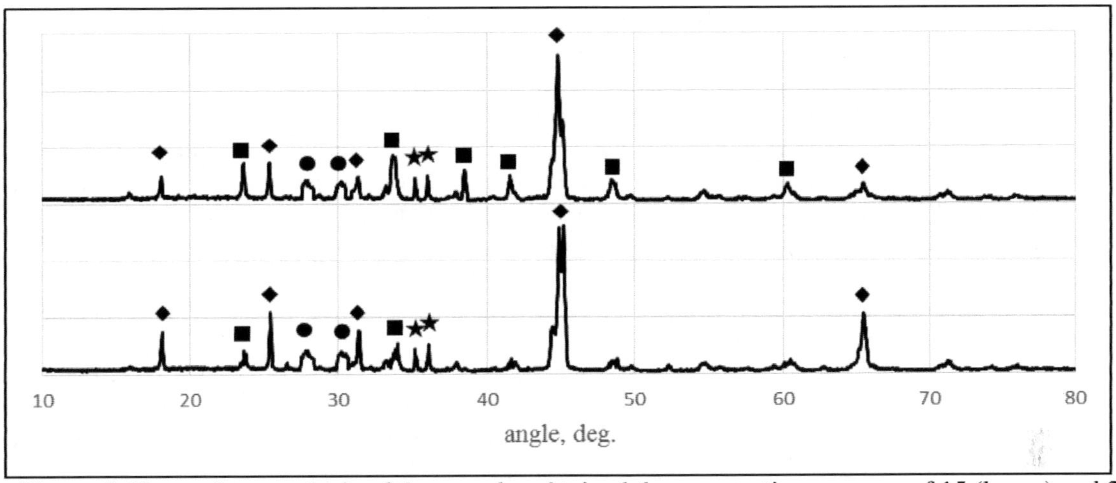

Figure 2. X-ray phase analysis of the samples obtained the compaction pressure of 15 (lower) and 25 MPa (upper), where ■ – $NdAlO_3$, ♦ – Ni_2Al_3, ▲– $NiAl_3$, ● – Nd_2O_3, ★ – Al.

As a result of the nickel additive use, most of unreacted aluminum entered into the exothermic reaction with the further the formation of the nickel aluminide phases, wherein phase formation shifted towards lower formation of perovskite phase. According to [15] aluminum perovskite formation depending on the stoichiometric mixture ratio starts from 1100 K, therefore, it can be assumed that the SHS does not involve higher temperatures.

Figure 3 is shows a X-ray pattern of the samples containing 25% wt. of additives:

- phase composition of the sample (% wt.) with the compaction pressure 20 MPa: 27.6% $NdAlO_3$, 36.9% Ni_2Al_3, 31.6% $NiAl_3$, 3.2% Nd_2O_3;
- phase composition of the sample (% wt.) with the compaction pressure 25 MPa: 39.9% $NdAlO_3$, 56.5% Ni_2Al_3, 3.6% Nd_2O_3;
- phase composition of the sample (% wt.) with the compaction pressure 30 MPa: 48.4% $NdAlO_3$, 46.8% Ni_2Al_3, 4.8% Nd_2O_3.

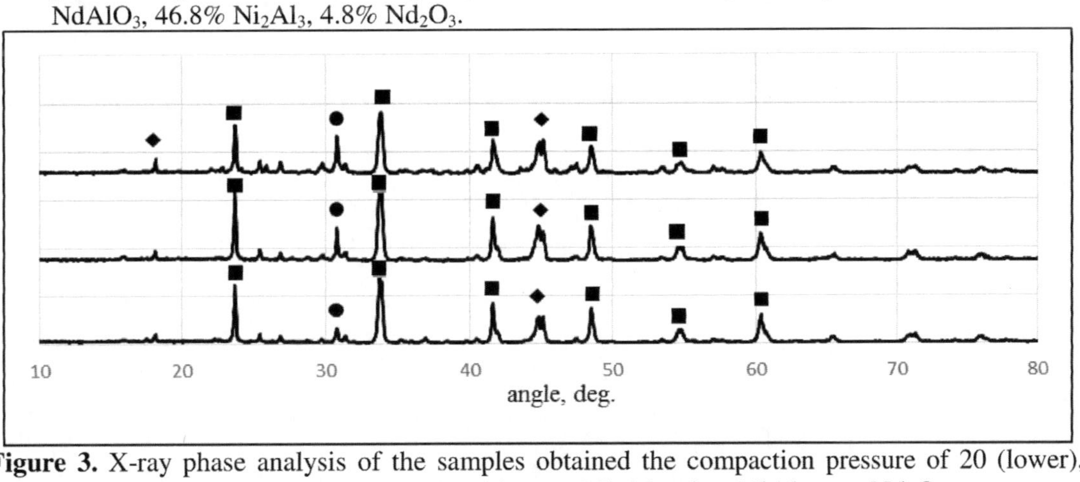

Figure 3. X-ray phase analysis of the samples obtained the compaction pressure of 20 (lower), 25 (middle) и 30 MPa (upper), where ■ – $NdAlO_3$, ♦ – Ni_2Al_3, ▲ – $NiAl_3$, ● – Nd_2O_3.

Thus, by increasing the content of additives amounts up to 25% wt. and compressing pressure of 30 MPa, the content of the modified aluminum-based perovskite based in the samples reaches almost 50 wt.%, and the amount of unreacted neodymium oxide is less than 5% wt.

4. Conclusions

Thus, results of the accomplished studies allowed to certify the following:

1) In the synthesis of matrix material, there is a lack of reaction energy yield for the $NdAlO_3$ formation according to reactions 1, therefore, it is necessary to use additives which increase the energy yield to ensure complete interaction between initial components
2) It is established that the samples obtained by compression of the mixture with the addition of Ni samples at the compaction pressure increased from 15 to 30 MPa experience the phase shift towards the final product formation due to the higher energy yield.
3) The maximum amount of aluminum perovskite is observed when the initial mixture compaction pressure equal to 30 MPa and 25% wt. nickel.

References

[1] International Atomic Energy Agency, Design and Operation of High Level Waste, Vitrification and Storage Facility, Technical report series No.176, IAEA, Vienna, 1977.
[2] M.I. Ojovan, W.E. Lee, An Introduction to Nuclear Waste Immobilization, Elsevier Ltd, Oxford, UK, 2005, p.213–267.
[3] D. Caurant, P. Loiseau, O. Majèrus, V. Aubin-Chevaldonnet, I. Bardez, A. Quintas, Glasses, Glass-Ceramics and Ceramics for Immobilization of Highly Radioactive Nuclear Wastes, Nova Science Publishers, New York, NY, 2009.
[4] Asuvathraman R, Joseph K, Madhavan R R, Sudha R, Prabhu R K, Govindan Kutty K V 2015 *J. Eur. Ceram. Soc.* **35** 4233–4239
[5] Mao X, Qin Z, Yuan X, Wang C, Cai X, Zhao W, Zhao K, Yang P and Fan X 2013 *J. Nucl. Mater.* **443** 428–31.
[6] Zhang K, Wen G, Zhang H and Teng Y 2015 *J. Nucl. Mater.* **465** 1-5.
[7] Wena G, Zhanga K, Zhangb H, Tenga Y, Zhoua Y 2015 *Ceram. Int.* **41** 6869-6875
[8] Liu M, Wang Z, Wu J, Li Q 2015 *J. Alloy. Compd.* **648** 116-121
[9] Yi H, Moore J 1990 *J. Mater. Sci. Lett.* **25** 1159-1168
[10] Demyanyuk D G, Dolmatov O Y, Isachenko D S, Kuznetsov M S, Semenov A O and Chursin S S 2015 *Adv. Mat. Res.* **1084** 728-732
[11] Sanchez-Rodríguez D, Wada H, Yamaguchi S, Farjas J, Yahiro H 2015 *J. Alloy. Compd.* 649 1291-1299
[12] Wen G, Zhang K, Zhang H, Teng Yu, Zhou Ye 2015 *Ceram. Int.* **41** 6869-6875
[13] Liqin Q, Kaituo W, Xuehang W, Wenwein W, Sen L, Gengming L 2014 *Ceram. Int.* **40** 3003-3009
[14] Nikbakht R, Assadi H 2012 *Acta. Mater.* **60** 4041-4053
[15] Couteres J 1985 *J. Am. Ceram. Soc.* **68** 105–107

IOP Conf. Series: Materials Science and Engineering **135** (2016) 012047 doi:10.1088/1757-899X/135/1/012047

The ability to create NTD silicon technology in the IRT-T reactor in a horizontal experimental channel with one-side access

V A Varlachev, A V Golovatsky, E G Emets and Ya A Butko
National Research Tomsk Polytechnic University, 30, Lenina Avenue, Tomsk, 634050, Russia

Abstract. The article shows the ability of creation of neutron transmutation doping (NTD) of monocrystalline silicon technology in the reactor's channel, which has a one-side access. In the article a distribution of thermal neutron flux through the length of channel and it's radius, neutron spectrum were obtained which confirmed that horizontal experimental channel HEC-1 is suitable for NTD.

1. Introduction

On the reactor IRT-T [1] of the Tomsk Polytechnic University the great attention among other radiation technologies is given to the neutron transmutation doping of the silicon (NTD). This technology was created in 1985 [2]. It realized on the horizontal experimental channel of the reactor HEC-4. Since then, it has continuously improved [3-5] from the point of view of quality and productivity of NTD. The main criteria of quality of NTD silicon are:

a) the dispersion of specific electric resistance (s.e.r.) on the ends of silicon ingots;
b) the deviation of the obtained values of s.e.r. from a given value;
c) the lifetime of minority charge carriers.

These indicators of the quality of the existing technology at reactor IRT-T are following:

a) the dispersion of s.e.r. on the ends of silicon ingots is not exceeds 2-4 %;
b) the deviation of s.e.r. from the given value, typically is less than 7%;
c) the lifetime of minority charge carriers, typically is 600-800 microseconds even when doping of silicon on low-resistivity values of s.e.r. (40-60 Ω·cm).

These quality indicators correspond to the best foreign analogs. The proof of this is the constant offers from domestic and foreign companies for cooperation. As for doping productivity, the possibilities of the reactor IRT-T are still realized no more than for 20% for different reasons. However, at the present time 4-5 tons of NTD silicon are produced on the reactor IRT-T per the year. As an example, China uses 20 tons of this material in a year at requirement of 40 tons.

One of opportunities of increase of doping productivity is as follows. On the IRT-T reactor there are two tangent experimental channels of big diameter (150 mm) – HEC-4 and HEC-1. However, the doping is realized only on channel – HEC-4. In this channel the axial uniformity of a doping is achieved due to the reciprocating movement of the containers with ingots through entire irradiation zone. Both the forward and reverse directions, the containers are moved so that they are completely left an irradiation zone. Therefore, at each such movement from one extreme position to another all elementary volumes of containers are irradiated with the same neutron fluence at any distribution of the neutron field along an axis of the channel. The length of HEC-4 channel has allowed to move containers with such amplitude, and the HEC-1 channel has limited length. In one of extreme positions it is impossible to completely remove containers out of irradiation zone limits. Therefore, so far HEC-1 is not used for the production of NTD silicon. This was not necessary, since all orders for a doping were covered by irradiation in HEC-4 channel.

Currently, there is a need to considerably increase of volume of NTD silicon. In this regard, for the tangent channels of the reactor limited on the one side, we have developed a method of irradiation [6] at which the axial uniformity of a doping is achieved. Its essence is illustrated in Fig. 1.

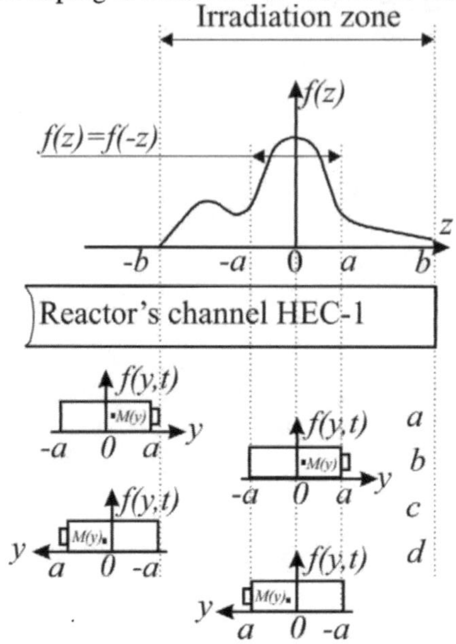

Figure 1. Position of container when using the channel with one-side access.

At implementation of this method the neutron field previously is formed so that in an irradiation zone ($-b \le z \le b$) there is area on which the distribution of density of a flow of thermal neutrons f(z) along an axis of the channel (axis z) would be even function ($f(z) = f(-z)$), and define this area $-a \le z \le a$. The length of the container must not exceed the length of the selected section (2a). Then the container once or repeatedly reciprocatingly is moved on the channel of the reactor from a position *a* to *b* and back. The irradiation is stopped when the fluence of neutrons average on container volume becomes equal to a half of the required. Then the container is deployed that its ends in the channel are reversed, and similarly is irradiated, moving it from a position *c* to *d* and back. At the same time in one extreme situation (positions *a* and *c*) the container is located outside an irradiation zone, and in the other (positions *b* and *d*) - combine the center of the container with the middle of a section (z=0), on which $f(z) = f(-z)$. The proof, that at such method of irradiation the axial uniformity of a doping is achieved, is provided in [6].

The main requirement to NTD silicon monocrystals is the high uniformity of distribution of s.e.r. on ingot volume. So, according to specifications 48-4-443-83, acting in Russia, the dispersion of s.e.r. at ends of an ingot shouldn't exceed 5%, and a deviation of average value of s.e.r. at each end from nominal value shouldn't exceed 10%. The strictest requirements are put in standards of Wacker firm: 5% and 7% respectively. Therefore when the performing of researches the main attention was paid to definition of conditions under which the uniform irradiation of silicon ingots is achieved by thermal neutrons. According to it spatial distribution of the field of thermal neutrons in HEC-1 channel was experimentally defined. According to the obtained distribution of thermal neutrons on HEC-1 channel

IOP Conf. Series: Materials Science and Engineering **135** (2016) 012047 doi:10.1088/1757-899X/135/1/012047

the optimum mode of irradiation of silicon ingots was chosen: velocities of rotation and linear movement of ingots on the channel of the reactor, the amplitude of reciprocating movement of ingots.

2. Experiment technique

At the measurements of the neutron field in HEC-1 the method of activation detectors [7] was used. As activation detectors, a copper foils and a copper wire were applied, which were fixed on the special device and by means of a bar were moved to the reactor channel. The reference points put on a bar allowed to install the device in the channel of the reactor with a margin error not worse than 3 mm, i.e. each activation detector was installed in the defined point of the irradiation zone with a margin error mm ±3, and their relative position – with a margin error about 1 mm. The irradiation of activation detectors was carried out at a power of reactor is 20 kW within 30 minutes. After the endurance of the irradiated samples within 24 hours, from irradiated foils and a wire the interesting areas were excised, weighed them and dissolved in nitric acid. The measurements of activity of solutions were conducted in the same geometry on the installation γ-γ coincidence. At the same time errors of relative measurements didn't exceed 3%.

As a result of measurements of activity of a copper wire the relative distribution of density of a neutron flow along an axis of the channel was obtained (Fig. 2).

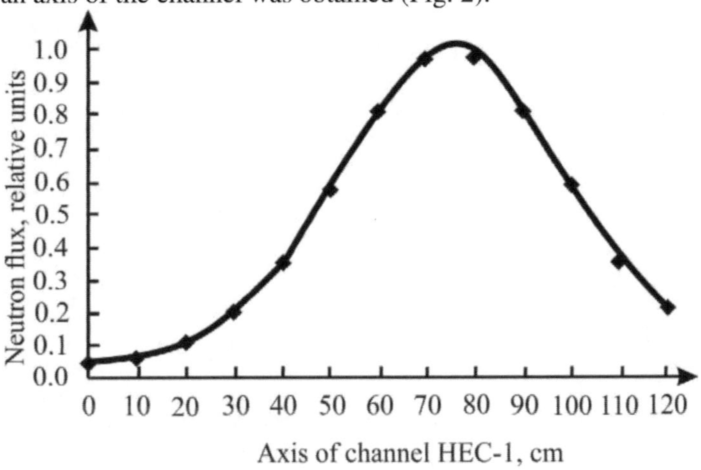

Figure 2. The distribution of the thermal neutron flux along axis of channel HEC-1.

In this Figure, the system of coordinates is as follows. The point 0 corresponds to border a gate of HEC-1 channel – a tank with water. The length of the channel in a tank is equal to 120 cm. The maximum density of a flow of thermal neutrons in the channel ($1{,}95\cdot10^{13}cm^{-2}s^{-1}$) at the coordinate of 75 cm. Figure 2 shows that distribution of thermal neutron flux in the range of 30 cm – 120 cm is even function relative to the coordinate of 75 cm. It means that in HEC-1 the silicon doping by the method [6] developed by us is possible. According to this method the irradiation of silicon ingots with a total length up to 90 cm is possible. At the same time it is necessary that due to the rotation of ingots the high radial uniformity of a doping was achieved. For research of parameters of radial uniformity of a doping the device with 6 copper foils in the form of the disks established one after another through 20 cm was used. Seven rings of different radius were cut from each disk and their activities were measured. Thereby, the rotation of a disk was imitated. For this purpose previously each ring was weighed and dissolved in nitric acid, and the results of measurements of activity were normalized on weight and the square of rings. For imitation of reciprocating movement along an axis of the reactor channel of activity of six rings with identical radiuses were summarized. As a result the distribution of activity on the radius (Fig. 3) was obtained. Thus, the average distribution of a fluence of neutrons was estimated due to rotation and reciprocating movement on the radius of the channel of the reactor. This figure shows in the limits of measurement error (±2 of %), the activity of a copper disk was radially uniform.

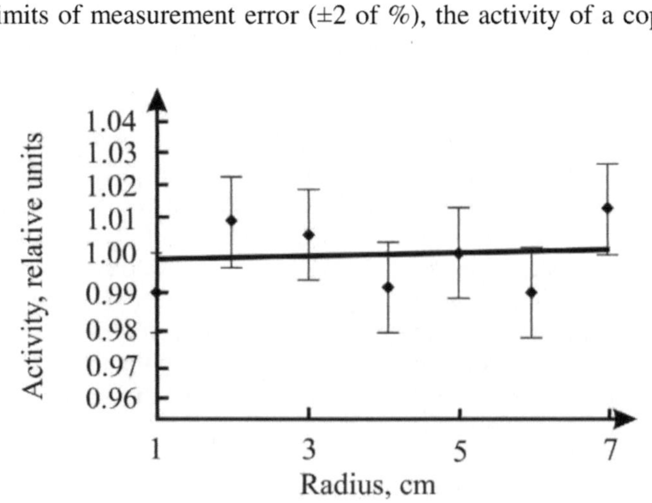

Figure 3. The neutron flux distribution through copper rings.

The neutron spectrum (Fig. 4) in HEC-1 was calculated with Monte-Carlo method.

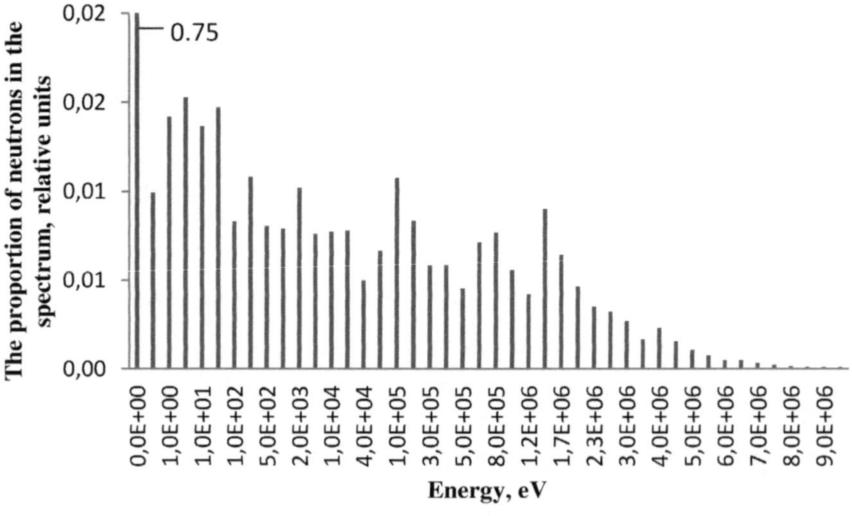

Figure 4. The spectrum of neutron flux in HEC-1.

3. Conclusion

Thus, the conducted measurements of neutron and physical parameters of the neutron field in HEC-1 lead to the conclusion that in HEC-1 channel the doping of silicon ingots up to 90 cm long with high axial and radial uniformity is possible.

Application of this method for moving containers is allows to create the NTD technology in the reactor's channel with one-side access. The resulting neutron spectrum, distribution of thermal neutron flux and deviation through channel's radius shows that the implementation of technology in the channel HEC-1 will allow to obtain the NTD silicon of high quality.

References

[1] Varlachev V A, Solodovnikov E S and Dudkin G N 2010 *Russian Physics Journal* **53** **11/2** 304-309 [In Russian]
[2] Varlachev V A, Kuzin A N, Lyhin S V, Solodovnikov E S, Usov Yu P, Fotin A V 1995 Atomic energy **79** 38-40 [In Russian]
[3] Varlachev V A, Zenkov A G, Solodovnikov E S *Research reactors: Science and High Technology* Dimitrovgrad vol 3. p. 173-186. [In Russian]
[4] Varlachev V A, Solodovnikov E S 2007 *Russian Physics Journal* **10/3** 261-265 [In Russian]
[5] Varlachev V A, Zenkov A G Solodovnikov E S 1998 *Russian Physics Journal* **4** 210-215 [In Russian]
[6] Varlachev V A, Solodovnikov E S 2007 *Russian Physics Journal* **53** **10/2** 313-316 [In Russian]

IOP Conf. Series: Materials Science and Engineering **135** (2016) 012048 doi:10.1088/1757-899X/135/1/012048

Effect of plasma suppression additives on electrodynamic characteristics of the torch discharge burning in the air

D Vidyaev, Yu Lutsenko and E Boretsky
National Research Tomsk Polytechnic University, Lenina Avenue, 30, Tomsk, 634050, Russia

E-mail: luts@mail.ru

Abstract. The paper shows the results of measurements of wave number of electromagnetic wave, which supports burning of high-frequency torch discharge in the mixture of air with water vapor and carbon dioxide. The nonmonotonic dependence of attenuation factor of electromagnetic waves is set on a concentration of water vapor. It is shown that the attenuation degree of electromagnetic field in the plasma with water vapor significantly exceeds the attenuation degree of electromagnetic field in the plasma with carbon dioxide.

1. Introduction

Burning of high-frequency torch discharge is carried out due to the dissipation of the electromagnetic wave energy. Therefore the characteristics of torch discharge are largely determined by the characteristics of its electromagnetic field. The most important quantity characterizing the propagation of electromagnetic waves in the discharge plasma is its wave number. The real part of the wave represents the damping coefficient of the electromagnetic wave in the discharge plasma; the imaginary part of the wave is a phase factor that characterizes the slowdown of the electromagnetic wave in the discharge plasma.

High-frequency discharge recently often used [1] as a plasma generator during various plasma-chemical processes. In particular, high-frequency plasma systems on the basis of torch discharge are used for sludge combustion, as well as for disposal of spent nuclear fuel. Conducting this process involves the formation of carbon dioxide and water vapor. Water vapor has a high heat capacity, which has a significant influence on the process of discharge burning. We should also note electronegative properties of water molecules as a result of which the electron density of the plasma discharge is significantly reduced. Changing the concentration of electrons of discharge plasma, in turn, has an effect on the propagation of electromagnetic waves, which supports the discharge. In particular, there is an increase of the attenuation coefficient of electromagnetic wave, and accordingly the changing density of energy sources in the discharge plasma. Carbon dioxide also has similar effect on the discharge burning process.

In this paper we were investigated the dependence of the wave number of the electromagnetic wave that supports the burning discharge, on the concentration of water vapor and carbon dioxide in the discharge chamber of plasma torch.

2. The scheme of experimental setup and methods of measurements

Determination of the wave number was conducted according to the method given in [2]. The scheme of experimental setup is given on the Fig. 1. Torch discharge excited in the air. Electromagnetic field frequency was 37 MHz. High-frequency torch discharge was excited in the cylindrical quartz chamber with the diameter of 28 mm. The discharge channel length ranged from 5 cm to 27 cm. It is depends on the water vapor concentration. The steam was fed into the discharge coaxial of plasma gas forming. Air humidity was controlled by a hygrometer PCE - 310. The air temperature was 30°C. The measurements of torch discharge characteristics of combustion in the mixture of air and carbon dioxide have been separately. Carbon dioxide concentration in the plasma-forming gas mixture was in a range 0...30%.

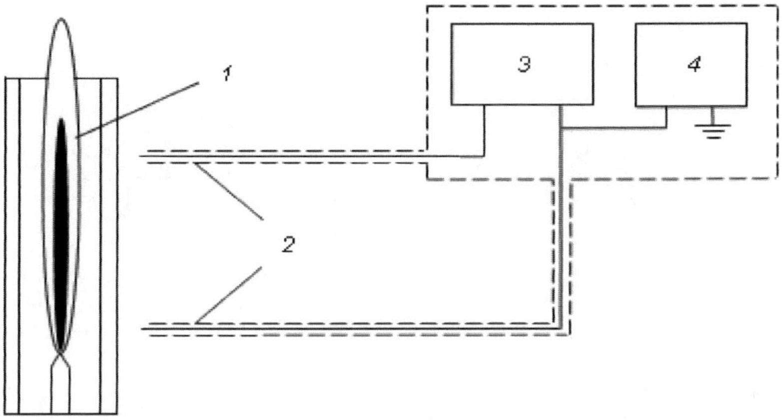

Figure 1. Scheme of the experimental setup. 1 - high-frequency torch discharge; 2 - capacitive and inductive probes; 3 - measuring device; 4 - generator "reference" signal.

Measurement of intensity of electric field of flame discharge was made by probes of capacity moved along plasma blob of discharge. Capacitive probe was a copper rod with length of 3 ... 5 mm. The signal from probes across the wire with double shielding was delivered on the input of measuring device. Depending on the type of the measured value as the measuring device used either oscilloscope or phase meter. When measuring the phase shift to increase the accuracy of the measurements was used a generator of "reference" signal. Defining the wave number of electromagnetic wave supporting burning of the discharge, was conducted by comparison calculated radial distribution of radial component of electrical field discharge with experimental data.

Note that, in the near zone of radiation of torch discharge, its electromagnetic field would be described by Hankel function. Namely, radial distribution of radial component of electric field might be presented by the [3] following equation:

$$E_r = AH_1^{(1)}\left(r\sqrt{k^2 - h^2}\right) \quad (1)$$

Where, A - constant; k - the constant of electromagnetic wave propagation in the air; h - the wave number of the electromagnetic wave; r - radial coordinate.

Thus, comparing the results of calculation of intensity value of radial component of the electric field E_r, conducted according to the formula (1) with the results of experimental measurements, the information about wave number might be gained and by doing so determine the mean of coefficient of phase and decreasing of electromagnetic wave in the discharge. As wave number from equation (1) is a complex value and is determined by two parameters α and β, so for it definite allocation it is required to conduct comparison of calculating and experimental data according to two parameters. In this case these parameters were in the form ratio of amplitude $E_r|_{r=30}/E_r|_{r=90}$ and difference of phases $\Psi_{Er}|_{r=30}-\Psi_{Er}|_{r=90}$ of radial component of electric field at a distance 30 mm and 90 mm from a discharge axis. The calculation data of the values of the amplitude ratio $E_r|_{r=30}/E_r|_{r=90}$ and phase difference

IOP Conf. Series: Materials Science and Engineering **135** (2016) 012048 doi:10.1088/1757-899X/135/1/012048

$\Psi_{Er}|_{r=30} - \Psi_{Er}|_{r=90}$ of radial component of the electrical field using the formula (1) with different meanings α and β, is presented on the Fig.2 and Fig.3.

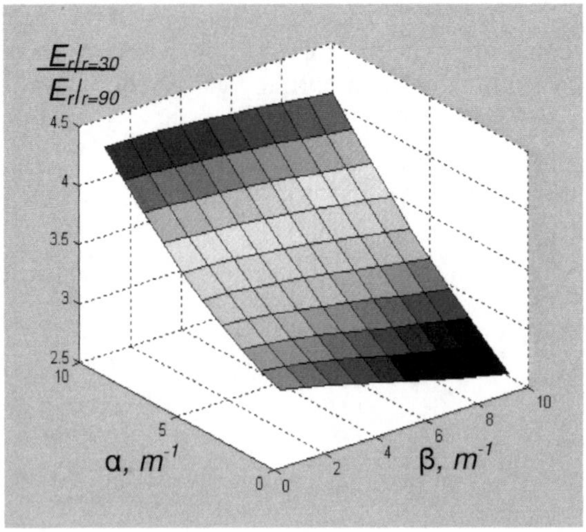

Figure 2. Dependence ratio between amplitude of electric field intensity from coefficient of phase and decrease of electromagnetic wave.

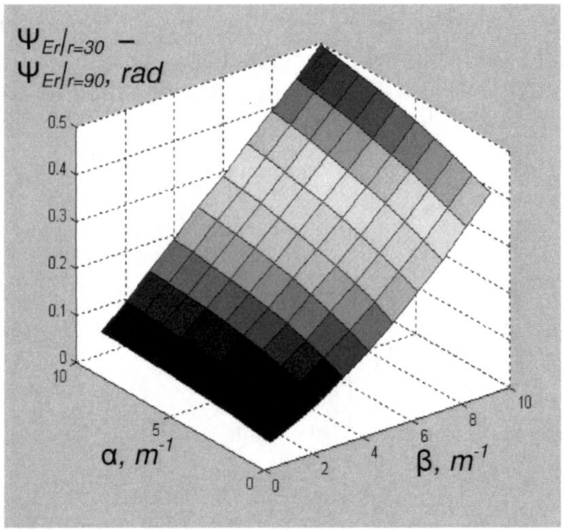

Figure 3. Dependence of phase shift of electrical field intensity versus coefficient of phase and decrease of electromagnetic wave.

As seen from these figures amplitude ratio depends mainly on the value of the phase coefficient, while the damping coefficient is determined primarily by the phase difference.

3. Results of measurements and their discussion

The results of measurements of the radial distribution of the radial component of the electric field in the case of torch discharge burning in the air with water vapor and having a power of 1 kW are presented in Figure 4. The dashed line shows the calculated data performed in accordance with equation (1). As can be seen from the data, the calculated and experimental results are practically identical, which in turn indicates the correctness of determining the value of the wave number by using the formula (1). When the concentration of carbon dioxide and water vapor in the plasma-forming gas are changing, then geometric characteristics of the discharge are also changes. Dependence the length of the channel of torch discharge from the concentration of water vapor and carbon dioxide for the discharge power of 1 kW is shown in Fig. 5. and Fig. 6, respectively. It is compared the experimental and calculated values. The values of the phase and attenuation coefficients of the electromagnetic wave that supports burning torch discharge were obtained. These values are determined depending on the concentration of water vapor and carbon dioxide in the plasma-forming gas. Results for torch discharge power of 1 kW, burning in the mixture of air and carbon dioxide are shown in Fig. 6. The figure shows that the increasing in carbon dioxide concentration in the air leads to increase in the attenuation coefficient and the coefficient of the phase of the electromagnetic wave. Resulting dependences has monotonic character. The value of the electromagnetic wave attenuation coefficient is slightly different from the value of the phase factor. The results of the measurement of the wave number of the electromagnetic wave propagating in the plasma torch discharge air with water vapor are shown in Fig. 7. The figure shows that in this case, with increasing concentrations of plasma-suppressing additive is increasing in the attenuation coefficient and phase coefficient of the electromagnetic wave. However, the dependence of the phase coefficient is nonmonotonic. By increasing the relative humidity of more than 90% there is a slight decrease in the phase factor.

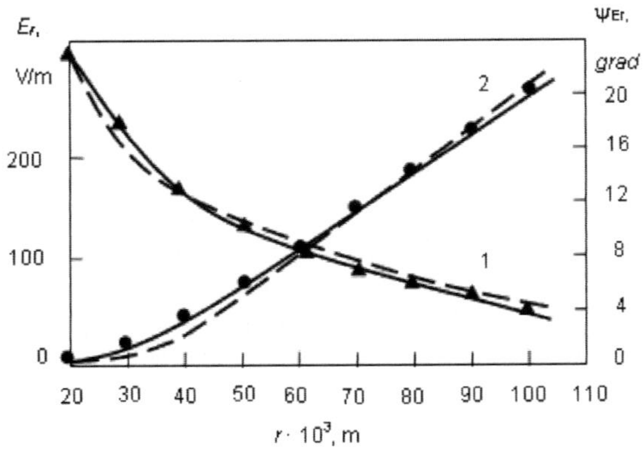

Figure 4. The radial distribution of amplitude and phase shift of the radial component of the electric field torch discharge.

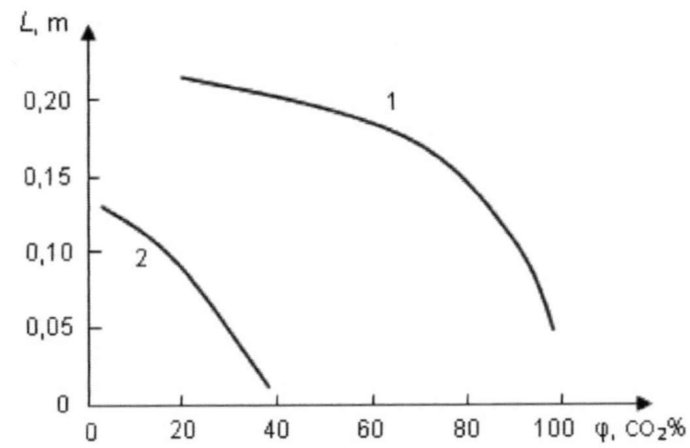

Figure 5. The dependence of channel length of the torch discharge power of 1 kW from plasma-suppression concentration of additives: 1 - the humidity φ; 2 - the concentration of CO_2 in the air.

This anomalous behavior of the electrodynamic characteristics of the torch discharge in this range humidity changes may be due to the changing nature of the interaction between the water molecules. Changing the conductivity of the discharge plasma due to an increase in the concentration of negative oxygen ions, it can also be accompanied by a change in the value of its dielectric constant. It is known that the magnitude of the wave number of the electromagnetic wave propagating along the channel of torch discharge is described [4] the following expression:

$$h = j\sqrt{\frac{2}{5}}\frac{1}{a}\left(\varepsilon_0 + j\frac{\sigma}{\omega}\right)^{-1/2} \qquad (2)$$

Where j - imaginary unit; a - the radius of the discharge channel; ε_0 - dielectric constant of the plasma discharge; σ - electrical conductivity of the plasma discharge; ω - frequency electromagnetic field.

Thus, according to the expression (2), the wave number and respectively the phase and attenuation coefficients of the electromagnetic waves will depend both on the conductivity of the discharge plasma, and on the value of its dielectric constant.

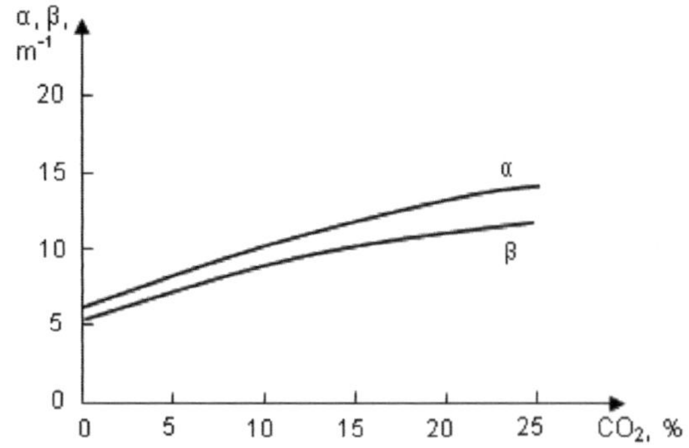

Figure 6. Dependence of the attenuation coefficient α and the coefficient β phase of the electromagnetic wave from the CO_2 concentrations in the plasma forming gas.

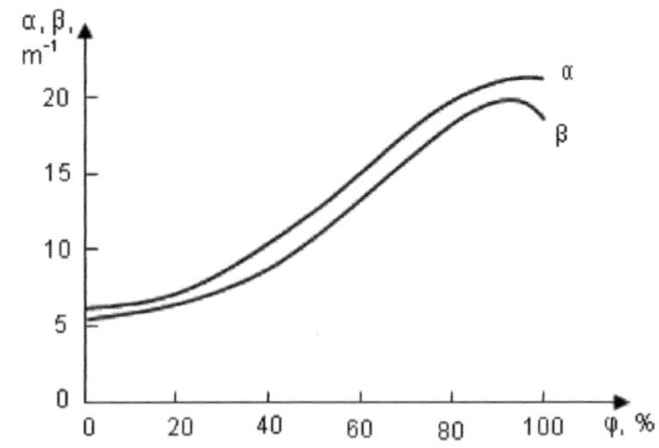

Figure 7. Dependence of the attenuation coefficient α and β phase coefficient of the electromagnetic waves from the humidity of the plasma gas.

From Fig. 7 also shows that the increase in air humidity reduction coefficient phase of the electromagnetic wave is less pronounced than the corresponding change in its damping coefficient.

Comparison of the results shown in Fig. 6 and Fig. 7 suggests a more pronounced effect of water vapor in the process of attenuation of the electromagnetic field in the plasma discharge, compared with carbon dioxide. The moisture content of the air is several times smaller than its relative humidity. At a temperature of 30° C and 30% humidity air the moisture content is 0.01 kg water vapor / kg dry air. Thus, to obtain the same damping effect of the electromagnetic field in the plasma discharge it should be added the water vapor into the plasma-forming gas in several times smaller by mass than carbon dioxide.

4. Conclusion

As a result of the measurement are defined phase growth coefficient and the damping coefficient of the electromagnetic wave that supports burning discharge with an increase in the plasma suppression concentration additives in the plasma forming gas. It is shown that the water vapor compared to carbon dioxide have a greater effect on the propagation of electromagnetic waves in a discharge plasma. It was found an anomalous reduction in attenuation of the electromagnetic field in the discharge plasma with increasing relative humidity of 90%. It is suggested that such a change in the electromagnetic wave attenuation coefficient caused by changing in the character the interaction between the water molecules. The experimental patterns can be used in the development of plasma chemical technologies based on the use of torch discharge burning a mixture of air with carbon dioxide and water vapor.

References

[1] Korolev Y. D., Frants O. B., Landl N. V., Geyman V. G., Karengin A. G., and others (2013) Plasma-Assisted Combustion System for Incineration of Oil Slimes. *IEEE Transactions on Plasma Science,* 41(12) 3214-3222.

[2] Vlasov V. A., Tikhomirov I. A., Lutsenko Y. Y. (2006). Determination of the wave number of an electromagnetic wave propagating in high-frequency torch discharge plasma. *Thermophysics and Aeromechanics*, 13(1), 131-134.

[3] Stratton J. A. *Electromagnetic theory* (1941) New York: McGraw-Hill.

Influence of pulse line switch inductance on output characteristics of high-current nanosecond accelerators

A I Mashchenko and I I Vintizenko
Institute of Physics and Technology, Tomsk polytechnic university, av. Lenina 30, Tomsk, 634050, Russia

E-mail: vintizenko@tpu.ru

Abstract. Various types of high-current nanosecond accelerators are simulated numerically using an equivalent circuit representation. The influence of pulse forming line switch inductance on the amplitude and waveform of output voltage and current pulses is analyzed.

1. Introduction

High-current (kiloampere level) nanosecond accelerators of charged particles [1] have a typical electrical scheme containing a pulse forming line (PFL), its switch, and charging unit. For PFL charging, various pulsed power supplies, Marx generators, or magnetic pulse generators are employed. Different types of PFL are possible: single line, double (Blumlein) line, in planar or coaxial geometry. Magnetic switches or spark gaps are used for PFL switching. For linear induction accelerators (LIAs) [2, 3], a set of ferromagnetic core inductors for voltage adding is included between a PFL and a load; there are also schemes with a high-voltage transformer [4].

It is implied that the inductance of the PFL switch and connections should be minimal to obtain the output pulse duration corresponding to the PFL electric length. The mentioned inductance, along with the PFL capacitance, determines the output pulse amplitude, full width, and rise-time. To reduce the inductance, multichannel spark gaps, spark gaps with small cathode-anode distance operating at a high gas pressure, or magnetic switches in the form of saturation chokes with single-turn windings and ferromagnetic cores having a rectangular hysteresis loop are employed. An analytical calculation of the inductance of the switch and connections is not always possible, so that such an important parameter usually is not indicated in the literature on the high-current nanosecond accelerators development. The analysis of the influence of PFL switch inductance on the amplitude and the waveform of the output pulses is possible using an equivalent circuit of the accelerator.

2. Equivalent circuit

The equivalent electric scheme of the accelerators is presented in figure 1. Its main elements are the power supply source PS, the double forming line DFL, the switch K, the high-voltage transformer T (ferromagnetic induction system if the LIA is simulated) and the ohmic load R_L. The scheme of the accelerator with the single PFL is obtained by excluding the cells of the second DFL section. Excluding the elements of the high-voltage transformer, one can reduce the scheme to that of usual accelerator, in which the PFL is directly connected to the load.

The following assumptions are accepted in the scheme: the distributed capacitances of the high-voltage transformer (induction system) and the load are replaced with the lumped ones; the line DFL is uniform (characteristic impedance is constant); ohmic losses and leakages in the line are neglected; the magnetization inductance of the transformer is a linear parameter; the electron diode, as the load, is replaced with the ohmic resistance.

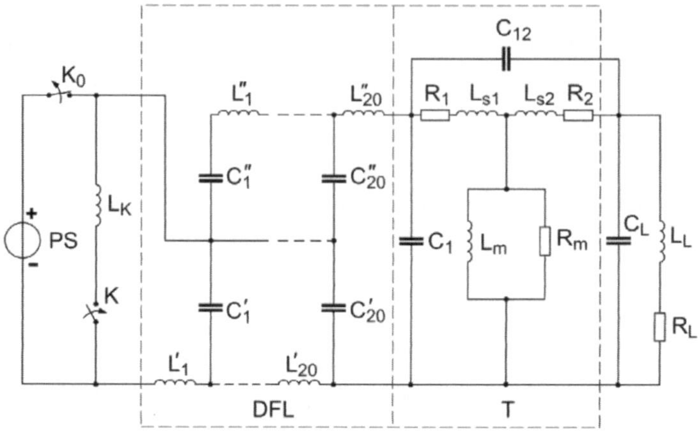

Figure 1. The equivalent electric scheme of accelerators.

The elements of the scheme are the following. L_K is the switch inductance, $L_1^{'} - L_{20}^{'}$, $L_1^{''} - L_{20}^{''}$, $C_1^{'} - C_{20}^{'}$, and $C_1^{''} - C_{20}^{''}$ are the inductances and capacitances per cell of the DFL sections. L_{S1}, L_{S2}, R_1, and R_2 are the leakage inductances and resistances of the transformer primary and secondary windings, respectively, L_m is the magnetization inductance of the transformer, R_m stands for the ohmic equivalent of losses in the transformer core, C_1 and C_{12} are, respectively, the input capacitance of the transformer and dynamic capacitance between its primary and secondary windings. C_L and L_L are the dynamic capacitance and the inductance of the load.

The resistances, inductances and capacitances per cell of the line, and dynamic capacitances and inductances can be easily calculated analytically using well-known expressions. For estimating the value of L_m, we use the theory of pulsed remagnetization of the ferromagnetic core [3]. The magnetization inductance L_m reflects the process of the transformer core magnetizing during discharging the PFL. The losses of energy in the core are determined as follows:

$$W_m = VH_0\Delta B + \frac{2VB_s}{\tau}\left(S_{\omega o}\lambda^2 + S_{\omega e}\lambda^3\right), \tag{1}$$

where V is the volume of the core, H_0 is the starting field strength, ΔB is the induction increase in the core, B_s is the saturation induction, $S_{\omega o}$ is the index of switching due to eddy currents, $S_{\omega e}$ is the index of switching due to magnetic viscosity, and $\lambda = \Delta B/2B_s$ is the index of the core loading by the flux linkage.

We determine the value of the magnetization inductance from the flux linkage of the transformer ψ_m:

$$L_m = \frac{\Psi_m}{\Delta I_m}, \tag{2}$$

where ΔI_m is the magnetization current. $\psi_m = lB_S k\omega(D_T - d_T)n$ where l is the width of the core steel, n is the number of cores, ω is the number of magnetization turns, $k \approx 0.8$ is the steel filling coefficient of the core for permalloy 50 NP (50% Fe, 50% Ni, electron beam vacuum melting, 0.01 mm thickness of rolling), D_T and d_T are the outer and inner diameter of the core, respectively.

Using the expression for pulsed remagnetization of the steel (1) and assuming $\lambda = 1$ (full remagnetization of the core from $-B_S$ up to $+B_S$), it is possible to write:

$$\Delta I_m = \frac{\pi(D_T + d_T)[H_0\tau + 2S_{\omega e} + S_{\omega o}]n}{2\tau\omega}. \tag{3}$$

From (2) and (3), one can obtain:

$$L_m = \frac{2\omega^2 B_S lk\tau(D_T - d_T)}{\pi(D_T + d_T)[H_0\tau + 2S_{\omega e} + S_{\omega o}]n}. \tag{4}$$

The charging of the first DFL section occurs directly from the pulsed source PS, and the second DFL section is charged through the primary winding of the transformer T, whose core at this time is demagnetized to a condition opposite to an operational one, and the load R_L. The switch K is closed and the switch K_0 is opened when DFL is charged up to the maximal value. After that, the first DFL section is recharged and the first and second sections connected in series are discharged through the transformer forming the high-voltage pulse on the load.

3. Simulation results

For simulations, the ohmic load R_L = 60 Ohm was taken and DFL parameters of the operating accelerators were chosen. The PS voltage of 60 kV was taken for the accelerator with the transformer (with the factor n = 12); the usual accelerator was simulated for the PS voltage of 720 kV.

First, the accelerator Tonus-NT [5] was simulated, when DFL was discharged, without the transformer, directly to the load. The scheme of the accelerator includes the high-voltage transformer with resonant contours (Tesla transformer) charging the double forming line filled with the transformer oil and the high-voltage diode. The gas-filled trigatron spark gap with up to 10 atm operating pressure is used for the double forming line switching. The main accelerator parameters are as follows: accelerating voltage range 0.4–1 MeV, line impedance 36.6 Ω, pulse duration 60 ns, pulse repetition rate up to 10 pps.

The voltage and current pulses on the load at 0.7 nF capacitance and 620 nH inductance of one DFL section and varied inductance L_K are shown in figure 2.

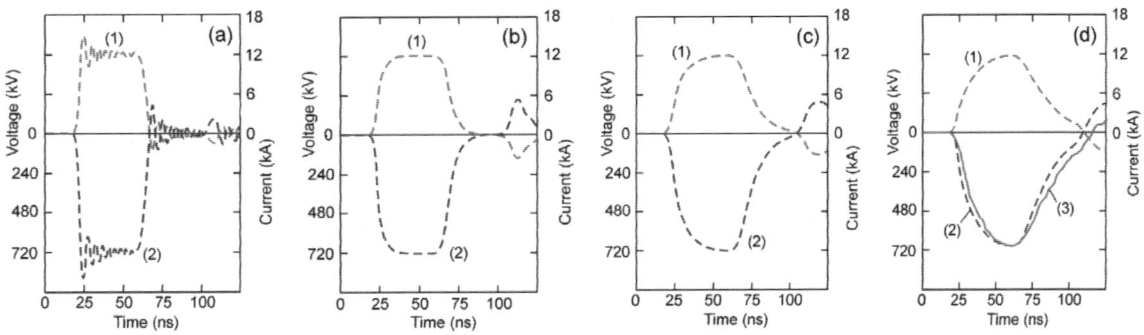

Figure 2. Simulated current (1) and voltage (2) pulses on the load R_L=60 Ohm. Inductance of the DFL switch: (a) L_K = 0 nH; (b) L_K = 100 nH; (c) L_K = 200 nH, (d) L_K = 320 nH. Tonus-NT observed voltage pulse (3).

The pulses have classical rectangular form (figures 2(a) and 2(b)) within a range of the switch inductance variation from 0 to 100 nH, the voltage value is U = 720 kV, the current is I = 12 kA and the duration is τ = 48 ns. At L_K = 200 nH, however, the rise and fall time of the pulses are lengthened, and their waveform becomes quasi-rectangular (figure 2(c)). The increase in the inductance L_K up to 320 nH (that corresponds to the inductance of the switch and connections in the Tonus-NT accelerator) results in more significant distortion of the pulse waveform, which approaches a trapezoidal one at U = 694 kV, I = 11.6 kA, and τ = 88 ns (figure 2(d)). The voltage reduction within the range of changing L_K from 0 up to 320 nH is small and caused by the increase in the pulse duration.

In the simulations of the full scheme of the accelerator (see figure 1) accounting for the parameters of the high-voltage transformer, the output pulses have a different waveform (figure 3). Even at L_K = 0 nH (figure 3(a)), the voltage and current pulses on the matched load have a triangular shape (simulations performed at 100 nF capacitance and 4.3 nH inductance of one DFL section). Such essential difference is due to the influence of the transformer, namely, the parasitic elements included to its equivalent circuit. The increase in the switch inductance from 5 nH (figure 3(b)) up to 18 nH (figure 3(d)) changes only the amplitude and duration of the pulses but not their waveform.

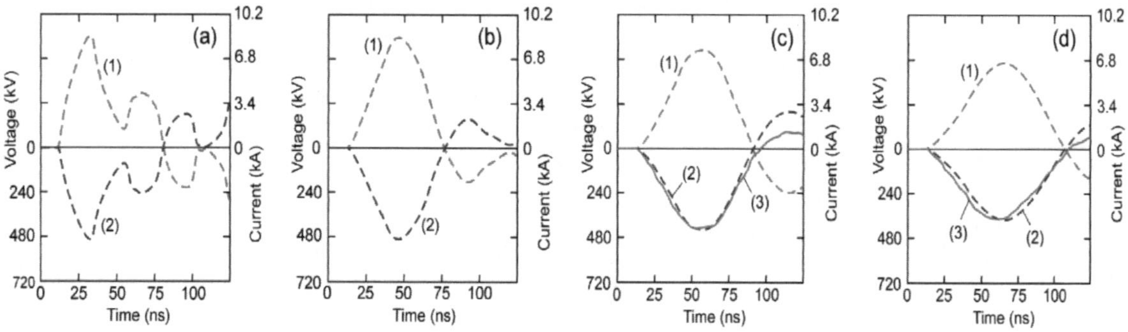

Figure 3. Simulated current (1) and voltage (2) pulses on the load R_L = 60 Ohm (R'_L = 0.416 Ohm reduced to the primary circuit) from the simulations of the scheme including the high-voltage transformer. The inductance of the DFL switch L_K is: (a) – 0 nH; (b) – 5 nH; (c) – 10 nH; (d) – 18 nH. LIA observed voltage pulses (3).

The amplitudes of the output voltage and current are following: at L_K = 5 nH, U = 495 kV, I = 8.26 kA, τ = 125 ns; at L_K = 10 nH, U = 439 kV, I = 7.32 kA, τ = 155 ns; at L_K = 18 nH, U = 405 kV, I = 6.75 kA, τ = 175 ns. Thus, the increase in the switch inductance from 0 up to 18 nH leads to the decrease in the accelerator output pulse power by ~34%; the pulse duration increases by ~11%.

Comparing the amplitudes and waveforms of the simulated pulses with the observed pulses of the accelerators Tonus-NT and LIA at Tomsk Polytechnic University, one can note the following. The difference in the amplitude and the duration of the simulated and observed pulses does not exceed 5%. The absence of the high-voltage transformer in the design of Tonus-NT allows forming the pulse trapezoidal on the load, even at the inductance of discharging contour as high as 320 nH.

The LIA with the multichannel spark gap has ~10 nH inductance of discharging contour (figure 3(c)) despite of multichannel switching (24 channels), synchronization of triggering the channels and forced division of current between the channels. The value of inductance is determined by the connections between the switch, DFL electrodes, and the induction system.

The LIA with the magnetic switch [6] has ~18 nH inductance of discharging contour (figure 3(d)). The higher inductance is caused by the processes of pulsed remagnetization of the switch cores, as the connections to the DFL and induction system is the same as for the LIA with spark gap switching. Thus, to decrease the inductance of discharging contour in the LIA with magnetic switching, one needs a core material with magnetization characteristics better than those of permalloy 50 NP.

IOP Conf. Series: Materials Science and Engineering **135** (2016) 012049 doi:10.1088/1757-899X/135/1/012049

However, common disadvantage of the accelerators with the transformer in the output circuit (the LIA is related to) is connected with the presence of parasitic reactive elements essentially affecting the output characteristics.

4. Conclusion

As a result of elaboration of the model including an analytical calculation of the circuit elements, a good agreement between the simulated pulses and the observed pulses of the working nanosecond accelerator (figure 2(d)) and linear induction accelerators (figures 3(c) and 3(d)) was achieved. It allows one to use the mentioned simulation model for the analysis of the influence of the element parameters of the equivalent circuit (in particular, PFL switch inductance changing in a wide range) on the amplitude and the waveform of the accelerator pulses on a load.

For the nanosecond accelerators decreasing inductance of the PFL switch and connections from 320 nH (Tonus-NT) down to 200 nH and below facilitates the formation of close-to-rectangular pulses. At the same time, within the 0–320 nH range of the switch inductance variation, the output pulse waveform depends on L_k significantly, and the amplitude changes slightly. Another situation takes place for the linear induction accelerators, with the spark gaps or with the magnetic commutators. Even the switch inductance and connections as low as several nH does not allow one to form a quasi-rectangular pulse; it looks rather triangular in accordance with simulations. With increasing inductance of the switch from 0 to 18 nH, the amplitude of the output pulse decreases and the pulse duration rises significantly but the waveform does not change. Such a difference between nanosecond accelerators and LIAs is connected with the presence of the induction system (high-voltage transformer) in LIAs; the elements of its equivalent circuit (dynamic capacitances, leakage and magnetization inductances, ohmic equivalents of losses in windings and magnetic system) determine the waveform and amplitude of the output pulses. The simulation has shown that the leakage inductance L_{S1} exerts the greatest influence. Therefore, for such accelerators, special attention must be devoted to designing the induction system (high-voltage transformer) and to a choice of the ferromagnetic core material (the parameters H_0, $S_{\omega o}$, $S_{\omega e}$) in order to achieve the maximal value of the magnetization inductance L_m. The simulation based on the presented equivalent circuit allows one to predict the characteristics of the output pulses of high-current nanosecond accelerators to a sufficient accuracy (~ 5%) and can be used for preliminary development of various types of accelerators with spark gaps, magnetic switches, Blumlein and single PFLs, with high-voltage transformers (induction systems) and without them in a discharging contour.

References

[1] Humphries S 2013 *Charged particle beams* (New York: Dover edition)
[2] Takayma K and Briggs R J 2011 *Induction accelerators* (New York: Springer science and business media)
[3] Vintizenko I I 2015 *Linear induction accelerator* (Moscow: Fizmatlit)
[4] Mazarakis M G et al 2010 *IEEE Transactions on Plasma Science* **38** 704
[5] Ryabchikov A I, Petrov A V, Karpov V B, Polkovnikova N M, Tolmacheva V G and Usov Yu P 2001 *Proceedings of the 5th Korea-Russia International Symposium on Science and Technology, (Tomsk)* vol 1 (Tomsk: TPU), pp372-375
[6] Chandra R et al 2014 *IEEE Transactions on Plasma Science* **42** 3491

Ways of improvement of methodological approaches to the assessment of the effectiveness of physical protection systems of nuclear facilities in consideration of modern requirements and threats

E A Vlasenko[1], A V Nikienko[1] and D G Demyanuk[2]

[1] Mining and chemical combine, Lenina street, 53, 662970, Zheleznogorsk, Krasnoyarsk region, Russia
[2] Physical Technical Institute of National research Tomsk Polytechnic University, Lenina avenue, 30, 634050, Tomsk, Russia

E-mail: atomlink@mcc.krasnoyarsk.su

Abstract. Methodological approaches to the assessment of the effectiveness of physical protection systems developed by Russian and foreign researchers are reviewed. Some ways of improvement of these approaches are offered. They consider tactics overview, application of two-person rule, aspects of inherent safety of nuclear materials, proposals on the use of test reliability data.

1. Introduction

Due to international agreements Russian authorities run state system of physical protection to prevent acts of nuclear sabotage and proliferation of nuclear materials. The operating organizations create physical protection systems (PPS) at nuclear sites including nuclear power plants. PPS is a set of organizational and technical measures taken by an administration of a nuclear site, its security department, guards forces with a help of technical means of physical protection. There is a set of state and departmental requirements to PPS. Fulfillment of these requirements and system design based on certain principles help to build an effective PPS which means that it can resist unauthorized actions of adversaries against objects of physical protection taking into account a design basis threat (DBT). DBT is usually defined during a vulnerability assessment of a nuclear facility. PPS effectiveness can be measured quantitatively as well as qualitatively. Quantitative measure of PPS effectiveness is called an efficiency indicator.

Efficiency assessment results are used to determine if PPS has to be improved or not. If yes, it helps to understand essentially important ways of improvement. When funding is limited it has a vital meaning.

2. Probabilistic-temporal analysis as a primary method of PPS efficiency assessment

Probabilistic-temporal analysis is a primary method of PPS efficiency assessment in Russia as well as in other countries. Researches all over the world develop and use similar techniques to perform it.

A.V. Boyarintsev, A.N. Brazhnik, A.G. Zuev say that efficiency indicator is a probability of suppression of unauthorized actions of intruders:

$$P = P_{\det} \times P_{del} \times P_{n} \tag{1}$$

where P_{det} is a probability of detection, P_{del} is a probability of delay, and P_n is a probability of neutralization. To define P_{det} it is offered to refer to technical specifications of intrusion sensors that are used at a site. At the same time, it is allowed to competently decrease a value of probability of detection compared to the one mentioned in manuals considering condition of communication lines and other equipment as well as a "human factor" caused by a need to assess an alarm by an operator [1].

S.V. Skryl, A.V. Dushkin and V.V. Gaifullin offer a probability of protection of a nuclear site against unauthorized actions of adversaries as an efficiency indicator:

$$P = P_{i} \cdot P_{fo} \tag{2}$$

where P_i is a probability of interception of adversaries by rapid response forces and P_{fo} is a probability that rapid response forces win an armed clash against adversaries.

$$P_{i} = P_{\det} \cdot P_{tr} \cdot P_{nfo} \cdot P_{dgrr} \tag{3}$$

where P_{det} is a probability of timely intrusion detection by an intrusion sensor system, P_{tr} is a probability of credible alarm signal transition to rapid response forces, P_{nfo} is a probability of no-failure operation of equipment, P_{dgrr} is a probability that rapid response forces deploy at a point of interception of adversaries before an armed clash after receiving an alarm signal.

It is stated that P_{tr} and P_{nfo} can be ignored because at a design stage they may be taken equal to 1. Authors say that P_{fo} can also be taken equal to 1 if a set of organizational measures is taken so that rapid response forces definitely win adversaries in an armed clash. Therefore, an efficiency indicator can be defined as:

$$P_{i} \approx P_{\det} \cdot P_{dgrr} \tag{4}$$

where P_{det} depends on characteristics, number of intrusion sensors and their location at a site. P_{dgrr} depends on tactics of rapid response forces as well as location and delay parameters of physical barriers that adversaries meet moving towards an object of physical protection [2].

M.L. Garcia offers two basic measures of PPS efficiency. The first one is based on comparison of minimal total delay time throughout the adversary path (TMIN) and response forces time (TG). PPS should provide sufficient delay so that response forces are able to intercept an adversary. The second measure is a total probability of detection of an adversary before he reaches his goal. To consider both measures during the PPS efficiency assessment, it is estimated in at a specific point of the adversary path which is called critical detection point (CDP). CDP is a point where an adversary delay time at the rest of the path (TR) just exceeds response forces time (TG).

$$P_{I} = 1 - \prod_{i=k}^{k-1} P_{NDi} \tag{5}$$

where P_I is a probability of interception of an adversary by response forces, k is a point where TR just exceeds TG, P_{NDi} is a probability that element i does not detect an adversary [3].

Taking into account a probability that response forces win an adversary in an armed clash an efficiency indicator may look like:

$$P_{E} = P_{I} * P_{N} \tag{6}$$

where P_N is a probability that an adversary is neutralized in case of interception. P_E characterizes vulnerability to a specific threat [4].

The methodical approach used by the State Corporation "Rosatom" is also based on a probability temporal analysis. Following indicators are used to assess the efficiency:

- Differential indicators that characterize probabilities of prevention of unauthorized actions against every object of physical protection.
- An integrated indicator that characterizes an ability of PPS to protect all the objects of physical protection [5].

An integrated indicator is calculated using differential indicators.

Efficiency assessment is carried out for various scenarios of actions of an outside adversary, an inside adversary and their combination.

A differential indicator of PPS efficiency against an outsider is defined as:

$$P_{out} = \max(f(P_{\det}, P^i_{inrf})P^n_{inrf}, f(P^i_{exrf}, P^n_{exrf})) \tag{7}$$

where f is a function that depends on tactics of response forces, P_{det} is a probability of detection of an adversary, P^i_{inrf} is a probability of interception of an adversary by internal response forces, P^n_{inrf} is a probability of neutralization of an adversary by internal response forces, P^i_{exrf} is a probability of interception of an adversary by external response forces, P^n_{exrf} is a probability of neutralization of an adversary by external response forces.

A probability of detection of an outsider overpassing a layer of physical protection is estimated as:

$$P_{\det} = 1-(1-P_{\det.before})\times(1-P_{\det.after}) \tag{8}$$

where $P_{\text{det.before}} = 1-\prod_{i=1}^{k}(1-P_i)$ is a probability of detection of an outsider before he overpasses a layer of physical protection, k is a total number of intrusion sensors which have a detection zone allowing to detect an adversary before he overpasses a layer of physical protection, P_i is a probability of detection of an intrusion sensor i that has a detection zone allowing to detect an adversary before he overpasses a layer of physical protection; $P_{\det.after} = 1-\prod_{j=1}^{m}(1-P_j)$ is a probability of detection of an outsider after he overpasses a layer of physical protection, m is a total number of intrusion sensors which have a detection zone allowing to detect an adversary after he overpasses a layer of physical protection, P_j is a probability of detection of an intrusion sensor j that has a detection zone allowing to detect an adversary before he overpasses a layer of physical protection.

A differential indicator of PPS effectiveness is taken equal to a minimal value of probability of prevention of unauthorized actions taken by an adversary using a variety of scenarios [6].

When assessing PPS effectiveness against an insider threat it is assumed that an adversary scenario consists of two parts: moving to a certain layer of physical protection using his own authorization and then a “force” breakthrough. In some cases second part of scenario may not be realized.

3. Ways of improvement of methodology

The assessment of PPS effectiveness based on probabilistic-temporal analysis helps to evaluate an ability to withstand unauthorized actions of different types of adversaries quantitatively. Along with this, a set of proposals to enhance methodology is formulated. Implementation of following suggestions may lead to better objectivity of results of assessment:

- When analyzing scenarios of actions of adversaries and guard forces it is advisable to model actions of several groups of guard forces (alarm group, reserve group) and potential adversaries (main group, cover group). Besides this, different tactics of actions should be considered. These tactics may vary depending on circumstances. Results of armed clashes should also vary depending on weapons and equipment of specific groups taking part in an action.

- Two-person rule should be considered when assessing an action in a zone or a vault where a rule is implemented. If a model of an insider consists of a single person it should be assumed that he is not allowed to such zones. As soon as he gets into a zone a single insider should be considered an outsider.
- When assessing an effectiveness of PPS against an insider that works with another person (colleague) in a zone following a two-person-rule a probability of generating an alarm by a partner should be taken into account.
- The inherent safety of an object of physical protection should be considered when assessing a probability of stealing. The main measure in this case is an equivalent dose level generated by an object of physical protection. If a time of an action is greater than a time required to achieve a certain level of dose a stealing may be considered impossible. E.g., for adversaries who are ready for self-sacrifice (according to a model) a level of dose can be extremely high. For those adversaries who are not ready to risk their health (according to a model) a level of dose should not exceed levels stated in radiation safety instructions.
- When assessing a stealing scenario it is advisable to overlook different paths of adversaries heading to an object of physical protection and back. For instance, when an insider approaches an object of physical protection he is most likely to go through an entry control point legally. However, he is not likely to go the same way back. To minimize a risk of being arrested, he may go back through an evacuation door escaping examination by a guard.
- During the calculation of a probability of detection it is advisable to take into account a probability that a complex of technical means of physical protection is in a state of operability at a moment of an attack. Such probability can be calculated using reliability data, information about periodicity and duration of maintenance, number of maintenance parties, maintenance prioritization, and spare parts delivery period. Reliability theory is a powerful instrument for calculation. State Corporation "Rosatom" collects data on reliability of technical means of physical protection [7]. However, collected data is not used for enhancing believability of results of assessment of the effectiveness of PPS. Some researches try to take into account a probability that a complex of physical protection devices is in a state of operability at a moment of an attack. For instance, P.G. Gorlach offers to use probability of successful data transition P_{dt} as well as availability ratio K when calculating probability of detection. He mentions that P_{dt} is a feature of a data processing system and $P_{dt} \rightarrow 1$ because of the modern protective coding methods and repeated look-up. To calculate K he suggests to use a formula:

$$K = \frac{T_f}{T_f + T_m} \tag{9}$$

where T_f is mean time between failures and T_m is mean time of restoration. It is mentioned that modern complexes of technical means of physical protection have $K \approx 0{,}99...0{,}999$ because of repeated look-ups, small time of restoration (several hours), and high mean time between failures of details and devices [8]. Similarly, other researchers suggest to use following multipliers when calculating probability of neutralization: probability of successful data transition to alarm forces, probability of no-failure operation of technical means of physical protection. They mention that these probabilities can be ignored at an early state of designing of PPS. Nevertheless, practice of exploitation of PPS in real conditions proves that values of probability of successful data transition to alarm forces and probability of no-failure operation of physical protection devices may vary drastically because of imperfection of maintenance, spare parts supply system, poor inspections at a site, run-out of equipment and mechanisms, and other reasons.

- During the calculation of a probability of detection it is advisable to take into account an efficiency of compensative measures that are taken for the period of restoration of main components. Usually allocation of guards at certain points, patrolling, buildup of temporary detection sensors and systems, buildup of temporary physical barriers are used at nuclear sites

IOP Conf. Series: Materials Science and Engineering **135** (2016) 012050 doi:10.1088/1757-899X/135/1/012050

as compensative measures. Compensative measures are especially actual when considering a mutual action of an insider and outside adversaries. Such threat is absolutely real. For instance, in October, 2014 Belgium authorities found that during 3 years a jihad devotee had been working in one of NPP departments as a security technician [9]. Famous assault on Pelindaba in 2007 could have been organized with a help of an insider [10]. An insider familiar with a state of an operability of PPS can transfer an information about location of a broken sensor to an outsider. The risk is too high that an outside adversary will try to attempt his assault before a broken device is restored.

4. Conclusion

Despite a wide set of instruments and algorithms, modern methods and computer programs designed to assess effectiveness of PPS have plenty of assumptions. Some of these assumptions forbid to consider individual features of a nuclear site (i.e., geographic position, technological features, etc.), guard force tactics, adversary tactics. Eventually, it affects result of an assessment significantly.

On this basis, development of methods of PPS efficiency assessment should be continued in order to provide objective results based on a whole set of individual features of a nuclear site.

References

[1] A V Boyarintsev, A N Brazhnik, A G Zuev 2006 *Issues of anti-terroristic activity: Categorization and vulnerability analysis of facilities* (Saint-Petersburg: JSC "NPP "ISTA-Systems") pp 98-101

[2] S V Skryl, A V Dushkin and V V Gaifullin 2011 Considering physical protection system efficiency evaluation *Information and security. Issue 2 293-96*

[3] Garcia M.L. 2001 *The design and evaluation of physical protection systems* (Burlington, MA, USA: Elsevier Butterworh-Heinemann) pp 243-246

[4] Garcia M.L. 2006 *Vulnerability assessment of physical protection systems* (Burlington, MA, USA: Elsevier Butterworh-Heinemann) p 382

[5] *Guidelines on efficiency evaluation of physical protection systems of nuclear objects* 2015 (Moscow: Rosatom) p 8

[6] *Guidelines on efficiency evaluation of physical protection systems of nuclear objects* 2015 (Moscow: Rosatom) p 18-9

[7] *Guidelines on collection, processing and distribution of information on reliability of technical means of physical protection* 2011 (Moscow: Rosatom)

[8] P G Gorlach 2009 Functional efficiency assessment of a complex of technical means of physical protection *T-Comm – Telecommunication and transport. Issue S2 56-7*

[9] http://rg.ru/2016/03/24/celiu-teraktov-v-belgii-dolzhny-byli-stat-aes.html

[10] https://www.washingtonpost.com/world/how-armed-intruders-stormed-their-way-into-a-south-african-nuclear-plant/2015/03/13/470fc8ba-579d-4dba-a0c0-f0a1ed332503_story.html

IOP Conf. Series: Materials Science and Engineering **135** (2016) 012051 doi:10.1088/1757-899X/135/1/012051

Analysis of physical processes via imaging vectors

V Volovodenko[1], N Efremova [2] and V Efremov[3]

[1]Department of Control System Optimization, Institute of Cybernetics, National Research Tomsk Polytechnic University, Tomsk, Russia
[2] Department of General Physics, Institute of Physics and Technology, National Research, Tomsk Polytechnic University, Tomsk, Russia
[3] Institute of Physics and Technology, National Research, Tomsk Polytechnic University, Tomsk, Russia

E-mail: ena@tpu.ru

Abstract Practically, all modeling processes in one way or another are random. The foremost formulated theoretical foundation embraces Markov processes, being represented in different forms. Markov processes are characterized as a random process that undergoes transitions from one state to another on a state space, whereas the probability distribution of the next state depends only on the current state and not on the sequence of events that preceded it. In the Markov processes the proposition (model) of the future by no means changes in the event of the expansion and/or strong information progression relative to preceding time. Basically, modeling physical fields involves process changing in time, i.e. non-stationay processes. In this case, the application of Laplace transformation provides unjustified description complications. Transition to other possibilities results in explicit simplification. The method of imaging vectors renders constructive mathematical models and necessary transition in the modeling process and analysis itself. The flexibility of the model itself using polynomial basis leads to the possible rapid transition of the mathematical model and further analysis acceleration. It should be noted that the mathematical description permits operator representation. Conversely, operator representation of the structures, algorithms and data processing procedures significantly improve the flexibility of the modeling process.

1. Introduction

The following research is devoted to the modeling of physical processes. Up-dated mathematical modeling of physical processes implies solving the following tasks: 1) mathematical model description of a specific physical process (or set of processes); 2) algorithm formulation in solving this or that problem; 3) numerical algorithmic mapping on computing system architecture. All these tasks are related. Before proceeding in examining closely any physical natural process via mathematical methods it is necessary to identify the basic principles and dynamics enabling to describe its quantitative and qualitative behavior adequately and simply i.e. produce a model. Real behavior of natural physical processes is actually much more complex comparable to primitive objects which are explorable by methods of modern theory. At the present development stage of Geoscience there are existing sets embracing numerous disordered elements. It is practically impossible to describe all these elements related to different Geoscience domains. To solve the tasks associated with ultra-long period processes, newly-developed methods in modeling and analysis of experimental data are required. Practically, all modeling processes in one way or another are random. Problem-solving of any random process is rather difficult. However, comparatively simple calculation methods are accessible only if general random processes are not considered, and, accordingly, only those processes having certain specific properties and being of some practical importance are considered. Such processes are termed as Markov processes [1]. Markov processes are one of the most important models for real existing natural processes, the apparatus of which is well-developed. The characteristic feature of Markov processes is the fact that the probability distribution of the next state depends only on the current state and not on the sequence of events that preceded it. Markov process of proposition (model) by no means changes in the event of the expansion and/or strong information progression relative to preceding time.

2. Theoretical background

The analysis of dynamic processes occurring under the influence of different fields could be based on various mathematical models. When modeling physical structures with time-evolving processes, the description of such processes should be correlated to an appropriate mathematical apparatus. Numerous characteristic features of the processes in a physical environment (system) could be attributed to the peculiarities of the stationary system itself. The parameters and structures of this system are all assumed to be constant in time. In this case, the direct model is applied based on the following statistic methods: regression, factorial analysis, argument-dependent method and etc. In the event of dynamic processes differential equations with constant coefficients are included. Nevertheless, there is on problem class which excludes these methods as inappropriate. These problems are attributed to the manifestation of explicit temporal and structured non-stationarity. In particular, the changing parameters of a physical environment (system) in time could result in the change of the structure itself furnishing the inapplicability of a mathematical apparatus, which, in its turn, is intended to describe a system with constant parameters and structures.

On the background of steady stationary state, such systems and phenomena develop slowly, and under insignificant exposure could sharply change. In this case, this development process evidently transforms to another stage, and, as a result, its parameters also change. As it has been mentioned all initially employed description methods, oriented on the stationary state, decrease the process adequacy and become inappropriate in describing any phenomena. To preserve the model adequacy of the process being described it is necessary to change the mathematical framework of basic unit model representation. The mathematical framework change poses a series of questions which could exclude further process development modeling.

One question is - the mathematical, algorithmic and software consistency level of modeling and data representation. It seems the modeling of any physical process, all data, including even initial data in modeling and/or design, is various, and, in this case, requires identically diversified structures for its representation. Such a diversity of structure data is generated not only by its source, but also by its possible integrated storage, processing and representation. To solve the whole complex of above-mentioned problems a properly selected mathematical apparatus could serve as a base which is completely adequate for the described physical processes, on the one hand, and could provide uniform procedure of process modeling and data representation, on the other hand.

In our opinion, applied functional analysis could be a mathematical apparatus. Based on this fact, the designed model system architecture possesses sufficient general and/or universal properties to establish a theoretical model, their algorithms and accompanying software. The rather extensive universality of this proposed method is based on the fact that the methods being generated within the functional analysis could be used to describe not only the modeling process itself, but also data processing in general. Conversely, all descriptions presume operator representation. Further, operator representation of structures, algorithms and data processing significantly improve the modeling flexibility. Moreover, due to the uniform representation of model components as array structures and signals as vectors, the modeling system becomes more reliable and universally accepting block-oriented description in the form of series-parallel block diagrams.

The most attractive examples of the above-described approach are such methods as spectrum estimation of non-stationary systems [2], interpolation method [3], and method of imaging vectors[4]. Data processing of both stationary and non-stationary cases are easily interpreted by the above-mentioned methods. Moreover, these methods can be applied in describing operations on discrete variables. It is possible to determinate precisely the problem formulation which is acceptable for using random variables within the spectrum method itself and its supporting methods. This aspect considers practical requirements when referring to random factors and processes which develop on the background of strictly specified regularities.

When considering physical fields modeling is of great importance due to the fact that processes changing in time, i.e. nonstationary processes are mainly involved. In this case, the application of Laplace transformation provides unjustified description complications. Transition to other possibilities results in explicit simplification. One possibility is the application of the functional analysis. The method of imagining vectors provides constructive mathematical models with subsequent changes in modeling and analysis. The flexibility of the model itself using polynomial basis leads to the possible rapid transition of the mathematical model and further analysis acceleration.

The method is based on the principle of the transition of mathematical operations to operation coefficients representing functions and their derivatives and integrals. This is particularly effective in linear cases where existing mathematical support and software could be applied. The operation is based on the representation of the function involving generalized Fourier constants in polynomial basis:

$$f(t) = \sum f_i P_i(t), \tag{1}$$

where, $(P_i(t))$ is function representation basis

Representing polynomial $(P_i(t))$ is included in the polynomial basis and is enormously flexible, independently and selectively, i.e. they are easily normalized and adapted to an orthogonality interval. Due to this property the analysis could be repeated, which, in its turn, is convenient in modeling.

Vector coefficients

$$F = \{f_0, f_1, \ldots, f_{n-1}\}, \tag{2}$$

are termed as imaging vector of the function in basis $(P_i(t))$..

In the transition to different operations involving differentiation and spatial integration of their equivalents, the coefficients are vector-matrix expressions [4]. Particularly, such expressions are various differential-integral equations. These expressions could be the problem basis in considering

IOP Conf. Series: Materials Science and Engineering **135** (2016) 012051 doi:10.1088/1757-899X/135/1/012051

the modeling of physical processes, including initial and boundary non-stationary conditions. The data representation is exceptionally similar to computerization and programming support (software support).

Practically, all modeling processes in one way or another are random. The foremost formulated theoretical foundation embraces Markov processes, being represented in different forms. It is critically important that this property is easily extended to software-based and mathematical model properties, which are considered with imagining vector method application.

Assuming that the operation is described by linear (the simplest examples) equation systems [2], including initial conditions:

$$\frac{dP}{dt} = \begin{bmatrix} P_{11}(t) & \ldots\ldots & P_{1n}(t) \\ \ldots\ldots\ldots & \ldots\ldots\ldots & \ldots\ldots\ldots \\ \ldots\ldots\ldots & \ldots\ldots\ldots & \ldots\ldots \\ P_{N1}(t) & \ldots\ldots\ldots & P_{NN}(t) \end{bmatrix} \begin{bmatrix} P_1(t) \\ P_2(t) \\ . \\ . \\ P_N(t) \end{bmatrix} \qquad P = \begin{bmatrix} P_1(t) \\ P_2(t) \\ . \\ . \\ P_N(t) \end{bmatrix} \tag{3}$$

$P_{1i}(t)$ – probability state, $P_{ij}(t)$- transition probability from the state S_i to the state S_j .

This system of linear differential equations is consistent to the algebraic equation system, where the argument t is the matrix J . Therefore, we have a block matrix of imagining matrix of coefficients:

$$\{P_{ij}(t)\} \rightarrow \{P_{ij}(J)\}\ , \tag{4}$$

while, vector function image is a block vector:

$$\left\{ \begin{array}{l} P_1(t) = \pi_{10}\overrightarrow{P_0}(t) + \pi_{11}\overrightarrow{P_1}(t) + \pi_{12}\overrightarrow{P_2}(t) + \ldots\ldots + \pi_{1N-1}\overrightarrow{P_{N-1}}(t) \\ P_2(t) = \pi_{20}\overrightarrow{P_0}(t) + \pi_{21}\overrightarrow{P_1}(t) + \pi_{22}\overrightarrow{P_2}(t) + \ldots\ldots + \pi_{2N-1}\overrightarrow{P_{N-1}}(t) \\ . \\ . \\ P_N(t) = \pi_{N0}\overrightarrow{P_0}(t) + \pi_{N1}\overrightarrow{P_1}(t) + \pi_{N2}\overrightarrow{P_2}(t) + \ldots\ldots + \pi_{NN-1}\overrightarrow{P_{N-1}}(t) \end{array} \right\}, \tag{5}$$

$$P_1(t)=\begin{pmatrix}\pi_{10}\\ \pi_{11}\\ \pi_{12}\\ \cdot\\ \cdot\\ \pi_{1N-1}\end{pmatrix}=\pi_1,\ P_2(t)=\begin{pmatrix}\pi_{20}\\ \pi_{21}\\ \pi_{22}\\ \cdot\\ \cdot\\ \pi_{2N-1}\end{pmatrix}=\pi_2 \dots P_N(t)=\begin{pmatrix}\pi_{N0}\\ \pi_{N1}\\ \pi_{N2}\\ \cdot\\ \cdot\\ \pi_{NN-1}\end{pmatrix}=\pi_N . \quad (6)$$

This algebraic system of equations could be solved.

Noting the linear equations, the spectrum method enables to write the expression of closed-form equation solutions using matrix symbolism. The application of the spectrum method results in the algebraic connection between the system attributes. This connection is expressed as an operator matrix form. In this case, it is possible to obtain an expression of transfer functions, system output signals and their explicit characteristics; perform conversion symbolically describing the investigated system to simplify the expressions; and in the final investigation stage of finding the variable characteristics to perform computations according to the matrix-operator formula by general-purpose subroutines and reference sources.

Spectrum method used for system solution of differential equations has the following advantages: 1.homogeneous representation for operations and procedures; 2.homogeneous representation of one-dimensional signals; and 3. possibility of structure determination.

The following fact should be highlighted- a set of conditions associated with the loosening of time-dependency models are reduced to simple parametric associations within the operator representation domain. As a result, there is the possibility of not only solving broad-class problems, but also accumulating information.

It is possible to determinate precisely the problem formulation which is acceptable for using random variables within the spectrum method itself and its supporting methods. This aspect considers practical requirements when referring to random factors and processes which develop on the background of strictly specified regularities.

Block diagrams are introduced to solve complicated non-stationary linear systems, as in the case of determination analysis. The first stage involves the differential equations connecting input and output systems [4]. In this case, the apparatus of linear differential operators is used. The second stage involves the transition to spectrum characteristics. In this case, the transition is based on spectrum (matrix) description of the systems and adapted to computer software.

3. Conclusion

The problem-solving of any random process is rather difficult. However, comparatively simple calculation methods are accessible only if general random processes are not considered, and, accordingly, only those processes having certain specific properties and being of some practical importance are considered. Such processes are as Markov processes [1]. The characteristic feature of Markov processes is the fact that the probability distribution of the next state depends only on the current state and not on the sequence of events that preceded it. Markov processes involve a well-developed theoretical foundation.

Considering Markov processes changing in time the possibility of state transition could indicate specific investigation methods: method of imaging vectors and spectrum method. Both methods are oriented in investigating orthogonal functional basis within an energy-limited function which corresponds to the physics of obtained results and introduces a special operator expression describing the physical phenomenon in finite time interval. The behavior of resulting relations is that matrix representation of linear operators is used as an information carrier of the processes. In these cases the introduction of numerical analysis modeling into up-dated computer programs is possible.

References

[1] Bharucha-Ried A 1969 *Elements of the theory of Markov processes and their application* McGraw Hill Book Company 511
[2] Solodovnikov V 1960 *Statisticheskaja dinamika linejnyh sistem avtomaticheskogo upravlenija* Moscow State Publishing House 655
[3] Dmitriev A 1967 *Teorija avtomaticheskogo regulirovanija. Analiz i sitnez nepreryvnyh i diskretnyh sistem avtomaticheskogo regulirovanija* Mechanical Engineering 679
[4] Solodovnikov V and Dmitriev A 1975 *Spektral'nye metody rascheta i proektirovanija sistem* Mechanical Engineering 439
[5] Solodovnikov V and Semenov V 1975 Spectral method of account of non-stationary control systems of flying means *Mechanical Engineering* 271
[6] Feller V 1984 *Introductions in a theory of probabilities and it of appendix* The World **2** 527

IOP Conf. Series: Materials Science and Engineering **135** (2016) 012052 doi:10.1088/1757-899X/135/1/012052

Error reduction in gamma-spectrometric measurements of nuclear materials enrichment

D Zaplatkina[1], A Semenov[2], E Tarasova[2], V Zakusilov[2] and M Kuznetsov[2]

[1] St. Petersburg Nuclear Physics Institute n.a. B.P. Konstantinov, Gathina, Russia
[2] National Research Tomsk Polytechnic University, Tomsk, Russia

E-mail: kms@tpu.ru

Abstract. The paper provides the analysis of the uncertainty in determining the uranium samples enrichment using non-destructive methods to ensure the functioning of the nuclear materials accounting and control system. The measurements were performed by a scintillation detector based on a sodium iodide crystal and the semiconductor germanium detector. Samples containing uranium oxide of different masses were used for the measurements. Statistical analysis of the results showed that the maximum enrichment error in a scintillation detector measurement can reach 82%. The bias correction, calculated from the data obtained by the semiconductor detector, reduces the error in the determination of uranium enrichment by 47.2% in average. Thus, the use of bias correction, calculated by the statistical methods, allows the use of scintillation detectors to account and control nuclear materials.

1. Introduction

One of the systems providing the required control level over the activities in the nuclear industry is a system of accounting and control of nuclear materials. The system is designated to ensure continuous systematic accuracy of information on nuclear material and to detect unauthorized actions. Nuclear materials accounting is based on the measurements results of quantitative characteristics and attribute properties of nuclear materials.

The development and improvement of non-destructive methods of isotopic composition analysis are directly related to the solution of practical problems in the field of accounting and control of nuclear materials, as well as detection and prevention of illicit trafficking. One of the main methods of non-destructive analysis is the gamma-ray spectrometry which is widely applied for the nuclear materials control. This is a key method in determining the isotopic composition of the nuclear materials and uranium enrichment [1].

However any measurement process is inevitably linked with the need to process the data. The most important aspects of the test results processing are accurate and correct calculations which include errors at all stages of the work and their correct interpretation [2].

Thus the research goal is to optimize the process of the uranium isotopic composition analysis with spectrometric channel based on NaI (Tl) scintillation detector by means of measurement error reduction with statistical methods.

2. Results and Discussion

To determine the accuracy of measurements it is necessary to run a series of experiments to define the amount of nuclear material. We have used a spectrometric channel on the basis of the NaI (Tl) scintillation detector (Canberra, model 802-2x2), spectra processing was performed in the Genie-2000 software, and uranium dioxide UO_2 with different mass of U^{235} enrichment by 2% was used as a sample. Nuclides were matched by the developed nuclide library. There were 3 measurements for each source spaced 0 cm from the detector at 1200 live time. The measurement results are shown in the Table 1 [3, 4].

Table 1. The results of the samples measurement with a scintillation detector

Sample mass, g.	4	6	8	10	12	14
	0.4557	0.3221	0.9272	0.4630	0.4886	0.4812
	0.3117	0.8896	0.9654	0.5847	0.9442	0.3015
Calculated enrichment, %	0.3037	0.2660	0.9389	0.6649	0.8236	0.5922
	0.8024	0.2670	0.5509	0.5418	0.8995	0.4091
	0.3157	0.8896	0.9674	0.6159	0.6552	0.4562
Moderate enrichment, %	0.3570	0.4926	0.9438	0.5709	0.7521	0.4583
Measured result dispersion	0.0574	0.4905	0.0004	0.0317	0.0985	0.1024
Absolute error of enrichment measurement	1.6430	1.5074	1.0562	1.4291	1.2479	1.5417
Relative error of enrichment measurement, %	82.1492	75.3710	52.8079	71.4569	62.3940	77.0843

As it can be seen from the table, the error of the results is significant and accounts for a prominent uncertainty in the measurements due to the plurality of the latter. To increase the accuracy of measurements the methods of mathematical statistics were applied.

The method for several samples was used to evaluate the systematic error and to introduce the bias correction. The high resolution semiconductor germanium detector Canberra in conjunction with the installed control environment Genie-2000 was used as a more accurate method [3].

To calculate the bias we performed measurements using the same set of uranium dioxide samples. There were 3 measurements for each source spaced 0 cm from the detector at 1200 live time. Average enrichment values measured by semiconductor detectors are considered as accepted ones.

The bias estimation was calculated by the formulas [5]:

$$\theta = \sum_{k=1}^{m} w_k \cdot (\bar{x}_k - \mu_k) \Big/ \sum_{k=1}^{m} w_k \quad (1)$$

$$w_k = (\sigma_k^2 + s^2/n_k)^{-1} \quad (2)$$

$$s^2 = \sum_{k=1}^{m} (n_k - 1) \cdot s_k^2 / (n - m) \quad (3)$$

Bias estimation results are shown in the Table 2.

Table 2. Bias estimation of the sample mass determination

Sample number, k	σ2	s2	w	θ
1	0.00008	0.05738	23.00663	
2	0.00366	0.49054	21.25564	
3	0.00181	0.00043	22.12301	-1.35436
4	0.00022	0.03171	22.92836	
5	0.00003	0.09854	23.02916	
6	0.00005	0.10243	23.02232	

The resulting bias estimate is introduced as a correction for every measured value with the opposite sign, thus compensating the calculated bias. Corrected results of uranium enrichment determination by scintillation detector are presented in Table 3.

Table 3. Corrected uranium enrichment determination by NaI(Tl) detector

Sample mass, g	4	6	8	10	12	14
Enrichment without correction, %	0.438	0.527	0.870	0.574	0.762	0.448
Relative enrichment error without correction	78.109	73.656	56.501	71.297	61.890	77.598
Enrichment with correction, %	1.792	1.881	2.224	1.928	2.117	1.802
Relative enrichment error with correction	10.390	5.938	11.217	3.579	5.828	9.880

According to the measurement results analysis, it was found that the measurement bias correction determined using a more accurate method of standard samples analysis as well as mathematical and statistical techniques can reduce the relative systematic error component in the measurements performed by the detector approximately by 47.2% in average. The main condition used for the bias estimation is that the samples' physical, chemical and material composition should match that of the samples investigated by the detector as close as possible. The studied experiment complies with this condition, because one set of uranium samples was used.

Since the systematic error is a main contributor to the measurement error, we need to consider the dependence of the unaccounted material amount (UMA) on the systematic error.

The dependence of the UMA value change, its variance, standard deviation and confidence intervals are expressed by the formulas [5]:

$$\Delta UMA = 0.472 \cdot \sqrt{S} \tag{4}$$

$$\Delta(\sigma_{UMA}^2) = S \cdot (\sigma_{\delta 2}^2 - \sigma_{\delta 1}^2) = -0.3244 \cdot S \tag{5}$$

$$\Delta(\sigma_{UMA}) = \sqrt{S} \cdot 0.0131 \tag{6}$$

$$\Delta Di = 2\sqrt{S} \cdot 0.472 \cdot (z_{1-\alpha/2} + 1) \tag{7}$$

According to the calculations by the formulas stated above, we plotted the graph showing the dependence of UMA value change on the stratum size and average mass of each stratum element.

With a decrease in the confidence interval the UMA value range decreases as well. This fact can be conditioned by the measurement errors, thus increasing the detection probability of loss or change of materials. The increase in the detection probability of material loss or change with the use of the studied method equals 14.6%.

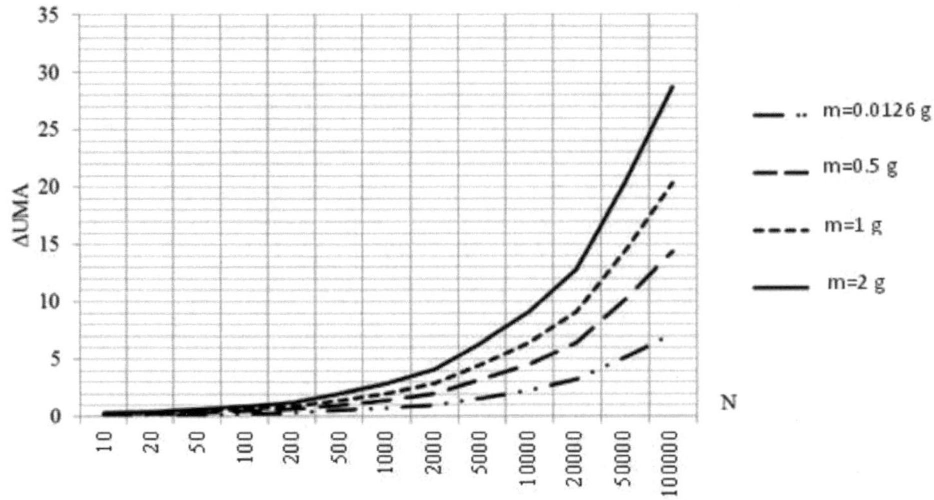

Figure 1. UMA value change depending on the stratum size and elements' mass.

3. Conclusion

The results analysis was used to estimate the bias correction estimation which equals 1.35%. The use of bias correction reduced the relative systematic error of measurements made by scintillation detector by 47.2%.

We have stated the dependence of the change in the unaccounted material amount, variance, standard deviation, confidence intervals of UMA, and the detection probability of nuclear materials change on the change in the measurement error. The described method allows for the increase in the detection probability of nuclear materials change by 14.6%.

Thus, the use of statistical methods to reduce the errors in uranium enrichment measurements made by the scintillation detector may be used to solve the accounting and control issues problems with acceptable accuracy.

References

[1] IAEA safeguards glossary. – 2001 ed. – Vienna : IAEA, 2002 – 230 p. – (International nuclear verification series, ISSN 1020–6205 ; no. 3)

[2] NP-030-12 "General guidelines for nuclear materials accounting and control": Federal regulations and guidelines, approved by Rostekhnadzor's decree of April 17, 2012, N 255 – Moscow: RF Ministry of Justice, 2012 – 23 p.

[3] Canberra Industries. Canberra Scintillation Detectors - Model 802/ – [Electronic resource] URL: http://www2.ph.ed.ac.uk/~td/SHlab/Projects/EPR/802.pdf, free. Accessed date: November 12, 2014.

[4] Saint Gobain Crystals. Physical Properties of Common Inorganic Scintillators./ – [Electronic resource] URL: http://www.hep.caltech.edu/~zhu/papers/08_tns_crystal.pdf, free. Accessed date: November 13, 2014.

[5] Kremer N. Sh. Probabilities theory and mathematical statistics / textbook. – Moscow: Unity, 2000. – 543 p.

[6] Dewji, S.A., Lee, D.L., Croft, S., Hertel, N.E., Chapman, J.A., McElroy, R.D., Jr., Cleveland, S. Validation of gamma-ray detection techniques for safeguards monitoring at natural uranium conversion facilities (2016) Nuclear Instruments and Methods in Physics Research, Section A: Accelerators, Spectrometers, Detectors and Associated Equipment, 823, pp. 135-148.

Model of mass transfer processes in the cascade of centrifugal extractors

E P Zelenetskaya[1] A G Goryunov[1] and N V Daneikina[2]
[1]Department of Electronics and Automation of Nuclear Pants, Institute of Physics and Technology, National Research Tomsk Polytechnic University, 2 Lenin Avenue, Tomsk 634050, Russia
[2]Department of Foreign Languages for Specialists in Physics and Technology, Institute of Physics and Technology, National Research Tomsk Polytechnic University, 2 Lenin Avenue, Tomsk 634050, Russia

E-mail: zeka@tpu.ru

Abstract. The paper describes a mathematical model of mass transfer processes in a cascade of reverse-flow centrifugal extractors. Model of operation of each extractor is given as tightly coupled system of mixing and separating chambers. All model units are represented by systems of differential equations. The article presents the results of testing of the developed model, which confirmed the validity of the assumptions made in the model. The authors assessed the impact of the overflow of dense phase level on the hydrostatic position of phase interface level in the extractor. The research showed that a change in the volume of dense and light phases occurs in each apparatus of a cascade even in the steady mode. Operation of the cascade consisting of 12 series-connected centrifugal extractors was simulated in order to verify the model. Computer simulation results confirm the adequacy of the developed model.

1. Introduction

Currently, there are various methods of chemical treatment, but extraction technologies are the most widespread because of their efficiency. Pulse column extractors connected by a cascade are used in production as the hardware implementation of the extraction technologies. In recent years, centrifugal extractors have been introduced into the technological process instead of column extractors. This is due to the factor that centrifugal extractors provide low contact time of phases, high coefficient of substances interface, increased nuclear safety. They also allow increasing production volume without expansion of workspaces. Such production aspects are important especially when dealing with radioactive materials. Although the extraction processes occurring in the pulse columns and centrifugal extractors are comparable, there are significant differences in the process flow diagram. These technological differences lead to the necessity for the development of extraction technologies. A number of technological schemes of extraction refining are being worked out within the framework of the project "PRORYV". These schemes include a large number of processing units with centrifugal extractor as the main hardware unit.

The search for optimal technological schemes of refining from the standpoint of safety and minimization of production costs is performed basically according to experimental research data. It is difficult to conduct experimental research at production site because of the reagents with high reactivity. In this regard, there is a need for mathematical and later computer models describing the operation of centrifugal cascade of extractors taking into account occurring basic physical and chemical processes. Moreover, computer simulation will evaluate the performance of centrifugal extractors, as separate devices or as a cascade of elementary units. This aspect will allow the introduction of safe technologies into chemically reactive production.

IOP Conf. Series: Materials Science and Engineering **135** (2016) 012053 doi:10.1088/1757-899X/135/1/012053

2. Mathematical model

The technological scheme of the extraction refining process contains blocks of cascades of centrifugal extractors with centrifugal extractor as elementary unit. Depending on industrial purposes, the number of extractors in the blocks varies. Since all the devices included in the blocks are identical structurally and by the principle of operation, we will describe the operation of one extractor.

2.1. Description of the simulation object

Operation of the centrifugal extractor is based on two processes. The first is the mixing of initial fluids in the mixing chamber of the apparatus, which provides mass transfer of the component. The other is the process of interface of the formed emulsion under the influence of centrifugal field on the light and dense phases in the separating chamber of the device. Therefore, the centrifugal extractor can be represented in the form of tightly coupled system of two parts, namely the mixing and separating during the simulation. The principle of operation of the extractor is as follows. After starting the drive initial mutually immiscible liquids are fed into the lower part of the mixing chamber where they are mixed with stirrer. After the liquids are mixed biphasic emulsion is formed in which the process of mass transfer occurs. Formed emulsion is fed from the mixing chamber into the separating chamber by bladed transfer unit. Emulsion is divided into dense and light phases in the mixing chamber under the influence of centrifugal field. At the same time the dense phase enters the manifold for collecting the dense phase through the water seal. Light phase is fed into the manifold for collecting light phase through the tubes of light phase output. Then, light and dense phases are removed from the machine through respective vertical channels.

Analyzing occurring technological processes in the centrifugal extractor, we concluded that the main input variables are the concentrations of the target components entering the extractor flow, volumetric flow and temperature of input streams, the number of rotations of stirrer and bladed transfer unit of extractor. Volumetric flow and temperature of the streams leaving the extractor, as well as the concentration of the target components in each of the streams are the output variables. Volumetric flow of resulting emulsion entering the separating chamber, the component wise concentration composition of emulsion refer to disturbing parameters. In addition, disturbing influences on the system should include the volumetric flow rate, temperature and concentration of the components of the poured solution, if it exists in the technological flow diagram.

Because the main processes taking place in the centrifugal extractor are mixing and interface processes, the mathematical model is based on the processes describing diffusion mass transfer and hydrodynamics. In addition, we will present the developed model as a cell model, where all the workspace is divided into n cells. The mechanism of the construction of this type of model is well represented in the [4].

2.2 The mathematical model of heat transfer processes in a centrifugal extractor as the basic unit of the cascade

According to [1-3] the extraction rate of the target components in the conditions of intensive mixing in a centrifugal extractor is determined by the rate of diffusion mass transfer. But the rate of the mass transfer process is influenced not only by physical and chemical processes occurring in the apparatus, but also by the rotation speed of internal moving parts of the machine, such as a stirrer or bladed transfer device. Therefore, it is possible to estimate the character of flowing of working media in the process of mixing using centrifugal similarity criteria such as Reynolds, Prandtl, Nusselt criteria presented in detail in [2, 5]. In practice, it is quite difficult to define the criteria, which include speed, because according to [1], the mode of developed free turbulence occurs in devices having internal rotating components. Velocity distribution in such devices is complex, therefore at the stage of design conventional velocity equal to the peripheral velocity of the stirrer blade end, excluding factor π=3.14 is used in centrifugal criteria [2].

Initial streams entering the extractor form a two phase emulsion during the mixing process in the mixing chamber. One liquid is dispersed as droplets and forms the dispersed phase in such a two-phase system. The other liquid forms a continuous phase in which the droplets of dispersed phase liquid move. According to [1, 3, 5] form of droplets depends on the rotational velocity of stirrer in apparatus. The greater the velocity, the greater the deformation of a drop. In this regard, the droplet diameter can only be determined experimentally. In the simulation, we will assume that the drops have a spherical shape and can only change its diameter. Dispersed phase will be denoted as X phase and continuous phase will be denoted as Y phase when developing the model. This denotation will be valid both for the mixing chamber unit and for the separating chamber unit. In the mixing chamber, due to the presence of the internal rotating mechanism, convective diffusion processes dominate over the processes of molecular diffusion. Therefore, the basic mass transfer equations for the block representing the mixing chamber can be represented as follows:

$$\begin{cases} V_i^x(t)\dfrac{\mathrm{d}\left[C_i^x(t)\right]}{\mathrm{d}t}+C_i^x(t)\dfrac{\mathrm{d}\left[V_i^x(t)\right]}{\mathrm{d}t}=Q_{\mathrm{in}}^x(\mathrm{t})C_{\mathrm{in},i}^x(t)-Q_{\mathrm{f}}^x(\mathrm{t})C_{\mathrm{f},i}^x(t)-\Delta m_i^x \\ V_i^y(t)\dfrac{\mathrm{d}\left[C_i^y(t)\right]}{\mathrm{d}t}+C_i^y(t)\dfrac{\mathrm{d}\left[V_i^y(t)\right]}{\mathrm{d}t}=Q_{\mathrm{in}}^y(\mathrm{t})C_{\mathrm{in},i}^y(t)-Q_{\mathrm{f}}^y(\mathrm{t})C_{\mathrm{f},i}^y(t)+\Delta m_i^x \end{cases} \quad (1)$$

where Q_{in}^x and Q_{in}^y – volumetric flow rates of phase X and phase Y respectively at the input of the mixing chamber; Q_{f}^x and Q_{f}^y – volumetric flow rates of phase X and phase Y respectively at the output of the mixing chamber; V_i^x and V_i^y – volumes of phase X and phase Y respectively; $C_{\mathrm{in},i}^x$ and $C_{\mathrm{in},i}^y$ – initial concentrations of i-th component in phase X and phase Y respectively; $C_{\mathrm{f},i}^x$ and $C_{\mathrm{f},i}^y$ – current value of concentration of i-th component in phase X and phase Y respectively; Δm_i^x and Δm_i^y – amount of a substance transferred from one phase to the other.

Volume of phase X and phase Y included in the system of equations (1) we will define as follows depending on the total volume of the chamber:

$$V_{\mathrm{m}}^{\ x}=\frac{Q_{\mathrm{in}}^x V_{\mathrm{m}}}{Q_{\mathrm{in}}^x+Q_{\mathrm{in}}^y} \qquad V_{\mathrm{m}}^{\ y}=V_{\mathrm{m}}-V_{\mathrm{m}}^x \quad (2)$$

where V_{m} – the working volume of the mixing chamber.

In the area of intensive mixing mass transfer occurs primarily by convective diffusion. Therefore, the amount of matter carried at each moment of time from one phase to the other presented in the system (1) can be defined as follows:

$$\begin{cases} \Delta m_i^x=6\cdot\beta_i^x\cdot\dfrac{V_{\mathrm{m}}^{\mathrm{d}}}{d_x}\left(C_i^x-{}^{*}C_i^x\right) \\ \Delta m_i^y=6\cdot\beta_i^y\cdot\dfrac{V_{\mathrm{m}}^{\mathrm{d}}}{d_x}\left({}^{*}C_i^y-C_i^y\right) \end{cases} \quad (3)$$

where $V_{\mathrm{m}}^{\mathrm{d}}$ – volume of dispersed phase; d_x – diameter of the disperse phase droplet; C_i^x и C_i^y – concentrations of i-th component in phase X and phase Y respectively; ${}^{*}C_i^x$ – concentration of i-th component in phase X equilibrium with the concentration C_i^y in the basic mass of phase Y; ${}^{*}C_i^y$ – concentration of i-th component in phase Y equilibrium with the concentration C_i^x in the basic mass of phase X; β_i^x and β_i^y – coefficient of mass transfer of i-th component in phase X and in phase Y respectively. Mass transfer coefficients in the system of equations (3) depend on the physical, geometrical and hydrodynamic factors and can be found mainly experimentally. But, according to [1], mass transfer coefficient is proportional to the product of stirrer rotation frequency to the diameter of blades of the stirrer and inversely proportional to the linear geometric size of the chamber in a turbulent flow mode of working environments, which is created in the mixing chamber. Based on this, and the fact that values of the equilibrium concentrations of target components in the set state are equal ${}^{*}C_i^x={}^{*}C_i^y$.

The values of the equilibrium concentrations in the system (3) are mainly determined by experimental data, curves of distribution are presented in literature. The disadvantage of such determination is that the reference data is presented according to the specific modes of operation of the device and composition of working environments, which often does not fully correspond to the investigated data. Therefore, based on the studies mentioned in [6], we will use a polynomial dependence for determination of the values of equilibrium concentrations of the target components in each phase.

The emulsion formed in the mixing chamber by means of a bladed transfer device is fed into the separating chamber. The process of interface of the emulsion is a complex hydrodynamic process, which is influenced by various factors, such as angular velocity of the rotor, emulsion supply rate, threshold level of phase interface, internal hydraulic resistance due to the geometry of the machine, change in the density of formed phases and distribution of the temperature gradient across the device in whole.

Based on the data in [1], we can assume that there are two liquid layers separated by phase interface level in separating chamber. The unblended layer is dispersed phase liquid layer and the second layer is the layer of continuous phase. Depending on the type and composition of emulsion entering the separating chamber, each of the layers can be either light or dense. Another assumption is that is a complete phase interface without mutual entrainment of components in the separating chamber. In this case, conical chamber can be replaced with a cylindrical one with equivalent volume. Phase volume ratio of X and Y will influence both the phase threshold position and value of the delay of matter concentrations in the chamber.

Volumes of light phase V_s^{l} and dense V_s^{h} phase in the separating chamber according to [1], are a function from the density of the light and dense phases, composition and concentration of the temperature gradient of the emulsion coming for interface, as well as from the structural features of the camera. However, our research showed that we could determine the amount of light and dense phases, without loss of accuracy in the following way:

$$V_{\mathrm{s}}^{\mathrm{l}} = \frac{V_{\mathrm{s}} \cdot V_{\mathrm{m}}^{x}}{V_{\mathrm{m}}^{y} + V_{\mathrm{m}}^{x}} \qquad V_{\mathrm{s}}^{\mathrm{h}} = V_{\mathrm{s}} - V_{\mathrm{s}}^{\mathrm{l}} \tag{4}$$

where V_{s} – working volume of the separating chamber, V_{m}^{x} – the volume of the dispersed phase of the mixing chamber, V_{m}^{y} – the volume of the continuous phase of the mixing chamber.

Since phases X and Y are continuous and incompressible, then we will put the model of ideal displacement as the basis of the separating chamber model. In general terms, taking into account all the chamber settings and the processes occurring in it, we receive a system of higher order equations. High order equations make the process of modeling and verification processes more complicated. Since concentrations of target components are significant parameters of the separating chamber, it is possible to reduce the order of the resulting model, for example up to the first. Accordingly, we will describe the concentration distribution within the separating chamber by the following system of differential equations:

$$\begin{cases} \dfrac{\mathrm{d}\left[C_i^x(t)\right]}{\mathrm{d}t} = \dfrac{C_{\mathrm{in},i}^x(t) - C_{\mathrm{f},i}^x(t)}{T_{\mathrm{s}}^x} \\ \dfrac{\mathrm{d}\left[C_i^y(t)\right]}{\mathrm{d}t} = \dfrac{C_{\mathrm{in},i}^{\mathrm{y}}(t) - C_{\mathrm{f},i}^{\mathrm{y}}(t)}{T_{\mathrm{s}}^{\mathrm{y}}} \end{cases} \tag{5}$$

where $C_{\mathrm{in},i}^x$ and $C_{\mathrm{in},i}^y$ – concentrations of i-th component in phase X and phase Y respectively in the input to separating chamber; $C_{\mathrm{f},i}^x$ and $C_{\mathrm{f},i}^y$ – concentrations of i-th component in phase X and in phase Y respectively at the output of separating chamber; T_{s}^x and T_{s}^y – concentration delay in phases X and Y respectively.

If the phase X at the output of mixing chamber is dispersed, then the concentration delay in phases X and Y we will estimate as follows:

$$T_s^x = k\cdot\frac{V_s^h}{Q^h};\; T_s^y = k\cdot\frac{V_s^l}{Q^l};\; Q^h = Q^x;\; Q^l = Q^y$$

where V_s^h and V_s^l – volumes of dense and light phases respectively, k – correction factor.

We will determine the change of the flow rate of phases in the output from separating chamber at the following way:

$$Q^l = \frac{V_s^l\cdot\left(Q_{s\_f}^x + Q_{s\_f}^y\right)}{k\cdot V_s}$$
$$Q^h = \frac{V_s^h\cdot\left(Q_{s\_f}^x + Q_{s\_f}^y\right)}{k\cdot V_s} \quad (6)$$

where k – correction factor, $Q_{s\_f}^x$ и $Q_{s\_f}^y$ – volumetric flow rates of phases X and Y at the output of the mixing chamber.

3. Simulation results

We have tested the developed model with the assumptions made using Matlab/Simulink software, since the software provides extensive instrumental basis for modeling physical systems. We will take the statement of the coincidence (with a possible deviation within 15%) of the experimental data with the simulation results for the target components, such as U and Pu as a criterion for the adequacy of the developed model.

When testing the developed model we have taken technological scheme of the extraction assembly of 12 consequently connected centrifugal extractors as a basic scheme. We have verified the model by the values of concentrations of target components both in the extract and in the raffinate at the cascade output.

The result of verification of the developed model is shown in Figure 1 and Figure 2 for the raffinate and the extract, respectively. For visual display of the results of simulation verification data are presented in relative units.

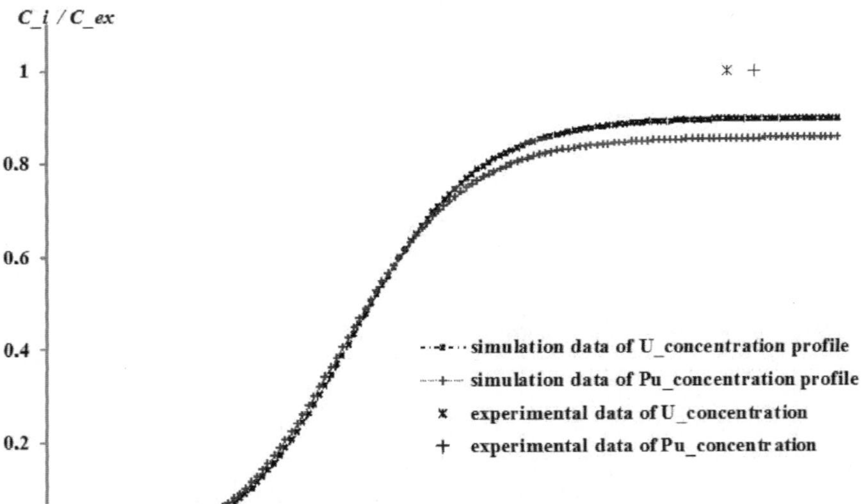

Figure 1. Distribution of the concentrations of U and Pu in the raffinate.

As we can see from the figure, the deviation of the distribution of the concentrations by target components, such as U and Pu in the raffinate does not exceed 15% from the experimental data, which meets production requirements.

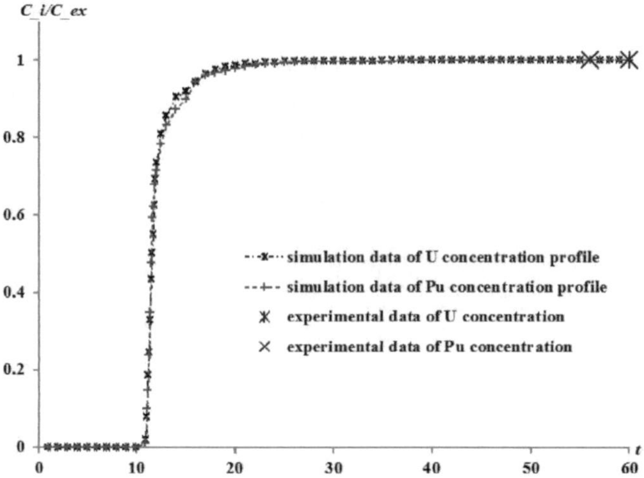

Figure 2. The distribution of concentrations of U and Pu in the extract

Figure 2 shows that the values of the concentrations of U and Pu in the extract obtained in the simulation are the same as the experimental data.

4. Conclusion

In the development of the mathematical model we presented the centrifugal extractor, the elementary unit of the cascade, in the form of tightly coupled system of two objects. We conducted analysis of technological process, on the basis of which we identified the main parameters influencing the process of mass transfer both in the extractor, and in the cascade. Preliminary testing of the model showed that an increase in radius of overflow of dense phase level in the water seal leads to reduce of the hydrostatic level position of phase interface in the separating chamber. If we change the density of the light phase, the volume of light phase also changes. This fact confirms the importance of considering changes in the volume of each phase in both chambers. The obtained verification data show that the concentration distribution of the target components at the cascade output both in the raffinate and in the extract, does not exceed the production limits. This fact confirms the admissibility of the assumptions made in the mathematical model.

References

[1] Kuznetsov G I, Pushkov A A, Kosogorov A.V 2000 *Centrifugal extractors CENTREX* (Moscow: RCHTU D.I. Mendeleev)

[2] Strenk F 1975 *Stirring and Stirred Apparatuses* [in Russian] (Leningrad: Khimiya)

[3] Goryunov A G 2011 Dynamic model of a multi-component nonequilibrium extraction process in the mixing chamber of centrifugal extractor *Scientific Technical Bulletin of Volga region* **4** pp 129-132

[4] Goryunov A G and Mikhaylov V S 2012 The automatic control system of a multi-component nonequilibrium extraction process in the pulse column *J. of Process Control* vol **22** pp. 1034-1043

[5] Vedantam S Wardle K E Tamhane T V Ranade V V and Joshi J B 2012 CFD Simulation of Annular Centrifugal Extractors *Int. J. of Chemical Engineering* vol **2012** 31 (doi:10.1155/2012/759397)

[6] Zilberman B Ya Fedorov Yu S Puzikov E A Goletskii N D Ryabkov D V Kudinov A S Blazheva I V and Kukharev D N 2014 Preliminary Studies of Dynamic Regimes in the Extraction–Scrubbing Groups of Units of the Experimental and Demonstration Center (EDC) Using Mathematical Modeling *Radiochemistry* vol **56** iss **6** pp 583-593

IOP Conf. Series: Materials Science and Engineering **135** (2016) 012054 doi:10.1088/1757-899X/135/1/012054

Experimental study of radiopharmaceuticals based on technetium-99m labeled derivative of glucose for tumor diagnosis

R. Zeltchan[1, 2], A. Medvedeva[1,2], I. Sinilkin[1,2], O. Bragina[2], V. Chernov[1,2], E. Stasyuk[2], A. Rogov[2], E. Il'ina[2], L. Larionova[2], V. Skuridin[2], A. Dergilev[3].

1 Tomsk Cancer Research Institute, Tomsk, Kooperativny Street, 5, Tomsk, Russia
2 Tomsk Polytechnic University, Tomsk, Lenin Avenue, 30, Tomsk, Russia
3Novosibirsk State Medical University, Krasny Prospect, 52, Novosibirsk, Russia

E-mail: r.zelchan@yandex.ru

Abstract. Purpose: to study the potential utility of 1-thio-D-glucose labeled with 99mTc for cancer imaging in laboratory animals. Materials and method: the study was carried out in cell cultures of normal CHO (Chinese hamster ovary cells CHO) and malignant tissues MCF-7 (human breast adenocarcinoma MCF-7). To evaluate the uptake of 99mTc-1-thio-D-glucose in normal and tumor tissue cells, 25 MBq of 1-thio-D-glucose labeled with 99mTc was added to the vials with 3 million cells and incubated for 30 minutes at room temperature. After centrifugation of the vials with cells, the supernatant was removed. Radioactivity in vials with normal and tumor cells was then measured. In addition, the study included 40 mice of C57B1/6j lines with tumor lesion of the right femur. For neoplastic lesions, Lewis lung carcinoma model was used. Following anesthesia, mice were injected intravenously with 25MBq of 99mTc-1-thio-D-glucose. Planar scintigraphy was performed 15 minutes later in a matrix of 512x512 pixels for 5 minutes. Results: when measuring the radioactivity of normal and malignant cells after incubation with 99mTc-1-thio-D-glucose, it was found that the radioactivity of malignant cells was higher than that of normal cells. The mean values of radioactivity levels in normal and malignant cells were 0.3±0.15MBq and 1.07±0.6MBq, respectively. All examined animals had increased accumulation of 99mTc-1-thio-D-glucose at the tumor site. The accumulation of 99mTc-1-thio-D-glucose in the tumor was on average twice as high as compared to the symmetric region. Conclusion: The present study demonstrated that 99mTc-1-thio-D-glucose is a prospective radiopharmaceutical for cancer visualization. In addition, high accumulation of 99mTc-1-thio-D-glucose in the culture of cancer cells and in tumor tissue of animals demonstrates tumor tropism of the radiopharmaceutical.

1. Introduction

Nuclear medicine techniques have become an indispensable tool for the diagnosis, treatment planning and management of cancer patients [[1], [2]]. Currently, there is a continuous search for new, more specific and efficient diagnostic and therapeutic radiopharmaceuticals. At the same time, the development of nuclear medicine imaging technologies in Russia is still far behind the world level. It is estimated that the demand for radiopharmaceuticals in Russia is satisfied by no more than 1-3%. A variety of positron-emitting nuclides is currently utilized to detect tumors and metastases as well as to monitor treatment response. The most commonly used PET radiopharmaceutical is a derivative of glucose, namely 2-(18F) fluoro-2-deoxy-glucose (18F-FDG) [[3]]. Despite the high diagnostic value of 18F-FDG-PET, the widespread use of this imaging modality in Russia is limited because of its high cost and lack of PET centers. In spite of advances in PET technologies worldwide, a large number of nuclear medicine laboratories in Russia are still equipped with gamma cameras. There are more than 200 SPECT (single photon emission computed tomography) centers in Russia. The most frequently used radionuclide for SPECT is technetium-99m, therefore, it seems relevant to develop innovative radiopharmaceuticals for molecular imaging based on technetium-99m-labeled glucose derivatives [[4], [5], [6],]. The main advantage of radiotracers based on technetium-99m-labeled glucose derivatives is the ability to image different tumor types using a gamma camera, thus significantly reducing the cost of the diagnostic procedure. A new tumor seeking agent, 99mTc-labeled 1-thio-D-glucose, is one of these radiotracers (Figure.1). It was developed at the Polytechnic University and Cancer Research Institute (Tomsk, Russia). It is currently under preclinical safety testing at the Tomsk Cancer Research Institute. Thus, the purpose of the present study was to evaluate the potential utility of 99mTc-1-thio-D-glucose for tumor tissue imaging in animals.

2. Materials and methods

The radiopharmaceutical (99mTc-1-thio-D-glucose) was prepared from the reagent (lyophilizate) by adding 4 ml of sodium pertechnetate (99mTc). The composition of reagent (lyophilizate) was a solution of 1-thio-D-glucose in the presence of tin (II) chloride dihydrate ($SnCl_2 \cdot 2\ H_2O$). After adding the reagent into a sterile vial, 99mTc sodium pertechnetate solution was incubated for 30 minutes at room temperature. The finished solution was a clear, colorless liquid. Description of the radiopharmaceutical: clear liquid, pH from 4.0 to 8.0 (6 ± 2); radiochemical impurity was not more than 8%. Radionuclide impurities: 99Mo - no more than $2\cdot10^{-3}$%, others - no more than 1·10-4%. The concentration of sodium chloride ranged from 8.0 to 10.0 mg/ml. The content of the derivative of 1-thio-D-glucose was not more than 0.625 mg / ml. The content of tin dichloride dehydrate was 0.044-0,052 mg / ml. Gamma spectrum 99mTc 1-thio-D-glucose had a characteristic line with the energy of 0.140 MeV.

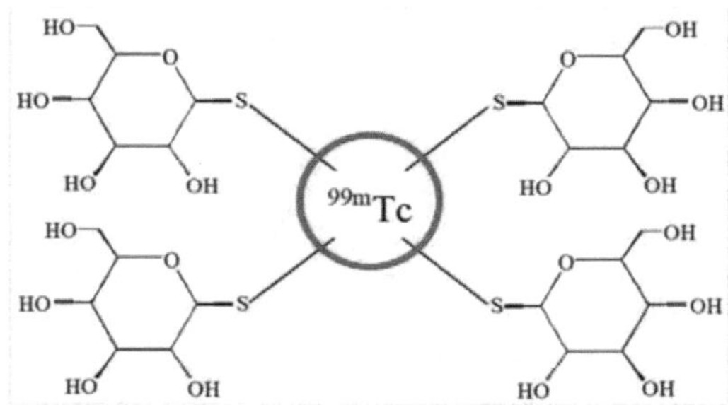

Figure 1. Chemical structure of 99mTc-1-thio-D-glucose.

The potential utility of 99mTc-1-thio-D-glucose for tumor imaging was studied in vitro and in vivo models. In vitro studies, cell cultures from normal (Chinese hamster ovary (CHO) cells) and cancer tissues (MCF 7–human breast adenocarcinoma cell line) were used. To evaluate the uptake of 99mTc-1-thio-D-glucose in normal and tumor tissue cells, 25 MBq in 200 μL of technetium-99m-labeled glucose derivative was added to vials with 3,000,000 cells/ml, 5 vials for each cell culture. The vials with cells were then incubated for 30 minutes at room temperature with periodic gentle shaking to avoid cell adhesion. After completion of incubation, vials were centrifuged at 300 rpm for 5 minutes. After sedimentation, the supernatant was removed carefully and the radioactivity measurement of the sediment and supernatant was made. Thereafter, cells were washed three times by adding 1 ml of buffer to each vial and centrifugation vials. PBS + 1% BCA were used as a buffer.

Radioactivity in the sediment and supernatant was measured after each washing. After the final cell washing, cell-labeling efficiency, a percentage of 99mTc-labeled 1-thio-D-glucose accumulation in normal and cancer tissue cells, was measured. In addition, planar scintigraphic imaging of the culture plates with vials was performed.

In vivo studies, the potential utility of 99mTc-labeled 1-thio-D-glucose for tumor visualization was studied using the intramuscularly implanted solid Lewis lung carcinoma model in C57BL/6J mice (Bank of Cell Lines, NN Blokhin Russian Cancer Research Center, Moscow). A total of 40 male and female mice, weighing 30±5g and 12 weeks of age, were used. The tumor size was 2± 0.2cm^3. The acquisition, care and housing of laboratory animals were in compliance with federal and local laws and regulations, as well as with ethical and legal principles of the European Convention for the Protection of Vertebrate Animals (Strasbourg, 1986). Mice were anesthetized by inhalation of ethyl ether. To study the 99mTc-labeled 1-thio-D-glucose accumulation in the body of C57BL/6J mice bearing Lewis lung carcinoma, the radiotracer was injected intravenously in diagnostic dose after anesthesia. Whole-body gamma scintigraphy using SiemensE.Cam 180 gamma camera was performed 10-15 min after intravenous injection of the radiotracer. We used the high-resolution, low-energy collimators for140keV energy. The mice were placed with their ventral surface to the gamma camera detector so that the imaging field of view could encompass the whole body of the animal. Each image was acquired for 5 minutes. About 500,000 counts were recorded in each image with matrix size of 256x256 pixels. Radioactivity in regions of interest drawn in various organs of the mice and in tumor tissue was measured. Statistical analysis was performed using E.soft Software Version 5.5 (Siemens).

3. Results

The radioactivity in normal and tumor tissue cells after the final washing was measured using a gamma counter. The mean radioactivity levels in normal and tumor tissue cells after the final washing were 0.3±0.15 MBq and 1.07±0.6 MBq, respectively. Thus, the 99mTc-labeled 1-thio-D-glucos uptake in cancer cells (MCF 7) was 3.5 times higher than that in normal tissue cells (CHO). Planar scintigraphic images of plates with culture vials showed more intense radiotracer accumulation in cancer tissue cells (Figure. 2).

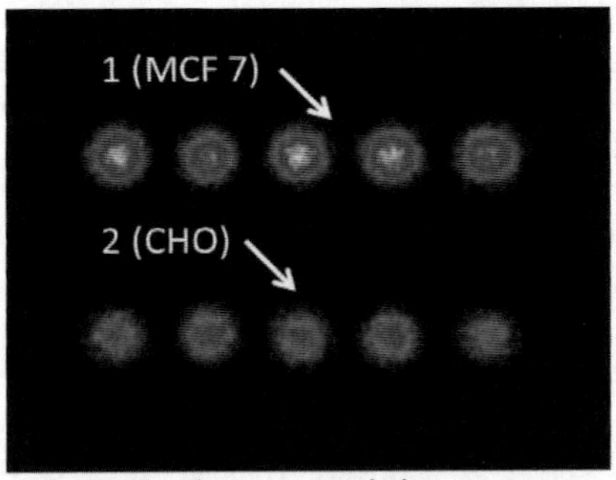

Figure 2. Scintigraphic imaging of culture plates containing cancer cells (1) and normal tissue cells (2) after incubation with 99mTc-labeled 1-thio-D-glucos and 3-time washing.

Planar scintigraphy was performed on tumor-bearing mice 10-15 minutes after intravenous injection of 25 MBq of 99mTc-labeled 1-thio-D-glucos. Tumor was visualized in all mice. Most scintigraphic images demonstrated the tumor as a site of extremely high radiotracer uptake with clear contours. The uptake of the radiotracer in the tumor was frequently heterogeneous, with different intensity (Figure.3).

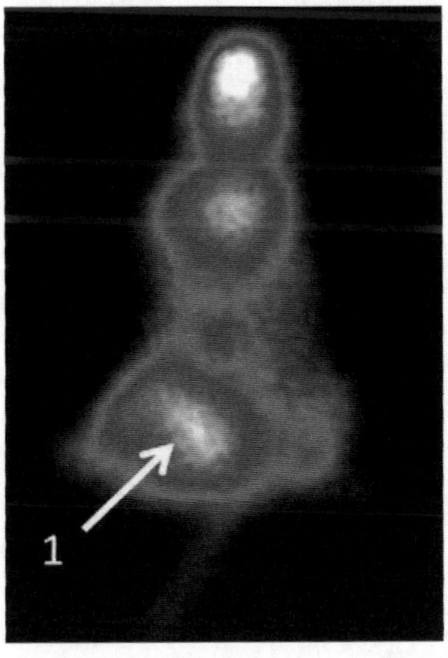

Figure 3. Scintigraphic image of tumor-bearing mice at 15 minutes after intravenous injection of 99mTc-labeled 1-thio-D-glucos. There is an extremely high radiotracer uptake in the tumor of the right femur (1).

Changes in 99mTc-labeled 1-thio-D-glucos accumulation in the tumor after intravenous injection is demonstrated in Figure 4. It should be noted that the maximum concentration of the 99mTc-labeled 1-thio-D-glucos in the tumor is recorded through 2 hours following intravenous administration.

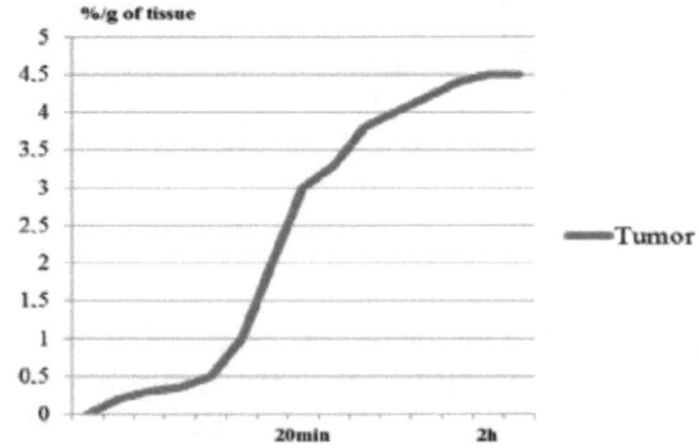

Figure 4. The dynamics of 99mTc-labeled 1-thio-D-glucos accumulation in tumors in mice after intravenous administration. The maximum concentration of the 99mTc-labeled 1-thio-D-glucos in the tumor is recorded through 2 hours following intravenous injection.

The images obtained showed that the mean radiotracer uptake intensity in the tumor was 2665.8±431 counts, while in the symmetrical tissue it was 878.3±134 counts. The relative 99mTc-labeled 1-thio-D-glucos uptake in the tumor was higher by 308.9±94% than that in the symmetrical tissue. The percentage of the radiotracer tumor uptake from the total injected radiotracer dose was calculated. At 10-15 minutes after intravenous injection, the percentage of 99mTc-labeled 1-thio-D-glucos uptake in the tumor appeared to be 8.14% of the total injected radiotracer dose. Thus, the use of 99mTc-labeled 1-thio-D-glucos radiotracer for tumor imaging by single photon emission scintigraphy allowed tumor imaging in 100% of the laboratory mice.

4. Conclusion. The results from this study indicate that 99mTc-labeled 1-thio-D-glucos, a new tumor-seeking agent based on technetium-99m-labeled glucose derivative, is a potentially useful tracer for tumor tissue imaging in the laboratory animals. Moreover, in vitro studies demonstrate highly selective accumulation of this radiotracer. The 99mTc-labeled 1-thio-D-glucos uptake in tumor tissue cells was 3 times higher than that in normal tissue cells. However, further pre-clinical studies of 99mTc-labeled 1-thio-D-glucos are necessary to investigate its pharmacokinetics and safety.

References

[1] Chernov V I, Medvedeva A A, Sinilkin I G, Zeltchan R V, Bragina O D, Skuridin V S 2015 *Siberian Journal of Oncology.* **2** 45-47
[2] Van Noorden R 2013 *Nature* **504** 202-4
[3] Jacobson M S, Hung J C, Mays T L, Mullan B P 2002 *Mol. Imaging Biol.* **4** 119-127
[4] Welling M M, Alberto R 2010 *Nucl. Med. Commun.* **31** 239-248
[5] Cheng D, Rusckowski M, Wang Y, Liu Y, Liu G, Liu X, Hnatowich D 2011 *Curr. Radiopharm.* **4** 5-9
[6] Chen Y, Huang Z W, He L, Zheng S L 2006 *Appl. Radiat. and Isot.* **64** 342-347

Deposition of the low resistive ITO-films by means of reactive magnetron sputtering of the In/Sn target on the cold substrate

Y S Zhidik[1], P E Troyan[1], E V Baturina[1], D V Korzhenko[2] and Y N Yurjev[2]
[1] Department of Physical Electronics, Faculty of Electronic Engineering, Tomsk State University of Control Systems and Radio-electronics, pr. Lenina, 40, Tomsk, 634050, Russia
[2] Department of Experimental Physics, Institute of Physics and Technology, National Research Tomsk Polytechnic University, pr. Lenina, 30, Tomsk, 634050, Russia

E-mail: zhidikyur@mail.ru

Abstract. Detailed information on the deposition technology of the low-resistive ITO-films in oxygen-containing media by magnetron reactive sputtering from the In(90%)/Sn(10%) target on the cold substrate is given. Developed technology allows deposition ITO-films with sheet resistance 2-3 Ω/□, transparency higher than 90%. Developed technology is notable for high reproducibility of results and is compatible with production technology of semiconductor devices of optoelectronics.

1. Introduction

Up-to-date electronics is substantially defined by the properties of low-dimensional semiconductor materials. At the present time a lot of research groups report about prospect researches of semiconductor transparent oxides, such as In_2O_3, ZnO, SnO_2, CdO, Ga_2O_3, TiO_2 and more complex double oxides and tertiary oxides [1-3]. Transition metal oxides in contrast with metals and semiconductors have broader spectrum of possible physical states and properties. Control of electron flow and light flux in semiconductor heterostructures, researched by J. Alferov, led to semiconductor revolution of the XX century. So, the next breakthrough in new functional materials is connected with control of electron flows and magnetism in low-dimensional heterostructures made of transition metal oxides [2].

There is no doubt that one of the most demand materials of such category is indium oxide films doped with tin (indium tin oxide, ITO), which found wide application as transparent electrodes in different devices [2, 4].

This paper is concerned with development of the technology of deposition of the low resistive ITO-films by means of reactive magnetron sputtering of the In(90%)/Sn(10%) metal target on the cold substrate with post-annealing.

2. Results and discussion

Magnetron sputtering of ITO-films was worked through and described in [6]. This technology provides value of sheet resistance as 10 Ω/□ with value of transparency 87% and implies sputtering of the metal target in the oxide-containing media, condensation of the oxidized atoms of tin and indium on the heated up to 200-250° C substrate, with post-annealing during 30-40 minutes at the temperature of 250° C.

There is need in industry to deposit such ITO-films with sheet resistance that is 10 Ω/□ on the cold substrate, since a lot of plants for magnetron sputtering do not have heating devices for substrate's heating. Furthermore, deposition of the low-resistive ITO-films is connected with its posterior integration to the technology of planar LED production, which implies deposition of ITO-films on the reverse mask of photoresist (explosive photolithography).

Features of ITO-films' deposition are described well in papers [6-7]. Main conclusion of these papers is that during ITO-films' deposition on the cold substrate the process of oxidation is nearly stopped on the substrate. Thus tolerance to the lack of oxygen is sufficiently smaller than the same deposition on the hot substrate. Content of oxygen in gas mixture should be sufficient for oxidation of all metal atoms to the moment of its condensation in the structural film grid. So, there is strong dependence between partial pressure of the oxygen in the chamber and film growth rate. Hence, the consequence of deposition technology of ITO-films on the cold substrate is the next.

2.1. Experiment I – "Measurement of ITO-film deposition rate"

Dependence between partial pressure of the oxygen in the chamber and film growth rate needs optimal film growth rate which is in the range of 2 – 3.5 Å/s [5-7]. Test samples were obtained by magnetron sputtering of metal target In/Sn in the gas mixture of Ar(50%)/O_2(50%) with different values of the discharge current (0.2 A; 0.3 A; 0.4 A) during the equal time (10 minutes). Distance from target to sample was 10 cm. Samples were made of glass ceramics. Thickness of the films was measured by means of interference microscope MII-4M. Results of the experiment are shown in the table 1.

Table 1. Measurement of the deposition rate of ITO-films.

Mode 1: discharge current – 0.2 A; deposition time - 10 minutes. Measured sputtering rate - 1Å/s	
• Before annealing: ρ_S =80 mΩ/□ d=0.06 μm Etching time in the cold HCl – 1.5 minutes	• After annealing in the N_2 media (450°C, 25 minutes): ρ_S=40 kΩ/□ d=0.057 μm Etching time in the cold HCl - 3 minutes
Mode 2: discharge current – 0.3 A; deposition time - 10 minutes. Measured sputtering rate – 1.9 Å/s	
• Before annealing: ρ_S =100 mΩ/□ d=0.11 μm Etching time in the cold HCl – 1.5 minutes	• After annealing in the N_2 media (450°C, 25 minutes): ρ_S =34 kΩ/□ d=0.09 μm Etching time in the cold HCl - 4 minutes
Mode 3: discharge current – 0.4 A; deposition time - 10 minutes. Measured sputtering rate 2.5 Å/s	
• Before annealing: ρ_S =100 mΩ/□ d=0.11 μm Etching time in the cold HCl – 3.5 minutes	• After annealing in the N_2 media (450°C, 25 minutes): ρ_S =26 kΩ/□ d=0.1 μm Etching time in the cold HCl - 8 minutes

Conducted experiment shows that structure of films with deposition rate higher than 2 Å/s is denser. That is proved by the time of films' etching – dense film needs more time for etching. Post-annealing make the film structure even denser.

IOP Conf. Series: Materials Science and Engineering **135** (2016) 012055 doi:10.1088/1757-899X/135/1/012055

Melting temperature of In/Sn alloy is relatively small, so optimal mode for ITO-films deposition was chosen to avoid target's melting: discharge current – 0.3 A; deposition time – 12 minutes. Approximate film thickness was 130 nm.

2.2. Experiment II "Determination of the optimal Ar/O_2 ratio in the operating media"

Range of oxygen percentage in the gas mixture is 10-50% [5-6]. It should be noted that optimal Ar/O_2 ratio in the gas mixture can vary from plant to plant due to different vacuum pumping systems and discharge power of the magnetrons sputtering system.

Determination of the optimal ratio of the partial pressures was made as follows. Series of experiments with optimal operating mode (discharge current 0.3 A; deposition time – 12 minutes) was made. Oxygen content was gradually reduced in series from 50 % by 3-5 % from deposition to deposition, until unoxidized atoms of Sn and In appeared in the growing film, i.e. transparent samples became brown colored. Conductivity and transparency of such films can be improved by annealing in the oxygen-containing media. In such annealing the structural grid of ITO-film is rebuild and not fully oxidized oxides become fully oxidized with simultaneous generation of oxygen vacancies.

Conductivity of the fully oxidized ITO-films can be improved by annealing in the vacuum or in the nitrogen media. This process is caused by diffusion of oxygen atoms, embedded in the crystal lattice or grain boundary from the volume of film during high-temperature annealing. Generated oxygen vacancies act like donor for electrons thus increasing electron concentration. So, annealing decrease sheet resistance both by increase in the concentration of electrons and by growth of their mobility.

Results of the conducted experiment of determination of the optimal Ar/O_2 ratio in the gas mixture are shown in the table 2.

Table 2. Determination of the optimal Ar/O_2 ratio in the gas mixture.

Mode 1: discharge current 0.3 A; deposition time - 12 minutes. Oxygen content - 16%. Films are not fully oxidized (black color).	
• Annealing in the N_2 (450°C; minutes): ρ_S =2.8 $\Omega/\square$ black colored films	• Annealing in the air (450°C; 25 minutes): ρ_S =7 $\Omega/\square$ brown colored films
Mode 2: discharge current 0.3 A; deposition time - 12 minutes. Oxygen content - 25%. Films are not fully oxidized (brown color).	
• Annealing in the N_2 (450°C; minutes): ρ_S =14 $\Omega/\square$ brown colored films	• Annealing in the air (450°C; 25 minutes): ρ_S =3 $\Omega/\square$ transparent films
Mode 3: discharge current 0.3 A; deposition time - 12 minutes. Oxygen content - 29%. Films are fully oxidized (transparent).	
• Annealing in the N_2 (450°C; minutes): ρ_S =24 $\Omega/\square$ transparent films	• Annealing in the air (450°C; 25 minutes): $\rho_S \rightarrow \infty$ transparent films
Mode 4: discharge current 0.3 A; deposition time - 12 minutes. Oxygen content - 33%. Films are fully oxidized (transparent).	
• Annealing in the N_2 (450°C; minutes): ρ_S =33 $\Omega/\square$ transparent films	• Annealing in the air (450°C; 25 minutes): $\rho_S \rightarrow \infty$ transparent films
Mode 5: discharge current 0.3 A; deposition time - 12 minutes. Oxygen content - 50%. Films are fully oxidized (transparent).	
• Annealing in the N_2 (450°C; minutes): ρ_S =98 $\Omega/\square$ transparent films	• Annealing in the air (450°C; 25 minutes): $\rho_S \rightarrow \infty$ transparent films

IOP Conf. Series: Materials Science and Engineering **135** (2016) 012055 doi:10.1088/1757-899X/135/1/012055

Conducted studies proved our theory of conductivity in films of transparent metal oxides, based on generation of oxygen vacancies that behave as donor centers after high-temperature annealing [8].

Optimal Ar/O_2 ratio for the gas mixture are the follows:

1) 25% O_2 + 75% Ar with post annealing in the air;
2) 29% O_2 + 71% Ar with post annealing in the nitrogen media.

After determination of the optimal Ar/O_2 ratio for the gas mixture correction of the deposition time was made to provide the deposited film with thickness 130 nm. It was necessary because deposition rate in the Ar/O_2 gas mixture depends on percentage of Ar. Deposition time was reduced to 10 minutes.

2.3. Experiment III "Testing of annealing mode of ITO-films"

Series of samples that were obtained during the experiment II with optimal gas mixture were annealed at different temperatures (400°C, 500°C and 600°C) within 15 and 25 minutes for determination of optimal high-temperature annealing mode of ITO-films.

Results of the conducted studies are shown in table 3. Transmission spectrums of samples obtained in gas mixtures with oxygen percentage of 25 % and 29 % before annealing and with minimal resistance after annealing are shown on the fig. 1 and fig. 2 respectively.

Table 3. Testing of annealing mode of ITO-films.

Mode 1 (cartridge A): discharge current 0.3 A; deposition time - 10 minutes; oxygen content - 25%; annealing in the air.	
Sample A6: without annealing ρ_S=26 Ω	Sample A1: 400°C, 25 minutes. ρ_S=1.7 Ω
Sample A2: 500°C, 15 minutes. ρ_S=6.6 Ω	Sample A3: 500°C, 25 minutes. ρ_S=6.3 Ω
Sample A4: 600°C, 15 minutes. ρ_S=2.6 Ω	Sample A5: 600°C, 25 minutes. ρ_S=2.3 Ω
Mode 2 (cartridge B): discharge current 0.3 A; deposition time - 10 minutes; oxygen content - 29%; annealing in the N_2.	
Sample B6: without annealing $\rho_S \to \infty$	Sample B1: 400°C, 25 minutes. ρ_S=24 mΩ
Sample B2: 500°C, 15 minutes. ρ_S=35 Ω	Sample B3: 500°C, 25 minutes. ρ_S=16 Ω
Sample B4: 600°C, 15 minutes. ρ_S=13 Ω	Sample B5: 600°C, 25 minutes. ρ_S=13 Ω

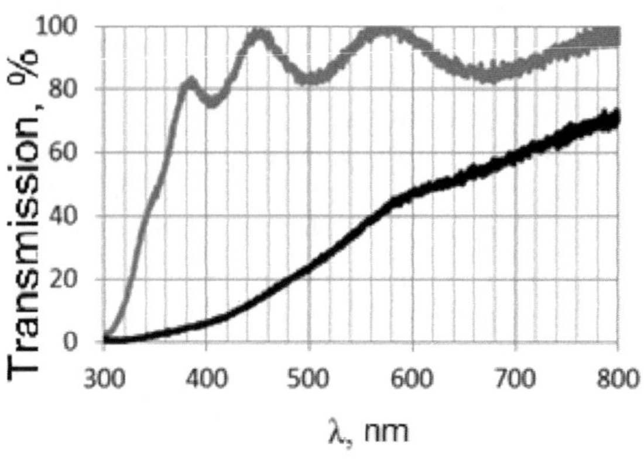

Figure 1. Transmission spectrum of A5 (grey curve) and A6 (black curve) samples

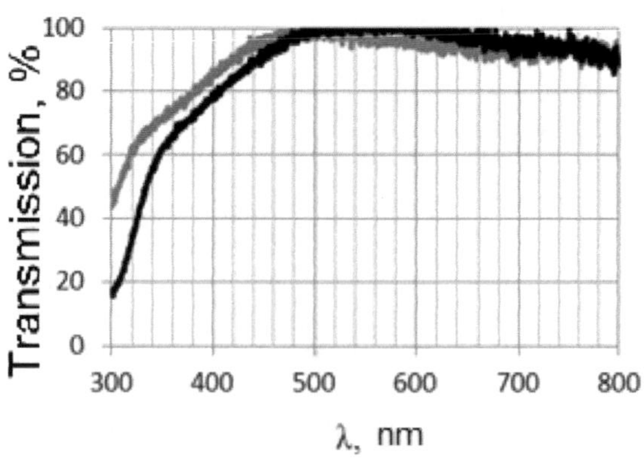

Figure 2. Transmission spectrum of B4 (grey curve) and B6 (black curve) samples

Thus, the best results in sheet resistance and transmittance in the range of blue spectral color (wavelength 440-485 nm) are obtained in the follows modes of deposition and annealing:

- deposition: discharge current – 0.3 A; deposition time - 10 минут; oxygen percentage in the gas mixture - 25%;

- annealing: in the air at the temperature 600°C; annealing time - 25 minutes.

3. Conclusion

Developed technology of the deposition of the low-resistive ITO-films in by magnetron reactive sputtering from the In(90%)/Sn(10%) metal target with post-annealing provide low-resistive ITO-films with sheet resistance from 2 to 3 Ω/□ on the cold substrate. Such technology provides temporal and temperature stability of ITO-properties after annealing as well as high electrical and optical properties. Densification of ITO-films after annealing leads to higher chemical stability.

References

[1] Granqvist C G and Hultaker A 2002 *Thin Solid Films* **411** 1-5
[2] Semikina T V, Komaschenko V N and Shmyreva L N 2010 *Electronics and Communications* **3(56)** 20-28
[3] Yurjev Y and Sidelev D 2013D *J. of Phys.:* Conference Series **479** 012018
[4] Untila G G, Kost T N, Chebotareva A B and Timofeyev M A 2012 *Semiconductors* **46** 984-90
[5] Zhidik Y S, Troyan P E and Saharov Y V 2014 *J. "Doklady Tomskogo gosudarstvennogo universiteta sistem upravleniya i radioelektroniki"* **1(31)** 99-102
[6] Amosova L P and Isaev M V 2014 *Technical Physics* **59**, 1545-9
[7] Amosova L P 2015 *Semiconductors* **49** 414-8
[8] Saharov Y V, Troyan P E and Zhidik Y S 2015 *J. "Doklady Tomskogo gosudarstvennogo universiteta sistem upravleniya i radioelektroniki"* **3(37)** 85-8

Institute of Physics
Dirac House, Temple Back
Bristol BS1 6BE UK

ISSN: 1757-8981
ISBN 978-1-5108-2818-6

CURRAN ASSOCIATES INC.
proceedings.com

International Conference on Aerospace Technology, Communications and Energy Systems (ATCES 2017)

IOP Conference Series: Materials Science and Engineering
Volume 302

Samara, Russia
28 - 30 September 2017

IOP
Institute of Physics